W9-DFH-650

SPICE for Circuits and Electronics Using PSpice®

Second Edition

Muhammad H. Rashid

Ph. D., Fellow IEE
Professor of Electrical Engineering
Purdue University at Fort Wayne

WITHDRAWN

Prentice Hall, Englewood Cliffs, New Jersey 07632

Library of Congress Cataloging-in-Publication Data

Rashid, M. H.
   SPICE for circuits and electronics using PSpice / Muhammad H.
Rashid.
      p.   cm.
   Includes bibliographical references and index.
   ISBN 0-13-124652-6
   1. SPICE (Computer file)   2. PSpice.   3. Electric circuit
analysis—Data processing.   I. Title.
TK454.R385   1995
621.319′2′0285—dc20   94-17481
                                                       CIP

Publisher: Alan Apt
Project Manager: Mona Pompili
Cover Designer: Wanda Lubelska Design
Copy Editor: Nick Murray
Buyer: Linda Behrens
Editorial Assistant: Shirley McGuire

 © 1995 by Prentice-Hall, Inc.
A Simon & Schuster Company
Englewood Cliffs, New Jersey 07632

The author and publisher of this book have used their best
efforts in preparing this book. These efforts include the
development, research, and testing of the theories and
programs to determine their effectiveness. The author and
publisher shall not be liable in any event for incidental or
consequential damages in connection with, or arising out of,
the furnishing, performance, or use of these programs.

All rights reserved. No part of this book may be reproduced,
in any form or by any means, without permission in writing
from the publisher.

Printed in the United States of America

10 9 8 7 6 5 4 3 2

ISBN 0-13-124652-6

PRENTICE-HALL INTERNATIONAL (UK) LIMITED, *London*
PRENTICE-HALL OF AUSTRALIA PTY. LIMITED, *Sydney*
PRENTICE-HALL CANADA, INC., *Toronto*
PRENTICE-HALL HISPANOAMERICANA, S.A., *Mexico*
PRENTICE-HALL OF INDIA PRIVATE LIMITED, *New Delhi*
PRENTICE-HALL OF JAPAN, INC., *Tokyo*
SIMON & SCHUSTER ASIA PTE. LTD., *Singapore*
EDITORA PRENTICE-HALL DO BRASIL, LTDA., *Rio de Janeiro*

TRADEMARK INFORMATION

PSpice and Probe are resistered
   trademarks of MicroSim
   Corporation.

IBM-PC is a registered trademark of
   International Business Machines
   Corporation.

Macintosh II is a registered
   trademark of Apple Computer Inc.

WordStar and WordStar 2000 are the
   registered trademarks of Micropro
   International Corporation.

Word is a registered trademark of
   Microsoft Corporation.

Program Editor is a registered
   trademark of WorkPerfect
   Corporation.

To my parents,
my wife Fatema,
and
my children, Fa-eza, Farzana, and Hasan

Preface

The Engineering Accreditation Commission of the Accreditation Board for Engineering and Technology (EAC/ABET) requirements specify the integration of computer-aided analysis and design in EE-curriculum. SPICE is very popular software for analyzing electrical and electronic circuits. Until recently, a mainframe or a VAX-class computer was required. Beside the cost, such a machine is not convenient for assignments in classes of 200-level circuits and electronics courses. MicroSim Corporation introduced the PSpice simulator that can run on personal computers (PCs) and is similar to the University of California (UC) Berkeley SPICE. The student version of PSpice, which is available free to students, is ideal for classroom use and for assignments requiring computer aided simulation and analysis. PSpice widens the scope for the integration of computer aided simulation to circuits and electronics courses for undergraduate students.

It may not be possible to add a 1-credit-hour course on SPICE to integrate computer aided analysis in circuits and electronics courses. However, the students require some basic knowledge of using SPICE. They are constantly under pressure with course loads and do not always have free time to read the details of SPICE from manuals and books of general nature.

This book is the outcome of difficulties faced by the author in integrating SPICE in circuits and electronics courses at the 200-level. The objective of this book is to introduce the SPICE simulator to the EE-curriculum at the sophmore or junior level with a minimum amount of time and effort. This book requires no prior knowledge about the SPICE simulator. A course on basic circuits should be a prerequisite or co-requisite. Once the students develop interests and appreciations in the applications of circuit simulator like SPICE, they can read advanced materials for the full utilization of SPICE or PSpice in solving complex circuits and systems.

This book can be divided into seven parts: (1) introduction to SPICE simulation—chapters 1, 2, and 3; (2) source and element modeling—chapters 4 and 5; (3) SPICE commands—chapter 6; (4) dc and ac circuits—chapters 7 and 8; (5) semiconductor devices modeling—chapters 9, 10, 11, and 12; (6) digital logic circuits—chapter 13; and (7) difficulties—chapter 14. Chapters 9, 10, 11, and 12 describe the simple equivalent circuits of transistors and op-amps, which are commonly used in analyzing electronic circuits. Chapter 13 introduces to the PSpice simulation of digital logic circuits using some off-the-shelf devices, and to the commands and techniques for developing subcircuits exhibiting the characteristics of user defined devices from basic digital primitives. Although SPICE generates the parameters of complex transistor models, the analysis with a simple circuit model exposes the students to the mechanism of computation by SPICE .MODEL commands. This approach has the advantage that the students can compare the results, which are obtained in a classroom environment with the simple circuit models of devices, to those obtained by using complex SPICE models.

The commands, models, and examples that are described for PSpice are also applicable to UC Berkeley SPICE with minor modifications. The changes for running a PSpice circuit file on SPICE and vice-versa are discussed in Chapter 11. The circuit files in this book are typed in uppercase so that the same file can be run on either the PSpice or the SPICE simulator.

Probe is a graphics post-processor and is very useful in plotting the results of simulation, especially with the capability of arithmetic operation it can be used to plot impedance, power, etc. Once the students have experience in programming on PSpice, they really appreciate the advantages of .Probe command. *Probe* is an option on PSpice, available with the student version. Running *Probe* does not require a math co-processor. The students can also get the normal printer output or printer plotting. The prints and plots are very helpful to the students in relating their theoretical understanding and making judgment on the merits of a circuit and its characteristics.

This book can be used as a text book on SPICE with a course on basic circuits being the prerequisite or co-requisite. It can also be a supplement to any standard text book on basic circuits or electronics. In the case of this option, the following sequence is recommended for the integration of SPICE at basic circuits level:

1. Supplement to a basic circuits course with 3 hours of lectures (or equivalent lab hours) and self-study assignments from chapters 1 to 8. Starting from chapter 2, the students should work with PCs.
2. Continue as a supplement to an electronics course with 2 hours of lectures (or equivalent lab hours) and self-study assignments from chapters 9 to 14.

For integrating SPICE at the electronics level, 3 hours of lectures (or equivalent lab hours) are recommended on chapters 1 to 6. Chapters 9 to 14 could be left for self-study assignments. From the author's experience in the class, it has been observed that after three lectures of 50 minutes duration, all students could solve assignments independently without any difficulty. The class could progress in a normal manner with one assignment per week on electronic circuits simulation and analysis with SPICE. Although the materials of this book have been tested in a basic circuits course for engineering students and in two electronics courses for EE-students, the book is also recommended for EET students.

Muhammad H. Rashid
Fort Wayne, Indiana

Note. The students are urged to print the contents of files "README.DOC" to set up for the right printer and monitor, and "EVAL.LIB" to obtain the list of subcircuits for semiconductor devices. These files come with the student version of PSpice software programs and can be printed by the DOS TYPE command.

Acknowledgments

I would like to thank the following persons for their comments and suggestions:

Frank H. Hielscher, Lehigh University
A. Zielinski, University of Victoria, Canada
Emil C. Neu, Stevens Institute of Technology

It has been a great pleasure working with the editorial staffs, Alan Apt, Mona Pompili, Sondra Chavez, and Shirley McGuire. Finally, I would thank my family for their love, patience, and understanding.

PSpice SOFTWARE

The PSpice student version software is available from Prentice Hall. To order the software, please see the card which is included in this book.

- PSpice student version disks (two $5\frac{1}{4}''$) IBM PC compatible (73476-4)
- PSpice student version disks (two $3\frac{1}{2}''$) IBM PC compatible (73475-6)
- PSpice student version disks (three $3\frac{1}{2}''$) MAC II compatible (73474-7)

The student version software can also be obtained directly from

MicroSim Corporation
20 Fairbanks
Irvine, CA 92718 USA
Tel: (800) 245-3022 (Toll Free)
     (714) 770-3022
Fax: (714) 455-0554

Any comments and suggestions regarding this book are welcomed and should be sent to the author.

Dr. Muhammad H. Rashid
Professor of Electrical Engineering
Indiana University—Purdue University Fort Wayne
Fort Wayne, IN 46805-1499 USA

Contents

Contents

Introduction

1-1 INTRODUCTION

Electronic circuit design requires accurate methods for evaluating circuit performance. Because of the enormous complexity of modern integrated circuits, computer-aided circuit analysis is essential and can provide information about circuit performance that is almost impossible to obtain with laboratory prototype measurements. Computer-aided analysis permits

1. Evaluating the effects of variations in elements, such as resistors, transistors, transformers, and so on.
2. The assessment of performance improvements or degradations
3. Evaluating the effects of noise and signal distortion without the need of expensive measuring instruments
4. Sensitivity analysis to determine the permissible bounds due to tolerances on each and every element value or parameter of active elements
5. Fourier analysis without expensive wave analyzers
6. Evaluating the effects of nonlinear elements on the circuit performance
7. Optimizing the design of electronic circuits in terms of circuit parameters.

SPICE is a general-purpose circuit program that simulates electronic circuits. SPICE can perform various analyses of electronic circuits: the operating (or the quiescent) points of transistors, a time-domain response, a small-signal frequency response, and so on. SPICE contains models for common circuit elements, active as well as passive, and it is capable of simulating most electronic circuits. It is a versatile program and is widely used both in industries and universities. The acronym SPICE stands for *Simulation Program with Integrated Circuit Emphasis*.

Until recently, SPICE was available only on mainframe computers. In addition to the initial cost of the computer system, such a machine can be expensive and inconvenient for classroom use. In 1984, MicroSim introduced the PSpice simulator, which is similar to the Berkeley SPICE and runs on an IBM-PC or compatible. It is available at no cost to students for classroom use. PSpice, therefore, widens the scope for the integration of computer-aided circuit analysis into electronic circuits courses at the undergraduate level. Other versions of PSpice that will run on computers such as the Macintosh II, VAX, SUN, and NEC are also available.

1-2 DESCRIPTIONS OF SPICE

The development of SPICE spans a period of about 30 years. During the mid-1960s, the program ECAP was developed at IBM [1]. Later ECAP served as the starting point for the development of program CANCER at the University of California (UC), Berkeley in the late 1960s. Based on CANCER, SPICE was developed at Berkeley in early 1970s. SPICE2, which is an improved version of SPICE, was developed during the mid-1970s at UC—Berkeley.

The algorithms of SPICE2 are general in nature but are robust and powerful for simulating electrical and electronics circuits, and SPICE2 has become a standard tool in the industry for circuit simulations. The development of SPICE2 was supported by public funds at UC—Berkeley, and the program is in the public domain. SPICE3, which is a variation of SPICE2, is designed especially to support the computer-aided design (CAD) research program at UC—Berkeley.

SPICE2 has become an industry standard and is now referred to simply as SPICE. The input syntax for SPICE is a free-format style; it does not require that data be entered in fixed column locations. SPICE assumes reasonable default values for unspecified circuit parameters. In addition, it performs a considerable amount of error checking to ensure that a circuit has been entered correctly.

PSpice, which uses the same algorithms as SPICE2 and is a member of the SPICE family, is equally useful for simulating all types of circuits in a wide range of applications. A circuit is described by statments that are stored in a file called the *circuit file*. The circuit file is read by the SPICE simulator. Each statement is self-contained and independent; the statements do not interact with each other. SPICE (or PSpice) statements are easy to learn and use.

1-3 TYPES OF SPICE

The commercially supported versions of SPICE2 can be divided into two types: mainframe versions and PC-based versions. Their methods of computation may differ, but their features are almost identical to those of SPICE2. However, some may include such additions as a pre-processor or shell program to manage input and provide interactive control, as well as a post-processor for refining the normal

SPICE output. A person who is familiar with one SPICE version (e.g., PSpice) should be able to work with other versions.

The mainframe versions are

HSPICE (Meta-Software), which is designed for integrated circuit design with special device models

RAD-SPICE (Meta-Software), which simulates circuits subjected to ionizing radiation

IG-SPICE (A.B. Associates)

I-SPICE (NCSS Time Sharing). IG-SPICE and I-SPICE are designed for interactive circuit simulation with graphic output.

Precise (Electronic Engineering Software)

PSpice (MicroSim), which is similar to PSpice

AccuSim (Mentor Graphics)

Cadence-SPICE (Cadence Design)

SPICE-Plus (Valid Logic)

The PC-versions are

AllSpice (Acotech)

IS-SPICE (Intusoft)

Z-SPICE (Z-Tech)

SPICE-Plus (Analog Design Tools)

DSPICE (Daisy Systems)

PSpice (MicroSim)

1-4 TYPES OF ANALYSIS

PSpice allows various types of analysis. Each analysis is invoked by including its command statement. For example, a statement beginning with the .DC command invokes the DC sweep. The types of analysis and their corresponding .(dot) commands are described below.

Dc Analysis is used for circuits with time-invariant sources (e.g., steady-state dc sources). It calculates all node voltages and branch currents over a range of values, and their quiescent (dc) values are the outputs.

Dc sweep of an input voltage/current source, a model parameter, or temperature over a range of values (.DC)

Determination of the linearized model parameters of nonlinear devices (.OP)

Dc operating point to obtain all node voltages (.OP)

Small-signal transfer function with small-signal gain, input resistance, and output resistance (Thévenin's equivalent) (.TF)

Dc small-signal sensitivities (.SENS)

Transient Analysis is used for circuits with time-variant sources (e.g., ac sources and switched dc sources). It calculates all node voltages and branch currents over a time interval, and their instantaneous values are the outputs.

Circuit behavior in response to time varying sources (.TRAN)

Dc and Fourier components of the transient analysis results (.FOUR)

Ac Analysis is used for small-signal analysis of circuits with sources of variable frequencies. It calculates all node voltages and branch currents over a range of frequencies, and their magnitudes and phase angles are the outputs.

Circuit response over a range of source frequencies (.AC)

Noise generation at an output node for every frequency (.NOISE)

1-5 LIMITATIONS OF PSpice

As a circuit simulator, PSpice has the following limitations:

1. The student version of PSpice is restricted to circuits with 10 transistors only. However, the professional DOS (or production) version can simulate a circuit with up to 200 bipolar transistors (or 150 MOSFETs).
2. The program is not interactive; that is, the circuit cannot be analyzed for various component values without editing the program statements.
3. PSpice does not support an iterative method of solution. If the elements of a circuit are specified, the output can be predicted. On the other hand, if the output is specified, PSpice cannot be used to synthesize the circuit elements.
4. The input impedance cannot be determined directly without running the graphic post-processor, Probe. The student version does not require a floating-point co-processor for running Probe, but the professional version does require such a co-processor.
5. The PC version needs 512 kilobytes of memory (RAM) to run.
6. Distortion analysis is not available in PSpice. SPICE2 allows distortion analysis, but it gives wrong answers.
7. The output impedance of a circuit cannot be printed or plotted directly.
8. The student version will run *with* or *without* the floating-point co-processor (8087, 80287, or 80387). If the co-processor is present, the program will run at full speed; otherwise it will run 5 to 15 times slower. The professional version requires a co-processor; it is not optional.

REFERENCES

1. R. W. Jensen and M. D. Liberman, *IBM Electronic Circuit Analysis Program and Applications*. Englewood Cliffs, N.J.: Prentice Hall, 1968.
2. R. W. Jensen and L. P. McNamee, *Handbook of Circuit Analysis Languages and Techniques*. Englewood Cliffs, N.J.: Prentice Hall, 1976.
3. *PSpice Manual*, Irvine, Calif.: MicroSim Corporation, 1992.

Circuit Descriptions

2-1 INTRODUCTION

SPICE is a general-purpose circuit program that can be applied to simulate and calculate the performance of electrical and electronic circuits. A circuit is described to a computer by using a file called the *circuit file*, which is normally typed in from a keyboard. The circuit file contains the circuit details of components and elements, the information about the sources, and the commands for what to calculate and what to provide as output. The circuit file is the input file to the SPICE (or PSpice) program, which, after executing the commands, produces the results in another file called the *output file*.

A circuit must be specified in terms of element names, element values, nodes, variable parameters, and sources. Consider the circuit in Fig. 2-1 that is to be simulated for calculating all node voltages and currents through R_2 and R_3. We shall show (1) how to describe this circuit to PSpice, (2) how to specify the type of analysis to be performed, and (3) how to define the required output variables. The description and analysis of a circuit require specifying the following:

Element values
Nodes
Circuit elements
Element models
Sources
Types of analysis
Output variables
PSpice output commands
Format of circuit files
Format of output files

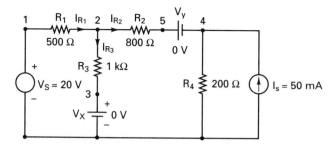

Figure 2-1 A dc circuit.

2-2 ELEMENT VALUES

The element values are written in standard floating-point notation with optional scale and units suffixes. Some values without suffixes that are allowable in PSpice are

$$5 \qquad 5. \qquad 5.0 \qquad 5E+3 \qquad 5.0E+3 \qquad 5.E3$$

There are two types of suffixes: the scale suffix and the units suffix. The scale suffix multiplies the number that it follows. The scale suffixes recognized by PSpice are

$$F = 1E{-}15$$
$$P = 1E{-}12$$
$$N = 1E{-}9$$
$$U = 1E{-}6$$
$$MIL = 25.4E{-}6$$
$$M = 1E{-}3$$
$$K = 1E3$$
$$MEG = 1E6$$
$$G = 1E9$$
$$T = 1E12$$

The unit suffixes that are normally used are

$$V = \text{volt}$$
$$A = \text{amp}$$
$$HZ = \text{hertz}$$
$$OHM = \text{ohm } (\Omega)$$

$$H = henry$$

$$F = farad$$

$$DEG = degree$$

The first suffix is always the scale suffix, and the unit suffix follows the scale suffix. In the absence of a scale suffix, the first suffix may be a units suffix, provided it is not the symbol of a scale suffix. The units suffixes are always ignored by PSpice. If the value of an inductor is 15 μH it is written as 15U or 15UH. In the absence of scale and units suffixes, the units of voltage, current, frequency, inductance, capacitance, and angle are, by default, volts, amps, hertz, henrys, farads, and degrees, respectively. PSpice ignores any units suffix, so the following values are equivalent:

| 25E–3 | 25.0E–3 | 25M | 25MA | 25MV | 25MOHM | 25MH |

Notes
1. The scale suffixes are all uppercase, but PSpice allows lowercase.
2. M means ''milli,'' not ''mega.'' 2 MΩ is written as 2MEG or 2MEGOHM.

2-3 NODES

The location of an element is identified by the node numbers. Each element is connected between two nodes. Node numbers are assigned to the circuit in Fig. 2-1. Node 0 is predefined as the ground. All nodes must be connected to at least two elements and should, therefore, appear at least twice. Node numbers must be integers from 0 to 9999 for SPICE2, but need not be sequential. PSpice allows any alphanumeric string up to 131 characters long. The node names shown in Table 2-1 are reserved and cannot be used.

The node numbers to which an element is connected are specified after the name of the element. All nodes must have a dc path to the ground node. This condition, which is not always satisfied in some circuits, is normally met by connecting very large resistors; it is discussed in Sec. 11-10.

2-4 CIRCUIT ELEMENTS

Circuit elements are identified by names. A name must start with a letter symbol corresponding to the element, but after that it can contain either letters or numbers. Names can be up to 8 characters long for SPICE2, and up to 131 characters long for PSpice. However, names longer than 8 characters are not normally necessary and not recommended. Table 2-2 shows the first letters of elements and sources. For example, the name of a resistor must start with R, an independent current source with I, and an independent voltage source with V, respectively.

TABLE 2-1. RESERVED NODE NAMES

Reserved node names	Value	Description
0	0 volts	Analog ground
$D_HI	1	Digital high level
$D_LO	0	Digital low level
$D_X	X	Digital unknown level

The format for describing passive elements is

⟨element name⟩ ⟨positive node⟩ ⟨negative node⟩ ⟨value⟩

where positive current is assumed to flow into positive node N+ and out of negative node N−. If the nodes are interchanged, the direction of the current through the element will be reversed. The formats for passive elements are described in Chapters 4, 5, and 6. The active elements are described in Chapters 7, 8, and 9.

The values of some circuit elements depend on other parameters, such as the resistance as a function of temperature, and the capacitance as a function of voltage. Models may be used to assign values to the various parameters of circuit elements. The techniques for specifying models of resistors are described in Sections 3-2 and 3-3, and those of inductors and capacitors in Section 4-2.

TABLE 2-2 SYMBOLS OF CIRCUIT ELEMENTS AND SOURCES

First letter	Circuit elements and sources
B	GaAs MES field-effect transistor
C	Capacitor
D	Diode
E	Voltage-controlled voltage source
F	Current-controlled current source
G	Voltage-controlled current source
H	Current-controlled voltage source
I	Independent current source
J	Junction field-effect transistor
K	Mutual inductors (transformer)
L	Inductor
M	MOS field-effect transistor
Q	Bipolar junction transistor
R	Resistor
S	Voltage-controlled switch
T	Transmission line
V	Independent voltage source
W	Current-controlled switch

Note: Voltage-controlled switch, current-controlled switch, and GaAs are not available in SPICE2, but they are available in SPICE3.

The passive elements in Fig. 2-1 are described as follows:

- The statement that R_1 has a value of 500 Ω and is connected between nodes 1 and 2 is

  R1   1   2   500

- The statement that R_2 has a value of 800 Ω and is connected between nodes 2 and 5 is

  R2   2   5   800

- The statement that R_3 has a value of 1 kΩ and is connected between nodes 2 and 3 is

  R3   2   3   1K

- The statement that R_4 has a value of 200 Ω and is connected between nodes 4 and 0 is

  R4   4   0   200

2-5 SOURCES

Voltage (or current) sources can be dependent or independent. The letter symbols for the names of sources are also listed in Table 2-2. The format for sources is

⟨source name⟩ ⟨positive node⟩ ⟨negative node⟩ ⟨source model⟩

where the voltage of node N+ is specified with respect to node N−. The positive current is assumed to flow out of the source (from positive) node N+ through the circuit to negative node N−. If the nodes are interchanged, the polarity of the source will be reversed. The order of nodes N+ and N− is important.

An independent voltage (or current) source can be dc, sinusoidal, pulse, exponential, polynomial, piecewise linear, or single-frequency frequency modulation. The techniques for specifying models of sources are described in Sections 3-5, 3-6, and 4-3.

The model for a simple dc source is

DC ⟨value⟩

- For $V_S = 20$ V and assuming node 1 is a higher potential with respect to node 0, the statement to specify the voltage source V_S that is connected between nodes 1 and 0 is

  VS   1   0   DC   20V

- For $I_S = 50$ mA, and assuming I_S flows from node 0 to node 4, the statement to specify the current source I_S that is connected between nodes 0 and 4 is

  IS   0   4   DC   50MA

2-6 TYPES OF ANALYSIS

SPICE and PSpice (hereafter called SPICE/PSpice) can perform various analyses, discussed in Section 1-4. Each analysis is invoked by including its command statement. Since Fig. 2-1 is a dc circuit, we are concerned with dc analysis only. Whenever a circuit file is run, SPICE/PSpice always calculates the dc bias point, which consists of all node voltages and the currents through all voltage sources.

The details of all node voltages as well as the current and power dissipation of all voltage sources can be sent to the output file by the .OP command (discussed in Section 3-9.1), whose format is

```
. OP
```

2-7 OUTPUT VARIABLES

SPICE/PSpice has some unique features in printing or plotting output voltages or currents. The various types of output variables that are permitted by PSpice are discussed in Sections 3-7, 4-5, and 5-2. The voltage of node 4 with respect to node 0 is specified by V(4,0) or V(4), and the voltage of node 2 with respect to node 3 is specified by V(2,3).

SPICE/PSpice can give the current of a voltage source as an output. A dc dummy voltage source of 0 V (say, $V_Z = 0$ V) is normally added and used as an ammeter to measure the current of that source, e.g., I(V_Z). Voltage sources V_X and V_Y in Fig. 2-1 act as ammeters and measure the currents of resistors R_2 and R_3, respectively. The statements for V_X and V_Y are

```
VX  3  0  DC  0V   ; Measures current through R3
VY  5  4  DC  0V   ; Measures current through R2
```

2-8 PSpice OUTPUT COMMANDS

The most common forms of output are print tables and plots, and they require output commands. However, with the .OP command, SPICE/PSpice automatically directs all node voltages and the current and power dissipation of all voltage sources to the output file, and therefore does not require any output command.

2-9 FORMAT OF CIRCUIT FILES

A circuit file that can be read by SPICE/PSpice may be divided into five parts: (1) the title, which describes the type of circuit or any comments; (2) the circuit description, which defines the circuit elements and the set of model parameters; (3) the analysis description, which defines the type of analysis; (4) the output description, which defines the way the output is to be presented; and (5) the end of the program (the .END command). The format for a circuit file is as follows:

```
Title
Circuit description
Analysis description
Output description
.END    (end-of-file statement)
```

Notes

1. The first line is the title line, and it may contain any type of text.
2. The last line must be the .END command.
3. The order of the remaining lines is not important and does not affect the results of simulations.
4. If a PSpice statement is more than one line, the statement can continue on the next line. A continuation line is identified by a plus sign (+) in the first column of the next line. The continuation lines must follow one another in the proper order.
5. A comment line may be included anywhere, preceded by an asterisk (*). Within a statement, a comment is preceded by a semicolon (;), for PSpice only.
6. The number of blanks between items is not significant (except in the title line). The tabs and commas are equivalent to blanks. For example, " " and " " and "," and " , " are all equivalent.
7. PSpice statements or comments can be in either upper- or lowercase.
8. SPICE2 statements must be in uppercase only. *It is advisable to type PSpice statements in uppercase, so that the same circuit file can also be run on SPICE2.*
9. If you are not sure of any command or statement, the best thing is to run the circuit file by using that command or statement and see what happens. SPICE/PSpice is user-friendly software; it gives an error message in the output file that identifies a problem.
10. In electrical circuits, subscripts are normally assigned to symbols for voltages, currents, and circuit elements. However, in SPICE the symbols are represented without subscripts. For example, V_s, I_s, and R_1 are represented by VS, IS, and R1, respectively. As a result, the SPICE symbols in a circuit description of voltages, currents, and circuit elements are often different from the normal circuit symbols.

2-10 FORMAT OF OUTPUT FILES

The results of simulation by SPICE/PSpice are stored in an output file. It is possible to control the type and amount of output by various commands. If there is any error in the circuit file, SPICE/PSpice will display a message on the screen indicating that there is an error and will suggest looking at the output file for details. The output falls into four types:

1. A description of the circuit itself that includes the net list, the device list, the model parameter list, and so on.

2. Direct output from some of the analyses without the .PLOT and .PRINT commands. This includes the output from .OP, .TF, .SENS, .NOISE, and .FOUR analyses.

3. Prints and plots by .PLOT and .PRINT commands. These include the output from the .DC, .AC, and .TRAN analyses.

4. Run statistics. These include the various kinds of summary information about the whole run, including times required by various analyses and the amount of memory used.

2-11 EXAMPLES OF SPICE SIMULATIONS

We have discussed all the details of describing the circuit of Fig. 2-1 as a SPICE/PSpice input file. We will illustrate the PSpice simulations by two examples.

Example 2-1

The circuit of Fig. 2-1 is to be simulated on PSpice to calculate and print all node voltages and the current and power of all voltage sources (V_S, V_X, and V_Y). The circuit file is to be stored in file EX2-1.CIR, and the outputs are to be stored in file EX2-1.OUT.

Solution   The circuit file contains the following statements:

Example 2-1   Simple dc circuit

```
▲ VS   1   0   DC   20V   ; Dc voltage source of 10 V
  IS   0   4   DC   50MA  ; Dc current source of 50 mA
▲▲ R1  1   2   500        ; Resistance of 500 ohms
   R2   2   5   800        ; Resistance of 800 ohms
   R3   2   3   1KOHM      ; Resistance of 1 kohms
   R4   4   0   200        ; Resistance of 200 ohms
   VX   3   0   DC   0V    ; Measures current through R3
   VY   5   4   DC   0V    ; Measures current through R2
▲▲▲ .OP                   ; Directs the bias point to the output file
.END                      ; End of circuit file
```

Performing a simulation with PSpice requires that you know how to install the PSpice software, create a circuit file, and run the circuit file. Running PSpice on PCs is described in Appendix A. Later versions of PSpice are menu-driven and don't require typing separate run commands. PSpice has a built-in text editor. The circuit file Ex2-1.CIR can be typed in from the File menu and then can be run from the Analysis menu. The menus of the PSpice control shell are shown in Fig. 2-2.

If the PSpice programs are loaded on a fixed disk and the circuit file is stored on a floppy diskette on drive A, the general command to run the circuit file is

```
PSPICE   a:⟨input file⟩ a:⟨output file⟩
```

For the input file EX2-1.CIR and the output file EX2-1.OUT, the command is

```
PSPICE a:EX2-1.CIR   a:EX2-1.OUT
```

If the output file name is omitted, the results are stored by default in an output file that has the same name as the input file and is on the same drive, but has the extension .OUT.

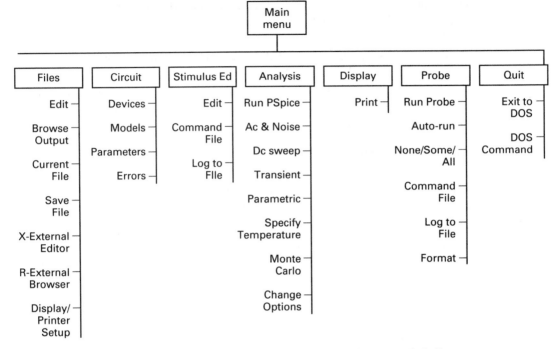

Figure 2-2 The menus of the PSpice control shell.

It is a good practice to use .CIR and .OUT extensions with all circuit files, so that a circuit file and its corresponding output file can be identified. Thus the command can simply be

```
PSPICE a:EX2-1.CIR
```

The output file EX2-1.OUT contains the results, which can be seen by using the Browse menu of PSpice. The results that appear in the output file EX2-1.OUT are shown below.

```
 ****      SMALL-SIGNAL BIAS SOLUTION        TEMPERATURE =    27.000 DEG C
 NODE    VOLTAGE      NODE    VOLTAGE      NODE    VOLTAGE      NODE     VOLTAGE
 (    1)   20.0000   (    2)   12.5000   (    3)    0.0000   (    4)    10.5000
 (    5)   10.5000
         VOLTAGE SOURCE CURRENTS
         NAME          CURRENT
         VS            -1.500E-02                       IR1 = 15 mA

         VX             1.250E-02                       IR3 = 12.5 mA

         VY             2.500E-03                       IR2 = 2.5 mA
         TOTAL POWER DISSIPATION    3.00E-01   WATTS
              JOB CONCLUDED
              TOTAL JOB TIME           1.04
```

Note. Since the type of analysis is not specified, the .OP command will not change the information about the bias point on the output file. Students are encouraged to run the circuit file EX2-1.CIR with and without the .OP command, and compare the output files.

Example 2-2

The circuit of Fig. 2-1 is to be simulated on PSpice to calculate and print the voltage at node 4, the current I_{R_2} and I_{R_3} for $V_S = 10$ V, 20 V, and 30 V. The circuit file is to be stored in the file EX2-2.CIR, and the outputs are to be stored in the file EX2-2.OUT.
Solution The circuit file is similar to that of Example 2-1, except that the statements for the analysis and the output are different. Since the dc voltage V_S is swept from 10 V to 30 V with a 10 V increment, the dc analysis (discussed in Section 3-9.3) can be invoked by the .DC command, whose format is

```
.DC  SWNAME  SSTART  SEND  SINC
```

where SWNAME = the sweep variable name, either voltage or current source
SSTART = the sweep start value
SEND = the sweep end value
SINC = the sweep increment value

For SSTART = 10 V, SEND = 30 V, and SINC = 10 V, the statement for the dc sweep is

```
.DC  VS  10V  30V  10V
```

The output variables for the dc analysis are discussed in Section 3-7, and the .PRINT command is discussed in Section 3-8.1. The statement to print V(4), I(VX), and I(VY) for the results of the .DC sweep is

```
.PRINT  DC  V(4)  I(VX)  I(VY)
```

If the .DC and .PRINT statements are included in the circuit file EX2-1.CIR, we will get the circuit file for EX2-2.CIR. The circuit file contains the following statements:

Example 2-2   Dc sweep of VS
```
▲ VS  1  0  DC   20V    ; Dc voltage source of 10 V
  IS  0  4  DC   50MA   ; Dc current source of 50 mA
▲▲ R1  1  2  500        ; Resistance of 500 ohms
   R2  2  5  800        ; Resistance of 800 ohms
   R3  2  3  1KOHM      ; Resistance of 1 kohms
   R4  4  0  200        ; Resistance of 200 ohms
   VX  3  0  DC   0V    ; Measures current through R3
   VY  5  4  DC   0V    ; Measures current through R2
▲▲▲ *  Dc sweep of VS from 10 V to 30 V in 10 V increment
     .DC   VS   10V   30V   10V
     .PRINT  DC   V(4)   I(VX)   I(VY) ; Prints the results of dc sweep
 .END                          ; End of circuit file
```

The output file EX2-2.OUT contains the results, which can be seen by using the Browse menu of PSpice. The results that appear in the output file EX2-2.OUT are shown below.

```
****      DC TRANSFER CURVES              TEMPERATURE =  27.000 DEG C
 VS           V(4)        I(VX)         I(VY)
  1.000E+01   9.500E+00   7.500E-03   -2.500E-03
  2.000E+01   1.050E+01   1.250E-02    2.500E-03
  3.000E+01   1.150E+01   1.750E-02    7.500E-03
            JOB CONCLUDED
            TOTAL JOB TIME          1.21
```

Notes
1. For the dc sweep in Example 2-2, PSpice calculates the dc bias point with $V_S = 20$ V, but will not direct the details of the bias point to the output file unless the .OP command is included.
2. Students are encouraged to run the circuit file EX2-2.CIR with and without the .OP command, and compare the output files.

Example 2-3

The circuit of Fig. 2-1 is to be simulated on PSpice to calculate and print the voltage at node 4, the current I_{R_2} and I_{R_3} for $V_S = 5$ V, 20 V, and 30 V for each value of $I_S = 50$ mA, 100 mA, and 150 mA. The current I_{R_2} is to be plotted. The circuit file is to be stored in file EX2-3.CIR and the outputs are to be stored in file EX2-3.OUT. The results should also be available for display and hard copy by the .PROBE command.

Solution   The circuit file is similar to that of Example 2-1, but the current source I_S has three values, and a plot of I_{R_2} versus V_S is required. For $V_S = 5$ V, 20 V, and 30 V, V_S cannot be swept linearly with an increment. We can simply list the values by using the key word LIST (discussed in Section 3-9.3). The format for the dc sweep with a list of values is

```
.DC  SWNAME  LIST ⟨value⟩*
```

where SWNAME = the sweep variable name, either voltage or current source.
        LIST = a key word
    ⟨value⟩* = list of sweep values

For $V_S = 5$ V, 20 V, and 30 V, the statement for the dc list sweep becomes

```
.DC  VS  LIST  5V  20V  30V
```

Since I_S has three values, there will be three sets of results for every value of V_S, for a total of nine sets. I_S can be regarded as the nested sweep within the sweep of V_S and can be swept from 50 mA to 150 mA with a 50 mA increment. The statement (discussed in Section 3-9.3) for the nested dc sweeps of V_S and I_S is

```
.DC  VS  LIST  5V  20V  30V  IS  50MA  150MA  50MA
```

which sweeps the current I_S linearly within the list sweep of V_S.

The command for output in the form of plots is .PLOT. The statement for the plots of I(VY) from the results of the .DC sweep is

```
.PLOT   DC   I(VY)
```

The outputs of .PRINT and .PLOT commands are stored in an output file created automatically by PSpice.

Probe is a *graphical waveform analyzer* for PSpice. The statement to invoke Probe is

```
.PROBE
```

This command makes the results of simulation available for graphical outputs on the display and on the hard copy. After executing the .PROBE command, Probe puts a menu on the screen that allows you to obtain graphical output. It is very easy to use Probe. The .PRINT command gives a table of data, and .PLOT generates the plot on the output file, while the .PROBE command provides graphical output on the monitor screen that can be dumped directly into a plotter and/or a printer. The .PLOT and .PRINT commands could generate a large amount of data in the output file and should be avoided if graphical output is available, as it is in PSpice with .PROBE. With the .PROBE command, there is *no* need for the .PLOT and/or .PRINT commands.

The circuit file contains the following statements:

Example 2-3   Dc sweep of VS and IS

```
▲ VS  1  0  DC   20V      ; Dc voltage source of 10 V
  IS  0  4  DC   50MA     ; Dc current source of 50 mA
▲▲ R1  1  2  500          ; Resistance of 500 ohms
  R2  2  5  800           ; Resistance of 800 ohms
  R3  2  3  1KOHM         ; Resistance of 1 kohms
  R4  4  0  200           ; Resistance of 200 ohms
  VX  3  0  DC   0V       ; Measures current through R3
  VY  5  4  DC   0V       ; Measures current through R2
▲▲▲ *  Dc sweep of VS from 10 V to 50 V in 20 V increment and
   *  Dc sweep of IS from 0 to 100 mA in 50 mA increment
  .DC    VS   LIST  5V   20V   30V   IS  50MA  150MA  50MA
  .PRINT DC  V(4)  I(VX)  I(VY) ; Prints the results of dc sweep
  .PLOT   DC  I(VY)             ; Plots I(VY) on the output file
  .PROBE                       ; Graphical waveform analyzer
.END                          ; End of circuit file
```

The output file EX2-3.OUT contains the results, which can be seen by using the Browse menu of PSpice. The results that appear in the output file EX2-3.OUT are shown below.

```
****       DC TRANSFER CURVES              TEMPERATURE =  27.000 DEG C
 VS           V(4)        I(VX)        I(VY)
  5.000E+00   9.000E+00   5.000E-03   -5.000E-03
  2.000E+01   1.050E+01   1.250E-02    2.500E-03
  3.000E+01   1.150E+01   1.750E-02    7.500E-03
```

5.000E+00	1.750E+01	7.500E-03	-1.250E-02
2.000E+01	1.900E+01	1.500E-02	-5.000E-03
3.000E+01	2.000E+01	2.000E-02	3.000E-12
5.000E+00	2.600E+01	1.000E-02	-2.000E-02
2.000E+01	2.750E+01	1.750E-02	-1.250E-02
3.000E+01	2.850E+01	2.250E-02	-7.500E-03

The plots of the dc sweep I_{R_2} versus V_S that are displayed on the monitor by .PROBE command and are dumped directly into a plotter are shown in Fig. 2-3. The plots of the dc sweep I_{R_2} versus V_S that are stored in the output file EX2-3.OUT by the .PLOT command and are printed on a printer (by pressing Print-Screen and then typing TYPE EX2-3.OUT), are shown in Fig. 2-4.

Note. Notice the quality of plots obtained with the .Probe command compared to those obtained with the .PLOT command. The .PLOT statement generates graphical plots in the output file. If the .PROBE command is included, there is no need for the .PLOT command.

2-12 GRAPHICAL INPUT FILES

The main disadvantage of the circuit (or input) file is that one has to draw the circuit and generate the net list of components and devices. The PSpice Design Center has a *Schematics editor* for circuit drawings that uses part symbols to represent devices and wire symbols for connections. The menus for the Schematics editor are shown in Fig. 2-5. The circuit diagram is saved with a .SCH extension and is called the *Schematics input file*. One can set up simulation parameters, run PSpice, and run Probe from the Schematics menu. One can also generate a circuit file directly from the circuit diagrams and simulation specifications. For many circuit simulations, working from a circuit diagram is much more convenient than typing in a circuit file. One could also interface *OrCADS*'s DRAFT Schematics editor to PSpice for generating the NETLIST.

For simulations requiring advanced commands and techniques, the generation of the circuit file is often convenient and necessary. One must often work from both the circuit file and the Schematics file, depending upon the applications and the types of simulation problems. However, the student version of PSpice allows only twenty components, so it is not complex enough for classroom examples in circuits and electronics courses.

2-13 PSpice ON THE MACINTOSH II

PSpice has a student version that runs on the Macintosh II. PSpice commands and features for IBM PCs are applicable to the Mac II. A circuit file developed for the IBM PC will run on the Mac II. The PSpice program for the Mac II comes on three floppy diskettes, and the steps for installing the PSpice program follow:

1. Place the first diskette into the floppy drive. Double click on the application **Install PSpice Part I.**

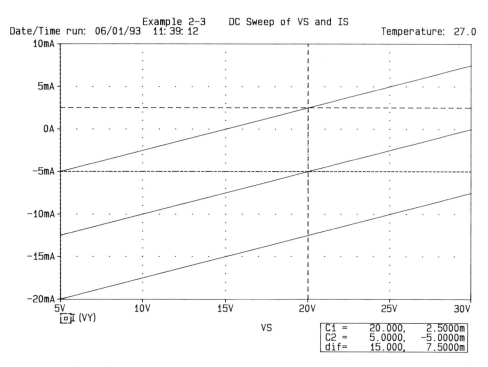

Figure 2-3 Plots of dc sweeps obtained by the .PROBE command.

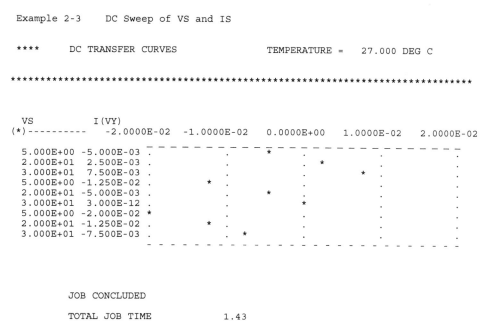

Figure 2-4 Plots of dc sweeps obtained by the .PLOT command.

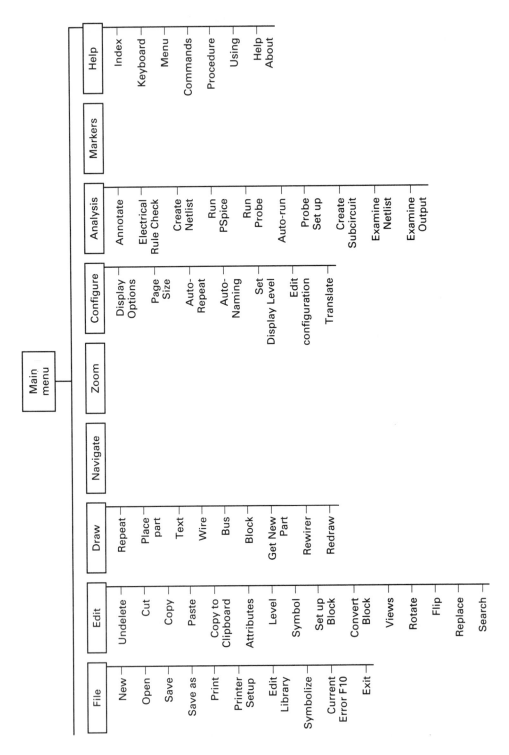

Figure 2-5 Menus for the Schematics editor of PSpice Design Center.

Circuit Descriptions   Chap. 2

2. Choose **Create folder** from the options.
3. Create a folder with a name, say **PSpice51,** and save it.
4. Repeat steps (1) to (3) and create folders **PSpice52** and **PSpice 53.**
5. Open the folder **PSpice52,** and drag all the files to **PSpice51.**
6. Open the folder **PSpice53,** and drag all the files to **PSpice51.**
7. Drag the folders **PSpice52** and **PSpice53** to the trash.
8. The following text files contain more information about PSpice:

   ReadMe.Word or ReadMe.Text
   RunningPSpice.Word or RunningPSpice.Text
   RunningProbe.Word or RunningProbe.Text
   RunningParts.Word or RunningParts.Text

To run a circuit file, follow these steps:

1. Double click on the PSpice icon.
2. Select **Open** from the **File Menu.**
3. Choose an input file (e.g., EX2-1.CIR).
4. Choose an output file name; the default is the circuit file with an .OUT extension.

To abort PSpice, follow these steps:

1. Click on the close box.
2. Select **Quit** from the **File Menu.**

REFERENCES

1. M. H. Rashid, *SPICE for Power Electronics and Electric Power.* Englewood Cliffs, N.J.: Prentice Hall, 1993.
2. Paul W. Tuinenga, *SPICE: A Guide to Circuit Simulation and Analysis Using PSPICE.* Englewood Cliffs, N.J.: Prentice Hall, 1992.
3. Herbert W. Jackson, *Introduction to Electric Circuits.* Englewood Cliffs, N.J.: Prentice Hall, 1986.

PROBLEMS

2-1. The circuit of Fig. P2-1 is to be simulated on PSpice to calculate and print **(a)** all node voltages, **(b)** the current and power dissipation of voltage sources V_A and V_B, and **(c)** the currents through all resistors.

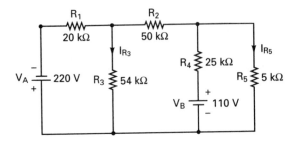

Figure P2-1

2-2. Repeat Problem 2-1 for V_A = 100 V, 120 V, and 150 V.

2-3. Repeat Problem 2-1 for V_A = 100 V, 120 V, and 150 V, and for V_B = 80 V, 110 V, and 140 V.

2-4. The circuit of Fig. P2-4 is to be simulated on PSpice to calculate and print **(a)** all node voltages, **(b)** the current and power dissipation of voltage sources V_A and V_B, and **(c)** the currents through all resistors.

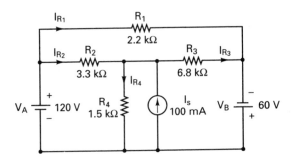

Figure P2-4

2-5. Repeat Problem 2-4 for V_A = 100 V, 120 V, and 140 V.

2-6. Repeat Problem 2-4 for V_A = 100 V, 120 V, and 140 V, and for V_B = 40 V, 60 V, and 80 V.

2-7. Repeat Problem 2-4 for V_A = 100 V, 120 V, and 140 V, for V_B = 40 V, 60 V, and 80 V, and I_S = 50 mA, 100 mA, and 150 mA.

Circuit Descriptions   Chap. 2

DC Circuit Analysis

3-1 INTRODUCTION

Sources in dc circuits are constant voltages or currents; that is, voltages and currents are invariant with time. Sources of this type are referred to as *direct current* or *dc* sources. In a circuit analysis course, dc circuits are covered first in order to introduce the basic circuit laws and techniques, for example, Kirchhoff's voltage and current laws, node-voltage and mesh-current methods, and Thévenin and Norton equivalents. Students can simulate dc circuits in SPICE to verify and reinforce their theoretical knowledge and to study the effects of parameter variation on voltages and currents in dc circuits. The simulation of dc circuits with passive elements requires the modeling of

Resistors
Model parameters
Operating temperature
Dc sources
Dc output variables
Type of dc analysis

We shall illustrate the SPICE simulations of dc circuits by examples. Students are encouraged to apply the basic circuit laws and to verify the SPICE results by hand calculation.

3-2 RESISTORS

The voltage and current relationships of resistors are shown in Fig. 3-1. The symbol for a resistor is *R*. The name of a resistor must start with *R*, and it takes

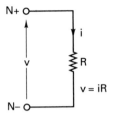

Figure 3-1 Voltage and current relationships.

the general form of

```
R (name) N+ N- RNAME RVALUE
```

A resistor does not have a polarity, and the order of the nodes does not matter. However, by defining N+ as the positive node and N− as the negative node, the current is assumed to flow from node N+ through the resistor to node N−. RNAME is the model name that defines the parameters of the resistor and is discussed in Section 3-3. RVALUE is the nominal value of the resistance.

Note. Some versions of PSpice or SPICE do not recognize the polarity of resistors and do not allow reference to currents through resistors, for example, $I(R_1)$.

The model parameters are shown in Table 3-1. It should be noted that the model parameter R is a resistance multiplier, rather than the value of the resistance. If RNAME is omitted, RVALUE is the resistance in ohms and can be positive or negative but *must* not be zero.

If RNAME is included and TCE is specified, the resistance as a function of temperature is calculated from

$$RES = RVALUE * R * 1.01^{TCE * (T - T0)}$$

If RNAME is included and TCE is not specified, the resistance as a function of temperature is calculated from

$$RES = RVALUE * R * [1 + TC1 * (T - T0) + TC2 * (T - T0)^2]$$

where T and T0 are the operating temperature and the room temperature, respectively, in degrees Celsius.

TABLE 3-1 MODEL PARAMETERS FOR RESISTORS

Name	Meaning	Units	Default
R	Resistance multiplier		1
TC1	Linear temperature coefficient	$°C^{-1}$	0
TC2	Quadratic temperature coefficient	$°C^{-2}$	0
TCE	Exponential temperature coefficient	$\%\ °C$	0

Some resistor statements

```
R1        6    5     10K
RLOAD    12   11     ARES   2MEG
.MODEL   ARES  RES   (R=1   TC1=0.02   TC2=0.005)
RINPUT   15   14     BRES   5K
.MODEL   BRES  RES   (R=0.8 TCE=2.5)
```

3-3 MODELING OF ELEMENTS

The models are necessary to take into account the parameter variations; that is, the value of a resistor depends on the operating temperature. A model that specifies a set of parameters for an element is specified in PSpice by the .MODEL command. The .(dot) is an integral part of the command. The same model can be used by one or more elements in the same circuit. The general form of the model statement is

```
.MODEL   MNAME   TYPE   (P1=A1 P2=A2 P3=A2   . . .   PN=AN)
```

MNAME is the name of the model and must start with a letter. Although not necessary, it is advisable to make the first letter the symbol of the element (e.g., R for resistor, L for inductor). The list of symbols for elements is shown in Table 2-1.

$P_1, P_2, \ldots$ are the element parameters (Table 3-1 for resistors) and $A_1, A_2, \ldots$ are their values, respectively. TYPE is the type name of the elements and must have the correct type, as shown in Table 3-2. An element must have the correct model type name. That is, a resistor must have the type name RES, not

TABLE 3-2 TYPE NAMES OF ELEMENTS

Type name	Element
RES	Resistor
CAP	Capacitor
D	Diode
IND	Inductor
NPN	NPN bipolar junction transistor
PNP	PNP bipolar junction transistor
NJF	N-channel junction FET
PJF	P-channel junction FET
NMOS	N-channel MOSFET
PMOS	P-channel MOSFET
GASFET	N-channel GaAs MOSFET
VSWITCH	Voltage-controlled switch
ISWITCH	Current-controlled switch
CORE	Nonlinear magnetic core (transformer)

the type IND or CAP. However, there can be more than one model of the same type in a circuit with different model names.

Some Model Statements

```
.MODEL  RMOD     RES   (R=1.1  TCE=0.001)
.MODEL  RLOAD    RES   (R=1    TC1=0.02  TC2=0.005)
.MODEL  CPASS    CAP   (C=1    VC1=0.01  VC2=0.002  TC1=0.02  TC2=0.005)
.MODEL  LFILTER  IND   (L=1    IL1=0.1   IL2=0.002  TC1=0.02  TC2=0.005)
.MODEL  DNOM     D     (IS=1E-6)
.MODEL  QOUT     NPN   (BF=50  IS=1E-9)
```

Note that the model parameter R is a resistance multiplier and scales the actual resistance value, RVALUE. $R = 1.1$ means that RVALUE is multiplied by 1.1, *not* that RVALUE is 1.1 Ω.

3-4 OPERATING TEMPERATURE

The operating temperature of an analysis can be set to any desired value by the .TEMP command. The general form of the statement is

```
.TEMP   ⟨(one or more temperature) values⟩
```

The temperatures are in degrees Celsius. If more than one temperature is specified, then the analysis is performed for each temperature.

The model parameters are assumed to be measured at a nominal temperature, which, by default, is 27°C. The default nominal temperature of 27°C can be changed by the TNOM option in the .OPTIONS statements that are discussed in Section 6-10.

Some Temperature Statements

```
.TEMP   50
.TEMP   25   50
.TEMP   0    25   50   100
```

3-5 INDEPENDENT DC SOURCES

The independent sources can be time-invariant or time-variant. They can be currents or voltages, as shown in Fig. 3-2.

3-5.1 Independent Dc Voltage Source

The symbol for an independent voltage source is V, and the general form is

```
V⟨name⟩ N+  N-  [DC ⟨value⟩]
```

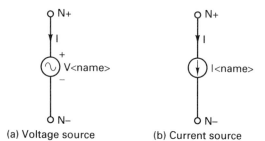

(a) Voltage source   (b) Current source

Figure 3-2 Voltage and current sources.

N+ is the positive node, and N− is the negative node, as shown in Fig. 3-2(a). Positive current flows from node N+ through the voltage source to the negative node N−. The voltage source need not be grounded.

The source is set to the dc value in dc analysis. A voltage source may be used as an *ammeter* in PSpice by inserting a zero-valued voltage source into the circuit for the purpose of measuring current. Since a zero-valued source behaves as a short circuit, there will be no effect on circuit operation.

Typical Statements

```
V1     15  0   6V      ; Dc voltage of 6 V, by default
V2     15  0   DC  6V ; Dc voltage of 6 V
```

3-5.2 Independent Dc Current Source

The symbol of an independent current source is *I*, and the general form is

```
I⟨name⟩ N+  N−  [DC ⟨value⟩]
```

N+ is the positive node, and N− is the negative node, as shown in Fig. 3-2(b). Positive current flows from node N+ through the current source to the negative node N−. The current source needs not be grounded. The source specifications are similar to those of independent voltage sources.

Typical Statements

```
I1     15  0   2.5MA      ; By default, dc current of 2.5 mA
I2     15  0   DC  2.5MA ; Dc current of 2.5 mA
```

3-6 DEPENDENT SOURCES

The four types of dependent sources that follow are shown in Fig. 3-4. They are

> Voltage-controlled voltage source
> Voltage-controlled current source

Current-controlled current source

Current-controlled voltage source

These sources can have either a fixed value or a polynomial expression.

3-6.1 Polynomial Source

The symbol for a polynomial or nonlinear source is POLY(n), where n is the number of dimensions of the polynomial. The default value of n is 1. The dimensions depend on the number of controlling sources. The general form is

```
POLY (n) ((controlling) nodes) ((coefficients) values)
```

The output sources or the controlling sources can be voltages or currents. For voltage-controlled sources, the number of controlling nodes must be twice the number of dimensions. For current-controlled sources, the number of controlling sources must be equal to the number of dimensions. The number of dimensions and the number of coefficients are arbitrary.

Let us call A, B, and C the three controlling variables, and Y the output source. Fig. 3-3 shows a source that is controlled by A, B, and C. The output source takes the form of

$$Y = f(A, B, C, \ldots)$$

where

Y can be a voltage or current

A, B, and C can be a voltage or current or any combination

For a polynomial of $n = 1$ with A as the only controlling variable, the source function takes the form of

$$Y = P_0 + P_1A + P_2A^2 + P_3A^3 + P_4A^4 + \cdots + P_nA^n$$

where $P_0, P_1, \ldots, P_n$ are the coefficient values; this is written in PSpice as

```
POLY  NC1+   NC1−   P₀   P₁  P₂   P₃   P₄   P₅ ... Pn
```

where NC1+ and NC1− are the positive and negative nodes, respectively, of controlling source A.

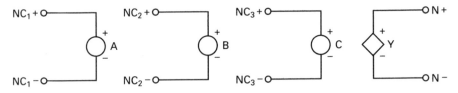

Figure 3-3 Polynomial source.

For a polynomial of degree $n = 2$ with A and B as the controlling sources, the source function takes the form of

$$Y = P_0 + P_1A + P_2B + P_3A^2 + P_4AB + P_5B^2 + P_6A^3$$
$$+ P_7A^2B + P_8AB^2 + P_9B^3 + \cdots$$

This is described in PSpice as

```
POLY(2) NC1+ NC1- NC2+ NC2-  P₀  P₁  P₂  P₃  P₄  P₅ ... Pₙ
```

where NC1+, NC2+ and NC1−, NC2− are the positive and negative nodes, respectively, of the controlling sources.

For a polynomial of degree $n = 3$ with A, B, and C as the controlling sources, the source function takes the form of

$$Y = P_0 + P_1A + P_2B + P_3C + P_4A^2 + P_5AB + P_6AC + P_7B^2 + P_8BC + P_9C^2$$
$$+ P_{10}A^3 + P_{11}A^2B + P_{12}A^2C + P_{13}AB^2 + P_{14}ABC + P_{15}AC^2 + P_{16}B^3$$
$$+ P_{17}B^2C + P_{18}BC^2 + P_{19}C^3 + P_{20}A^4 + \cdots$$

This is written in PSpice as

```
POLY(3) NC1+ NC1- NC2+ NC2- NC3+ NC3-  P₀  P₁  P₂  P₃  P₄  P₅ ... Pₙ
```

where NC1+, NC2+, NC3+ and NC1−, NC2−, NC3− are the positive and negative nodes, respectively, of the controlling sources.

Typical model statements.   For $Y = 2 V(10)$, the model is

```
POLY    10  0   2.0
```

For $Y = V(5) + 2[V(5)]^2 + 3[V(5)]^3 + 4[V(5)]^4$, the model is

```
POLY    5  0   0.0  1.0  2.0  3.0  4.0
```

For $Y = 0.5 + V(3) + 2V(5) + 3[V(3)]^2 + 4V(3) V(5)$, the model is

```
POLY(2)  3  0  5  0  0.5  1.0  2.0  3.0  4.0
```

For $Y = V(3) + 2V(5) + 3V(10) + 4[V(3)]^2$, the model is

```
POLY(3)  3  0  5  0  10  0  0.0  1.0  2.0  3.0  4.0
```

If $I(VN)$ is the controlling current through voltage source V_N and $Y = I(VN) + 2[I(VN)]^2 + 3[I(VN)]^3 + 4[I(VN)]^4$, the model is

```
POLY    VN  0.0  1.0  2.0  3.0  4.0
```

If $I(VN)$ and $I(VX)$ are the controlling currents and $Y = I(VN) + 2I(VX) + 3[I(VN)]^2 + 4I(VN)I(VX)$.

```
POLY(2)  VN  VX  0.0  1.0  2.0  3.0  4.0
```

Note. If the source is of one dimension and only one coefficient is specified, as in the first example, PSpice assumes $P_0 = 0$ and the specified value as P_1. That is, $Y = 2A$.

3-6.2 Voltage-Controlled Voltage Source

The symbol of a voltage-controlled voltage source in Fig. 3-4(a) is E; it takes a linear form, as in

```
E⟨name⟩ N+ N−  NC+  NC−  ⟨⟨voltage gain⟩ value⟩
```

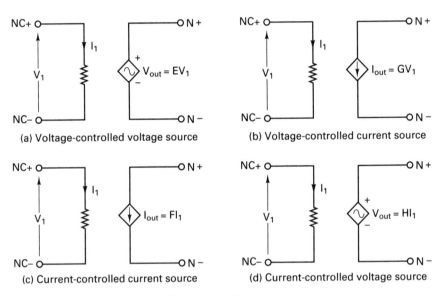

(a) Voltage-controlled voltage source

(b) Voltage-controlled current source

(c) Current-controlled current source

(d) Current-controlled voltage source

Figure 3-4 Dependent sources.

N+ and N− are the positive and negative output nodes, respectively, and NC+ and NC− are the positive and negative nodes, respectively, of the controlling voltage.

The nonlinear form is

```
E⟨name⟩  N+ N−  [POLY(⟨value⟩)
+            ⟨⟨⟨+ controlling⟩ node⟩ ⟨⟨− controlling⟩ node⟩⟩ (pairs)
+            ⟨⟨polynomial coefficients⟩ values⟩]
```

The POLY source was described in Section 3-6.1. The number of controlling nodes is twice the number of dimensions. A particular node may appear more than once, and the output and controlling nodes could be the same.

Typical Statements

```
EAB       1  2   4   6    1.0
EVOLT     4  7   20  22   2E5
ENONLIN   25 40  POLY(2) 3  0  5  0  0.0  1.0  1.5  1.2  1.7
E2        10 12  POLY  5  0  0.0  1.0  1.5  1.2  1.7
```

Notes

1. The source ENONLIN that specifies a polynomial voltage source between nodes 25 and 40 is controlled by V(3), and V(5). Its value is given by

$$Y = V(3) + 1.5V(5) + 1.2[V(3)]^2 + 1.7V(3)V(5)$$

The source E2 that specifies a polynomial voltage source between nodes 10 and 12 is controlled by V(5, 0) and is given by

$$Y = V(5) + 1.5[V(5)]^2 + 1.2[V(5)]^3 + 1.7[V(5)]^4$$

3-6.3 Voltage-Controlled Current Source

The symbol of a voltage-controlled current source, as shown in Fig. 3-4(b), is *G*; its linear form is

```
G⟨name⟩   N+ N−   NC+  NC−   ⟨(transconductance) value⟩
```

N+ and N− are the positive and negative output nodes, respectively. NC+ and NC− are the positive and negative nodes, respectively, of the controlling voltage. The nonlinear form is

```
G⟨name⟩   N+ N−   [POLY(⟨value⟩)
+              ⟨⟨(+ controlling) node⟩ ⟨(− controlling) node⟩⟩ (pairs)
+              ⟨(polynomial coefficients) values⟩]
```

Typical Statements

```
GAB       1  2   4   6    1.0
GVOLT     4  7   20  22   2E5
GNONLIN   25 40  POLY(2) 3  0  5  0  0.0  1.0  1.5  1.2  1.7
G2        10 12  POLY  5  0  0.0  1.0  1.5  1.2  1.7
```

Notes

1. The source GNONLIN that specifies a polynomial current source from node 25 to node 40 is controlled by V(3), and V(5) and is given by

$$I = V(3) + 1.5V(5) + 1.2[V(3)]^2 + 1.7V(3)V(5)$$

The source G2 that specifies a polynomial current source from node 10 to node 12 is controlled by V(5); it is given by

$$I = V(5) + 1.5[V(5)]^2 + 1.2[V(5)]^3 + 1.7[V(5)]^4$$

2. A nonlinear conductance can be simulated by a voltage-controlled current source. A linear voltage-controlled current source is the same as a conductance if the controlling nodes are the same as the output nodes.

```
GRES  4  6  4  6  0.1
```

is a conductance of 0.1 mhos with a resistance of $1/0.1 = 10\ \Omega$.

```
GHMO  1  2   POLY  1  2  0.0 1.5M  1.7M
```

represents

$$I = 1.5 \times 1^{-3}V(1,2) + 1.7 \times 10^{-3}[V(1,2)]^2$$

and is a nonlinear conductance in mhos.

3-6.4 Current-Controlled Current Source

The symbol of the current-controlled current source, as shown in Fig. 3-4(c), is F, and its linear form is

```
F⟨name⟩  N+ N−  VN  ⟨(current gain) value⟩
```

N+ and N− are the positive and negative nodes, respectively, of the current source. VN is a voltage source through which the controlling current flows. The controlling current is assumed to flow from the positive node of VN through the voltage source VN to the negative node of VN. The current through the controlling voltage source I(VN) determines the output current. The voltage source VN that monitors the controlling current must be an independent voltage source, and it can have a **zero** or finite value. If the current through a resistor controls the source, a dummy voltage source of 0 V should be connected in series with the resistor to monitor the controlling current.

The nonlinear form is

```
F⟨name⟩  N+ N−  [POLY(⟨value⟩)
+        VN1, VN2, VN3.....
+        ⟨(polynomial coefficients) values⟩]
```

The POLY source was described in Section 3-6.1. The number of controlling current sources must be equal to the number of dimensions.

Typical Statements

```
FAB      1   2  VIN  10
FAMP     13  4  VCC  50
FNONLIN  25 40 POLY  VN  0.0  1.0  1.5  1.2  1.7
```

Note. The source FNONLIN that specifies a polynomial current source from node 25 to node 40 is given by

$$I = I(VN) + 1.5[I(VN)]^2 + 1.2[I(VN)]^3 + 1.7[I(VN)]^4$$

3-6.5 Current-Controlled Voltage Source

The symbol of a current-controlled voltage source, as shown in Fig. 3-4(d), is *H*, and its linear form is

```
H⟨name⟩   N+ N−   VN   ⟨(transresistance) value⟩
```

N+ and N− are the positive and negative nodes, respectively, of the voltage source. VN is a voltage source through which the controlling current flows, and its specification is similar to that for a current-controlled current source.
    The nonlinear form is

```
H⟨name⟩   N+ N−   [POLY(⟨value⟩)
+         VN1, VN2, VN3 ...
+         ⟨(polynomial coefficients) values)]
```

Typical Statements

```
HAB       1  2  VIN   10
HAMP      13 4  VCC   50
HNONLIN   25 40 POLY  VN  0.0  1.0  1.5  1.2  1.7
```

Notes

1. The source HNONLIN that specifies a polynomial voltage source between nodes 25 and 40 is controlled by I(VN) and is given by

$$V = I(VN) + 1.5[I(VN)]^2 + 1.2[(VN)]^3 + 1.7[I(VN)]^4$$

2. A nonlinear resistance can be simulated by a current-controlled voltage source. A linear current-controlled voltage source is the same as a resistor if the controlling current is the same as the current through the voltage between output nodes.

```
HRES  4  6  VN  10
```

is a resistance of 10 Ω.

```
HMHO  1  2  POLY  VN  0.0  1.5M  1.7M
```

represents

$$H = 1.5 \times 1^{-3}I(VN) + 1.7 \times 10^{-3}[I(VN)]^2$$

and is a nonlinear resistance in ohms.

PSpice has some unique features for printing or plotting output voltages or currents. The output variables can be divided into two types: voltage output and current output. An output variable can be assigned the symbol of a device (or element) to identify whether the output is the voltage across the device or current through the device (or element). Table 3-3 shows the symbols of two-terminal elements. Table 3-4 shows the symbols and terminal symbols of three- or four-terminal devices.

TABLE 3-3 SYMBOLS FOR TWO-TERMINAL ELEMENTS

First letter	Element
C	Capacitor
D	Diode
E	Voltage-controlled voltage source
F	Current-controlled current source
G	Voltage-controlled current source
H	Current-controlled voltage source
I	Independent current source
L	Inductor
R	Resistor
V	Independent voltage source

TABLE 3-4 SYMBOLS AND TERMINAL SYMBOLS FOR THREE- OR FOUR-TERMINAL DEVICES

First letter	Device	Terminals
B	GaAs MESFET	D (Drain)
		G (Gate)
		S (Source)
J	JFET	D (Drain)
		G (Gate)
		S (Source)
M	MOSFET	D (Drain)
		G (Gate)
		S (Source)
		B (Bulk, substrate)
Q	BJT	C (Collector)
		B (Base)
		E (Emitter)
		S (Substrate)
T	Transmission line	A (Input port)
		B Output port

3-7.1 Voltage Output

For dc sweep and transient analysis (discussed in Chapter 4), the output voltages can be obtained by the following statements:

V($\langle$node$\rangle$)	Voltage at $\langle$node$\rangle$ with respect to ground
V(N1, N2)	Voltage at node N_1 with respect to node N_2
V($\langle$name$\rangle$)	Voltage across two-terminal device, $\langle$name$\rangle$
Vx($\langle$name$\rangle$)	Voltage at terminal x of three-terminal device, $\langle$name$\rangle$
Vxy($\langle$name$\rangle$)	Voltage across terminals x and y of three-terminal device, $\langle$name$\rangle$
Vz($\langle$name$\rangle$)	Voltage at port z of transmission line, $\langle$name$\rangle$

VARIABLES	MEANING
V(5)	Voltage at node 5 with respect to ground
V(4,2)	Voltage of node 4 with respect to node 2
V(R1)	Voltage of resistor R_1, where the first node (as defined in the circuit file) is positive with respect to the second node
V(L1)	Voltage of inductor L_1, where the first node (as defined in the circuit file) is positive with respect to the second node
V(C1)	Voltage of capacitor C_1, where the first node (as defined in the circuit file) is positive with respect to the second node
V(D1)	Voltage across diode D_1 where the anode is positive with respect to cathode
VC(Q3)	Voltage at the collector of transistor Q_3 with respect to ground
VDS(M6)	Drain-source voltage of MOSFET M_6
VB(T1)	Voltage at port B of transmission line T_1

Note. SPICE and some versions of PSpice do not permit measuring voltage across a resistor, an inductor, and a capacitor, for example, V(R1), V(L1), and V(C1). This type of statement is used only for outputs by .PLOT and .PRINT commands.

3-7.2 Current Output

For dc sweep and transient analysis (discussed in Chapter 4), the output currents can be obtained by the following statements:

I($\langle$name$\rangle$)	Current through $\langle$name$\rangle$
Ix($\langle$name$\rangle$)	Current into terminal x of $\langle$name$\rangle$
Iz($\langle$name$\rangle$)	Current at port z of transmission line, $\langle$name$\rangle$

VARIABLES	MEANING
I(VS)	Current flowing into dc source V_S
I(R5)	Current flowing into resistor R_5, where the current is assumed to flow from the first node (as defined in the circuit file) through R_5 to the second node
I(D1)	Current into diode D_1
IC(Q4)	Current into the collector of transistor Q_4
IG(J1)	Current into gate of JFET J_1
ID(M5)	Current into drain of MOSFET M_5
IA(T1)	Current at port A of transmission line T_1

Note. SPICE and some versions of PSpice do not permit measuring the current through a resistor, for example, I(R5). The easiest way is to add a dummy voltage source of 0 V (say, $V_X = 0$ V) and to measure the current through that source, for example, I(VX).

Example 3-1

For the bipolar junction transistor (BJT) circuit in Fig. 3-5, write the various currents and voltages in forms that are allowed by PSpice. The dc sources of 0 V are introduced to measure currents I_1 and I_2.

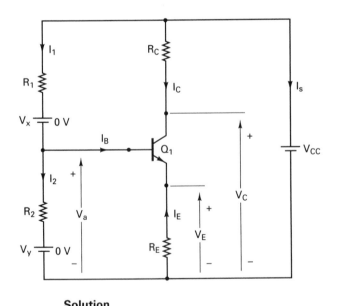

Figure 3-5 Bipolar junction transistor circuit.

Solution

PSpice	VARIABLES	
I_B	IB(Q1)	The base current of transistor Q_1
I_C	IC(Q1)	The collector current of transistor Q_1
I_E	IE(Q1)	The emitter current of transistor Q_1

PSpice	VARIABLES	
I_S	I(VCC)	The current through voltage source V_{CC}
I_1	I(VX)	The current through voltage source V_X
I_2	I(VY)	The current through voltage source V_Y
V_B	VB(Q1)	The voltage at the base of transistor Q_1
V_C	VC(Q1)	The voltage at the collector of transistor Q_1
V_E	VE(Q1)	The voltage at the emitter of transistor Q_1
V_{CE}	VCE(Q1)	The collector-emitter voltage of transistor Q_1
V_{BE}	VBE(Q1)	The base-emitter voltage of transistor Q_1

3-8 TYPES OF OUTPUT

The commands that are available to get output from the results of simulations are

```
.PRINT    Print
.PLOT     Plot
.PROBE    Probe
Probe Output
.WIDTH    Width
```

3-8.1 .PRINT (Print Statement)

The results from dc analysis can be obtained in the form of tables. The print statement for dc outputs takes the form

```
.PRINT DC [output variables]
```

The maximum number of output variables is eight in any .PRINT statement. However, more than one .PRINT statement can be used to print all the desired output variables.

The values of the output variables are printed as a table with each column corresponding to one output variable. The number of digits for output values can be changed by the NUMDGT option on the .OPTIONS statement (see Section 6-10). The results of the .PRINT statement are stored in the output file. An example of a print statement is

```
.PRINT DC V(2), V(3,5), V(R1), VCE(Q2), I(VIN), I(R1), IC(Q2)
```

Note. Having two .PRINT statements for the same variables will not produce two tables. PSpice will ignore the first statement and produce output for the second statement.

3-8.2 .PLOT (Plot Statement)

The results from dc analysis can also be obtained in the form of line printer plots. The plots are drawn by using characters, and the results can be obtained from any kind of printer. The plot statement for dc outputs takes the following form:

```
.PLOT DC ⟨output variables⟩
     + [⟨(lower limit) value⟩, ⟨(upper limit) value⟩]
```

The maximum number of output variables is eight in any .PLOT statement. More than one .PLOT statement can be used to plot all the desired output variables.

The range and increment of the x-axis is fixed by the dc analysis command. The range of the y-axis is set by adding [⟨(lower limit) value⟩, ⟨(upper limit) value⟩] at the end of a .PLOT statement. The y-axis range, [⟨(lower limit) value⟩, ⟨(upper limit) value⟩] can be placed in the middle of a set of output variables. The output variables will follow the specified range that comes immediately to the right.

If the y-axis range is omitted, PSpice assigns a default range determined by the range of the output variable. If the ranges of output variables vary widely, PSpice assigns the ranges corresponding to the different output variables.

Plot Statements

```
.PLOT DC V(2), V(3,5), V(R1), VCE(Q2), I(VIN), I(R1), IC(Q2)
.PLOT DC V(5) V(4,7) (0,10V) IB(Q1) (0, 50MA) IC(Q1) (−50MA, 50MA)
```

Notes. In the first statement, the y-axis is by default. In the second statement, the range for voltages V(5) and V(4,7) is 0 V to 10 V, that for current IB(Q1) is 0 MA to 50 MA, and that for the current IC(Q1) is −50 MA to 50 MA.

3-8.3 .PROBE (Probe Statement)

Probe is a graphics post-processor/waveform analyzer for PSpice and is available as an option for the professional version of PSpice. However, Probe comes with the student version of PSpice. The simulation results cannot be used directly by Probe. First, the results have to be processed by the .PROBE command, which writes the processed data on a file, PROBE.DAT, for use by Probe. The command takes one of these forms:

```
.PROBE
.PROBE ⟨one or more output variables⟩
```

In the first form, where no output variable is specified, the .PROBE command writes all the node voltages and all the element currents into the PROBE.DAT file. The element currents are written in the forms that are permitted as output variables (discussed in Section 3-7).

In the second form, where the output variables are specified, PSpice writes only the specified output variables to the PROBE.DAT file. This form is suitable for users without a fixed disk to limit the size of the PROBE.DAT file.

Probe Statements

```
.PROBE
.PROBE V(5), V(4,3), V(C1), VM(2), I(R2), IB(Q1), VBE(Q1)
```

3-8.4 Probe Output

Once the results of the simulations are processed by the .PROBE command, the results are available for graphical displays and can be further manipulated through expressions. Probe comes with a first menu, as shown in Fig. 3-6, that allows one to choose the type of analysis. After the first choice, the second level is the choice for the plots and coordinates of output variables, as shown in Fig. 3-7. After the

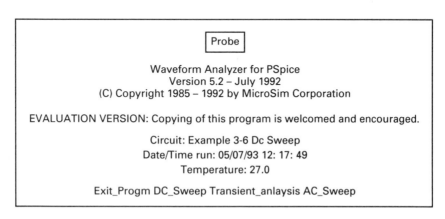

Figure 3-6 Select analysis menu for Probe.

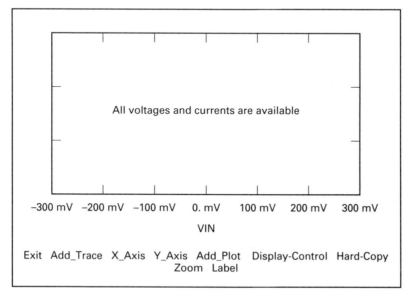

Figure 3-7 Select plot/graphics output.

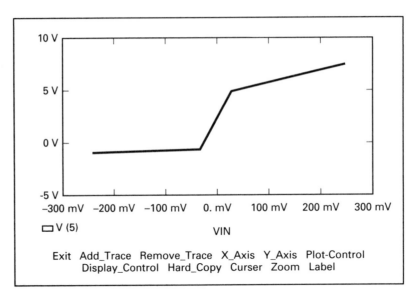

Figure 3-8 Output display.

choices are made, the output is displayed as shown in Fig. 3-8. It is very easy to use Probe.

Probe disregards case for letters; that is, V(8) and v(8) are equivalent. However, there is an exception with the scale suffix *m*, and one should be careful about "milli" and "mega"; *m* means "milli" (1E−3), but *M* means "mega" (1E+6). The suffixes MEG and MIL are not acceptable. The following units are recognized by Probe:

V	Volts
A	Amps
W	Watts
d	Degrees (of phase)
s	Seconds
H	Hertz

If units are omitted, Probe recognizes that W = V ∗ A, V = W/A, and A = W/V, and gives the results in appropriate units.

The arithmetic expressions of output variables use the following functions and the operators +, −, ∗, /, along with parentheses.

PROBE FUNCTION	MEANING
ABS(x)	$|x|$ (absolute value)
B(Kxy)	Flux density of coupled inductor *Kxy*
H(Kxy)	Magnetization of coupled inductor *Kxy*

PROBE FUNCTION	MEANING
SGN(x)	+1 (if $x > 0$), 0 (if $x = 0$), −1 (if $x < 0$)
EXP(x)	e^x
DB(x)	$20 \log(|x|)$ (log of base 10)
LOG(x)	$\ln(x)$ (log of base e)
LOG10(x)	$\log(x)$ (log of base 10)
PWR(x,y)	$|x|^y$
SQRT(x)	$x^{1/2}$
SIN(x)	$\sin(x)$ (x in radians)
COS(x)	$\cos(x)$ (x in radians)
TAN(x)	$\tan(x)$ (x in radians)
ARCTAN(x)	$\tan^{-1}(x)$ (result in radians)
d(y)	Derivative of y with respect to the x-axis variable
s(y)	Integral of y over the x-axis variable
AVG(x)	Running average of x
RMS(x)	Running RMS of x

If a dc voltage source VS is connected between nodes 2 and 3, the plot of the trace (or curve)

```
V(2,3) * I(VS)
```

will give the power delivered by the source VS.

Two or more traces can be added to the same plot by typing the variables, for example

```
V(8)  dV(8)  s(V(8))  AVG(V(8))  RMS(V(8))  V(2,3) * I(VS)
```

This is faster than adding one plot at a time.

If the analysis is done for more than one parameter or sweep variable, the PROBE.DAT file can contain all the results. If V(8) is calculated for three parameter values, the expression

```
V(8))
```

will give three curves instead of the usual one curve. One could specify a particular trace by the expression

```
V(8)@n
```

which will give the curve of V(8) for the nth transient analysis. The difference between mth curve and nth curve can be obtained by the expression

```
v(8)@m−V(8)@n
```

Notes

1. The .PROBE command requires a math co-processor for the professional version of PSpice, but not for the student version.

2. Probe is not available on SPICE. However, the newest version of SPICE (SPICE3) has a post-processor similar to Probe called Nutmeg.

3. It is required that the type of display and the type of hard-copy devices in the PROBE.DEV file be specified as follows:

```
Display = ⟨display name⟩
Hard copy = ⟨port name⟩, ⟨device name⟩
```

The details of names for display, port, and device (printer) can be found in the README.DOC file that comes with the PSpice program or in the PSpice manual.

4. The display and hard-copy devices can be set from the Display/Printer Setup menu.

3-8.5 .WIDTH (Width Statement)

The width of the output in columns can be set by the .WIDTH statement, which has the general form of

```
.WIDTH OUT=⟨value⟩
```

The ⟨value⟩ is in columns and must be either 80 or 132. The default value is 80.

3-9 TYPES OF DC ANALYSIS

In dc analysis, all the independent and dependent sources are dc types. If inductors and capacitors are present in a circuit, they are considered as short circuits and open circuits, respectively, because at zero frequency, the impedance represented by an inductor is zero and that of a capacitor is infinite. The commands that are commonly used for dc analysis are

.OP	Dc operating point
.TF	Small-signal transfer function
.DC	Dc sweep

3-9.1 .OP (Operating Point)

Electronic and electrical circuits contain nonlinear devices (e.g., diodes, transistors), whose parameters depend on the *operating point*. The operating point is also known as a *bias point* or *quiescent point*.

Let us consider the circuit of Fig. 3-9(a), where the voltage across the

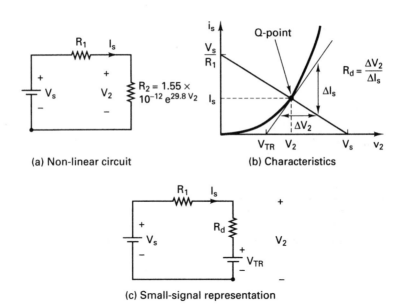

(a) Non-linear circuit

(b) Characteristics

(c) Small-signal representation

Figure 3-9 Dc circuit with a nonlinear resistance.

resistor R_2 depends upon its current. Using Kirchhoff's voltage law (KVL), the current I_S can be expressed as

$$V_S = R_1 I_S + V_2$$

or

$$I_S = -\frac{V_2}{R_1} + \frac{V_S}{R_1} \qquad (3-1)$$

which is the equation of a straight line.

Let us assume that the current I_S is related to the voltage V_2 by

$$I_S = 1.55 \times 10^{-12} e^{29.8V_2} \quad \text{for} \quad V_2 > 0 \qquad (3-2)$$

I_S depends on V_2, which in turn depends on I_S. The intersection of the two curves defined by Eqs. (3-1) and (3-2) gives the operating point or quiescent point (Q-point) as shown in Fig. 3-9(b). If V_S changes by a small amount, then the operating point will also change. The slope of the tangent at the Q-point will be a measure of this change, and will represent a small-signal resistance R_d, so that $R_d = \Delta V_2/\Delta I_S$. R_2 can be represented by a small-signal resistance R_d and a resistance that corresponds to the fixed threshold voltage V_{TR}. This type of representation is known as a *piecewise linear model* and is shown in Fig. 3-9(c).

The operating point is always calculated by PSpice in calculating the small-signal parameters of nonlinear devices during the dc sweep and transfer-function analysis. The command takes the form

.OP

The .OP command controls the output of the bias point but not the method of bias analysis. If the .OP command is omitted, PSpice prints only a list of the node voltages.

If the .OP command is present, PSpice prints the currents and power dissipations of all the voltage sources. The small-signal parameters of all nonlinear controlled sources and all the semiconductor devices are also printed.

The .OP command has no effect in circuits with linear elements. However, in circuits with nonlinear devices (e.g., diodes, transistors), the .OP command prints the small-signal model parameters of the devices.

Example 3-2

Repeat Example 2-1 with the values of R_1 and R_2 increased by $+5\%$ and those of R_3 and R_4 decreased by 10%.

Solution Since there are two types of resistances, we will use two models: RMOD1 with $R = 1.05$ and RMOD2 with $R = 0.9$.

The listing of the circuit file follows.

Example 3-2 Simple dc circuit

```
▲ VS   1   0   DC   20V      ; Dc voltage source of 20 V
  IS   0   4   DC   50MA     ; Dc current source of 2 mA
▲▲ R1   1   2     RMOD1  500      ; Resistance of 500 ohms with model RMOD1
   R2   2   5     RMOD1  800      ; Resistance of 800 ohms with model RMOD1
   .MODEL   RMOD1   RES (R = 1.05) ; Model for R1 and R2
   R3   2   3     RMOD2  1KOHM    ; Resistance of 1 kilohms with model RMOD2
   R4   4   0     RMOD2  200      ; Resistance of 200 ohms with model RMOD2
   .MODEL   RMOD2   RES (R = .9)  ; Model for R1 and R2
   VX   3   0   DC    0V     ; Measures current through R3
   VY   5   4   DC    0V     ; Measures current through R2
▲▲▲ *.OP                     ; Prints small-signal model parameters
.END                          ; End of circuit file
```

The results that are obtained by printing the contents of output file EX3-2.OUT follow.

```
****      SMALL-SIGNAL BIAS SOLUTION       TEMPERATURE =   27.000 DEG C
NODE    VOLTAGE     NODE   VOLTAGE     NODE    VOLTAGE     NODE    VOLTAGE
(    1)   20.0000  (    2)   11.7410  (    3)    0.0000  (    4)     9.4836
(    5)    9.4836
  VOLTAGE SOURCE CURRENTS
  NAME           CURRENT
  VS            -1.573E-02
  VX             1.305E-02
  VY             2.687E-03
  TOTAL POWER DISSIPATION   3.15E-01   WATTS
```

Note. Students are encouraged to run Example 3-2 with the .OP command in order to see its effect.

3-9.2 .TF (Small-Signal Transfer Function)

The small-signal transfer function capability of PSpice can be used to compute the small-signal dc gain, the input resistance, and the output resistance of a circuit. If V(1) and V(4) are the input and output variables, respectively, PSpice will calculate the small-signal dc gain between nodes 1 and 4, defined by

$$A_v = \frac{\Delta V_{\text{out}}}{\Delta V_{\text{in}}} = \frac{V(4)}{V(1)}$$

as well as the input resistance between nodes 1 and 0 and the small-signal dc output resistance between nodes 4 and 0.

PSpice calculates the small-signal dc transfer function by linearizing the circuit around the operating point. The statement for the transfer function has one of the following forms:

```
.TF VOUT VIN
.TF IOUT IIN
```

where VIN is the input voltage, and VOUT (or IOUT) is the output voltage (or current). If the output is a current, then that current must be through a voltage source. The output variable, VOUT or IOUT, has the same format and meaning as in a .PRINT statement. If there are inductors and capacitors in a circuit, the inductors are treated as short circuits and the capacitors as open circuits.

The .TF command calculates the parameters of Thévenin's (or Norton's) equivalent circuit for the circuit file. It automatically prints the output and does not require .PRINT, .PLOT, or .PROBE statements.

Statements for Transfer-Function Analysis

```
.TF  V(2,4)  VIN  ; VIN is the input, and V(2,4) is the output.
.TF  V(10)   IIN  ; IIN is the input, and V(10) is the output.
.TF  I(VX)   IIN  ; IIN is the input, and the current through VX is the output.
.TF  I(VX)   VIN  ; VIN is the input, and the current through VX is the output.
```

Example 3-3

A dc circuit is shown in Fig. 3-10. Use PSpice to calculate and print (a) the voltage gain $A_v = $ V(2,4)$/V_{\text{in}}$, (b) the input resistance $R_{\text{in}} = V_{\text{in}}/I_{\text{in}}$, (c) Thévenin's (output) resistance $R_{\text{out}} = R_{\text{Th}}$ between nodes 2 and 4, and (d) Thévenin's voltage V_{Th} between nodes 2 and 4.

Solution The output voltage V(2,4) is between nodes 2 and 4. The .TF command can calculate and print the dc gain, the input resistance, and the output resistance. The voltage source V_x in Fig. 3-9 acts as an ammeter, and an independent source of 0 V is normally connected to measure a current. The listing of the circuit file follows.

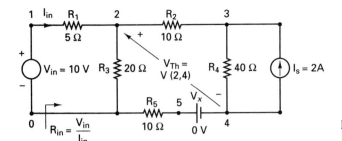

Figure 3-10 A dc circuit for determining Thévenin's equivalent.

Example 3-3 Thévenin's analysis

```
▲ VIN  1    0    DC    10V   ; Voltage source of 10 V dc
  IS   4    3    DC    2A    ; Current source of 2 A dc
▲▲ VX  4    5    DC    0V    ; Measures the current through R5
   R1  1    2    5
   R2  2    3    10
   R3  2    0    20
   R4  3    4    40
   R5  5    0    10
▲▲▲ .TF   V(2,4)   VIN      ; Transfer-function analysis
  .END                      ; End of circuit file
```

The results are obtained by printing the contents of output file EX3-3.OUT. PSpice always prints the small-signal bias solutions to the output file, and they are as follows:

```
****      SMALL-SIGNAL BIAS SOLUTION      TEMPERATURE =   27.000 DEG C
 NODE    VOLTAGE    NODE   VOLTAGE    NODE    VOLTAGE    NODE    VOLTAGE
(   1)    10.0000 (    2)   12.5000 (    3)   23.7500 (    4)  −11.2500
(   5)   −11.2500
        VOLTAGE SOURCE CURRENTS
        NAME            CURRENT
        VIN            5.000E−01
        VX            −1.125E+00
        TOTAL POWER DISSIPATION  −5.00E+00  WATTS
```

PSpice prints the results of .TF command to the output file, and they are as follows:

```
****     SMALL-SIGNAL CHARACTERISTICS
        V(2,4)/VIN =  6.250E−01              Av  = 0.625
        INPUT RESISTANCE AT VIN =  2.000E+01  Rin = 20 Ω
        OUTPUT RESISTANCE AT V(2,4) =  1.094E+01  RTh = 10.94 Ω
        JOB CONCLUDED
        TOTAL JOB TIME          1.05
```

Thus, Thévenin's voltage V_{Th} is $A_v V_{in} = 0.625 \times 10 = 6.25$ V.

Example 3-4

An amplifier circuit is shown in Fig. 3-11. Calculate and print (a) the voltage gain $A_v = V(5)/V_{in}$, (b) the input resistance, R_{in}, and (c) the output resistance, R_{out}. **Solution** The .TF command can calculate and print the dc gain, the input resistance, and the output resistance. The listing of the circuit file follows.

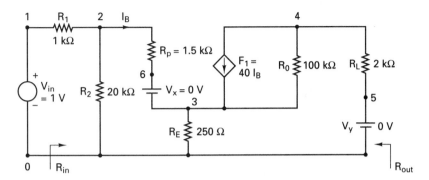

Figure 3-11 An amplifier circuit for determining Thévenin's equivalent.

Example 3-4 Transfer-function analysis

```
▲ VIN  1  0  DC   1V  ; Dc input voltage of 1 V
▲▲ R1  1  2  1K
   R2   2  0  20K
   RP   2  6  1.5K
   RE   3  0  250
   F1   4  3  VX    40  ; Current-controlled current source
   RO   4  3  100K
   RL   4  5  2K
   VX   6  3  DC   0V  ; Measures the current through Rp
   VY   5  0  DC   0V  ; Measures the current through RL
▲▲▲ .TF  V(4)  VIN    ; Transfer-function analysis
 .END                 ; End of circuit file
```

The results are obtained by printing the contents of output file EX3-4.OUT. PSpice always prints automatically the small-signal bias solutions to the output file, which are as follows:

```
****      SMALL-SIGNAL BIAS SOLUTION       TEMPERATURE =   27.000 DEG C
  NODE    VOLTAGE     NODE   VOLTAGE    NODE   VOLTAGE    NODE   VOLTAGE
(   1)    1.0000  (    2)    .8797  (    3)    .7653  (    4)   -5.9695
(   5)    0.0000  (    6)    .7653
      VOLTAGE SOURCE CURRENTS
      NAME           CURRENT
      VIN           -1.203E-04
      VX             7.630E-05
      VY            -2.985E-03
      TOTAL POWER DISSIPATION  1.20E-04  WATTS
```

The .TF command calculates the small-signal characteristics and sends the results automatically to the output file. The results of .TF command are as follows:

```
****    SMALL-SIGNAL CHARACTERISTICS
        V(4)/VIN = -5.969E+00                      A_v  = -5.969
        INPUT RESISTANCE AT VIN =  8.313E+03       R_in = 8.313 kΩ
        OUTPUT RESISTANCE AT V(4) =  1.992E+03     R_Th = 1.992 kV
            JOB CONCLUDED
            TOTAL JOB TIME              .99
```

Thus, Thévenin's voltage V_{Th} is $A_v V_{in} = -5.969 \times 1 = -5.969$ V.

3-9.3 .DC (Dc Sweep)

The dc sweep is also known as the *dc transfer characteristic*. The input variable is varied over a range of values. For each value of the input variable, the dc operating point and the small-signal dc gain are computed by calling the small-signal transfer function capability of PSpice. The dc sweep (or dc transfer characteristic) is obtained by repeating the calculations of the small-signal transfer function for a set of values. The statement for performing the dc sweep takes one of the following general forms:

```
.DC   LIN   SWNAME    SSTART    SEND    SINC
+               [(nested sweep specification)]
.DC   OCT   SWNAME    SSTART    SEND    NP
+               [(nested sweep specification)]
.DC   DEC   SWNAME    SSTART    SEND    NP
+               [(nested sweep specification)]
.DC   SWNAME     LIST ⟨value⟩
+               [(nested sweep specification)]
```

SWNAME is the sweep variable name and could be either a voltage or current source. SSTART, SEND, and SINC are the start value, the end value, and the increment value of the sweep variable, respectively. The sweep increment SINC must be positive; it must not be zero or negative. NP is the number of steps. LIN, OCT, or DEC specifies the type of sweep, as follows:

LIN (Linear sweep): SWNAME is swept linearly from SSTART to SEND. SINC is the step size. If LIN is omitted, SPICE assumes a linear sweep by default.

OCT (Sweep by octave): SWNAME is swept logarithmically by octave, and NP becomes the number of steps per octave. The next variable is generated by multiplying the present value by a constant larger than unity. OCT is used if the variable range is wide.

DEC (Sweep by decade): SWNAME is swept logarithmically by decade, and NP becomes the number of steps per decade. The next variable is generated by multiplying the present value by a constant larger than unity. DEC is used if the variable range is the widest.

LIST (*List of values*): There are no start and end values. The values of the sweep variables are listed after the keyword LIST.

The SWNAME can be one of the following types:

Source: The name of an independent voltage or current source. During the sweep, the source's voltage or current is set to the sweep value.

Model Parameter: The model name type and model name followed by a model parameter name in parentheses. The parameter in the model is set to the sweep value. The model parameters L and W for a MOS device and any temperature parameters such as TC1 and TC2 for the resistor *cannot* be swept.

Temperature: The keyword TEMP followed by the keyword LIST. The temperature is set to the sweep value. For each value of the sweep, the model parameters of all circuit components are updated to that temperature.

Global Parameter: The keyword PARAM followed by a parameter name. The parameter is sweeped. During the sweep, the global parameter's value is set to the sweep value, and all expressions are evaluated.

The dc sweep can be nested, similar to a DO loop within a DO loop in FORTRAN programming. The first sweep is the inner loop, and the second sweep is the outer loop. The first sweep is done for each value of the second sweep. The nested sweep specification follows the same rules as the main sweep variable.

PSpice does not print or plot any output by itself for the dc sweep; the results of dc sweep are obtained by .PRINT, .PLOT, or .PROBE statements. Probe allows nested sweeps to be displayed as a family of curves.

Statements for the Dc Sweep

```
.DC  VIN  -5V  10V  0.25V          ; Sweeps the voltage VIN linearly.
.DC  LIN  IIN  50MA  -50MA  1MA    ; Sweeps the current IIN linearly.
.DC  VA  0  15V  0.5V  IA  0  1MA  0.05MA ; Sweeps the current IA linearly within
                                     the linear sweep of VA.
.DC  RES  RMOD(R)  0.9  1.1  0.001 ; Sweeps linearly the model parameter R
                                     of the resistor model RMOD.
.DC  DEC  NPN  QM(IS)  1E-18  1E-14  10 ; Sweeps with a decade increment the
                                     parameter IS of the NPN transistor.
.DC  TEMP  LIST  0  50  80  100  150 ; Sweeps the temperature TEMP using the
                                     listed values.
.DC  PARAM  Vsupply  -15V  15V  0.5V ; Sweeps linearly the parameter PARAM
                                     Vsupply.
```

Notes
1. If the source has a dc value, its value is set by the sweep, overriding the dc value.
2. In the third statement, the current source IA is the inner loop, and the voltage source VA is the outer loop. PSpice will vary the value of the current source IA from 0 to 1 mA with an increment of 0.05 mA for each value of voltage source VA, and generate an entire print table or plot for each value of voltage sweep.

3. The sweep-start value SSTART may be greater than or less than the sweep-end value SEND.
4. The sweep increment SINC must be greater than zero.
5. The number of points NP must be greater than zero.
6. After the dc sweep is finished, the sweep variable is set back to the value it had before the sweep started.

Example 3-5

A dc circuit with controlled sources is shown in Fig. 3-12. Use PSpice to calculate all node voltages and branch currents, and Thévenin's equivalent circuit between nodes 2 and 5.

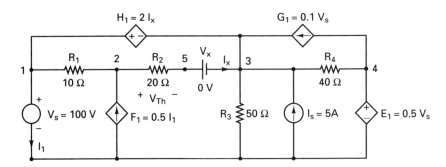

Figure 3-12 Dc circuit with controlled dc sources.

Solution We can use the .DC command to calculate all voltages and currents, and the .PRINT command to print the results of the dc analysis. A .TF command will calculate and print Thévenin's equivalent voltage and resistances along with all node voltages and currents through all independent voltage sources. The circuit file follows:

Example 3-5   Dc circuit with controlled dc sources

```
▲ VS  1  0  DC   100V  ; Voltage source of 100 V dc
  IS  0  3  DC   5A    ; Current source of 5 A dc
▲▲ R1 1  2  10
  R2  2  5  20
  R3  3  0  50
  R4  3  4  40
```

```
       VX  5  3   DC  0V      ; Measures current through R2
       E1  4  0   1   0   0.5 ; Voltage-controlled voltage source
       F1  0  2   VS      0.5 ; Current-controlled current source
       G1  4  3   1   0   0.1 ; Voltage-controlled current source
       H1  1  3   VX      2   ; Current-controlled voltage source
▲ ▲ ▲ .DC   VS   100V   100V   5V          ; Only one dc sweep value
      .PRINT  DC  I(R1)  I(R2)  I(R3)  I(R4)     ; Prints branch currents
      .PRINT  DC  V(1)   V(2)   V(3)   V(4)   V(5) ; Prints node voltages
       .TF  V(2,5)   VS           ; Transfer-function analysis
.END                              ; End of circuit file
```

Note that PSpice will sound a warning because there is only one item of data from dc sweep, and Probe can't plot the transfer characteristic. You can ignore this warning and look for the results on the output file.

PSpice always prints automatically the small-signal bias solutions to the output file, and they are as follows:

```
 ****       SMALL-SIGNAL BIAS SOLUTION        TEMPERATURE =   27.000 DEG C
 NODE     VOLTAGE     NODE    VOLTAGE     NODE    VOLTAGE    NODE    VOLTAGE
 (   1)   100.0000   (   2)  178.0400   (   3)   91.3280   (   4)  50.0000
 (   5)    91.3280
     VOLTAGE SOURCE CURRENTS
     NAME        CURRENT
     VS          2.428E+01
     VX          4.336E+00
     TOTAL POWER DISSIPATION  -2.43E+03   WATTS
```

The results of the dc sweep, which are obtained from the output file EX3-5.OUT, are shown below:

```
 ****       DC TRANSFER CURVES           TEMPERATURE =   27.000 DEG C
  VS          I(R1)       I(R2)       I(R3)       I(R4)
   1.000E+02  -7.804E+00   4.336E+00   1.827E+00   1.033E+00
  VS          V(1)        V(2)        V(3)        V(4)        V(5)
   1.000E+02   1.000E+02   1.780E+02   9.133E+01   5.000E+01   9.133E+01
```

The results of the .TF command, which are also obtained from the output file EX3-5.OUT, are shown below:

```
 ****       SMALL-SIGNAL CHARACTERISTICS
     V(2,5)/VS =   4.982E-01
     INPUT RESISTANCE AT VS =  -7.169E+00
     OUTPUT RESISTANCE AT V(2,5) =   5.240E+00
       JOB CONCLUDED
       TOTAL JOB TIME           1.27
```

Note. The .TF command calculates the small-signal characteristics and automatically sends the results to the output file. The dc sweep calculates the various voltages and currents for dc transfer curves, but printing them requires a PRINT statement.

Example 3-6

For the amplifier circuit of Fig. 3-11, calculate and plot the dc transfer characteristics, V_{out} versus V_{in}. The input voltage is varied from 0 to 1 V with an increment of 0.5 V. The resistance R_E changes by ±25%. Plot the results with the .PLOT command, print with the .PRINT command, and display them by using Probe.

Solution The .DC command will sweep the input voltage V_{in}. We can vary the resistance R_E by varying the model parameter R by ±25%.

The circuit file follows:

Example 3-6   Dc sweep

```
▲ VIN   1   0   DC    1V     ; Dc input voltage of 1 V
▲▲ R1   1   2   1K
   R2    2   0   20K
   RP    2   6   1.5K
   RE    3   0   RMOD  250   ; Resistance with model RMOD
   .MODEL  RMOD  RES (R = 1.0) ; Model statement for R_E
   F1    4   3   VX    40    ; Current-controlled current source
   RO    4   3   100K
   RL    4   5   2K
   VX    6   3   DC    0V    ; Measures the current through Rp
   VY    5   0   DC    0V    ; Measures the current through RL
▲▲▲ *  Dc sweep for VIN from 0 to 1 V with 0.5 V increment and
    *  use the listed values of parameter R in model RMOD
   .DC  VIN   0   1.5   0.5  RES  RMOD(R)  LIST   0.75  1.0  1.25
   .PRINT  DC  V(1)   V(4)      ; Prints a table on the output file
   .PLOT   DC   V(0,4)          ; Plots V(0,4) on the output file
   .PROBE                       ; Graphical waveform analyzer
.END                           ; End of circuit file
```

The transfer characteristics that are obtained by Probe are shown in Fig. 3-13. The plot produced by the .PLOT command is shown in Fig. 3-14. The results of the .PRINT command, which are obtained from the output file EX3-6.OUT, follow.

```
▲▲▲ ****     DC TRANSFER CURVES          TEMPERATURE =  27.000 DEG C
    VIN          V(1)          V(4)
    0.000E+00   0.000E+00   0.000E+00
    5.000E-01   5.000E-01   -3.736E+00
    1.000E+00   1.000E+00   -7.471E+00


    0.000E+00   0.000E+00   0.000E+00
    5.000E-01   5.000E-01   -2.985E+00
    1.000E+00   1.000E+00   -5.969E+00


    0.000E+00   0.000E+00   0.000E+00
    5.000E-01   5.000E-01   -2.485E+00
    1.000E+00   1.000E+00   -4.970E+00
           JOB CONCLUDED
           TOTAL JOB TIME          1.70
```

Note. The .PRINT command gives a table of data, and .PLOT generates the plot on the output file, while the .PROBE command gives graphical output on the monitor

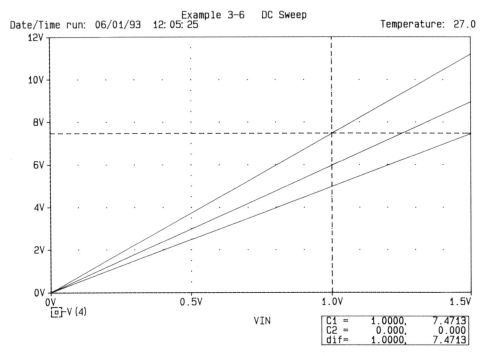

Figure 3-13   Dc transfer characteristics obtained by Probe for Example 3-6.

```
**** 06/01/93 12:16:38 ********** Evaluation PSpice (July 1992) *************

Example 3-6   DC Sweep


****     DC TRANSFER CURVES              TEMPERATURE =    27.000 DEG C


*****************************************************************************
```

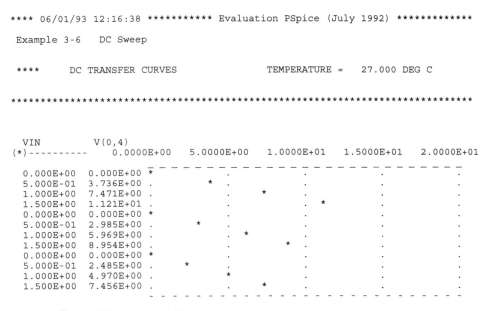

Figure 3-14   Dc transfer characteristics obtained by the .PLOT command for Example 3-6.

screen that can be dumped directly into a plotter and/or a printer. The .PLOT and .PRINT commands could generate a large amount of data in the output file and should be avoided if graphical output is available, as in PSpice, by .PROBE. With the .PROBE command, there is *no* need for the .PLOT and .PRINT commands.

SUMMARY

The symbol and statement for resistors are

R   Resistor
         R⟨name⟩ N+   N−   RNAME   RVALUE

The symbols and statements for dc sources are

E   Voltage-controlled voltage source
         E⟨name⟩ N+ N− NC+ NC− ⟨(voltage gain) value⟩
F   Current-controlled current source
         F⟨name⟩ N+ N− VN ⟨(current gain) value⟩
G   Voltage-controlled current source
         G⟨name⟩ N+ N− NC+ NC− ⟨(transconductance) value⟩
H   Current-controlled voltage source
         H⟨name⟩ N+ N− VN ⟨(transresistance) value⟩
I   Independent current source
         I⟨name⟩ N+ N− [dc ⟨value⟩]
V   Independent voltage source
         V⟨name⟩ N+ N− [dc ⟨value⟩]

The commands that are generally used for dc analysis are

.DC   Dc analysis
.END   End of circuit
.MODEL   Model
.OP   Operating point
.PLOT   Plot
.PRINT   Print
.PROBE   Probe
.TEMP   Temperature
.TF   Transfer function
.WIDTH   Width

REFERENCES

1. R. L. Boylestad, *Introductory Circuit Analysis*. New York: Macmillan, 1990.
2. A. E. Fitzerald, D. E. Higginbotham, and A. Grabel, *Basic Electrical Engineering*. New York: McGraw-Hill, 1981.
3. W. H. Hayt and J. E. Kemmerly, *Engineering Circuit Analysis*. New York: McGraw-Hill, 1990.
4. J. David Irwin, *Basic Engineering Circuit Analysis*. New York: Macmillan, 1989.
5. D. E. Johnson, J. Hilburn, and J. R. Johnson, *Basic Electric Circuit Analysis*. Englewood Cliffs, N.J.: Prentice Hall, 1990.
6. J. W. Nilson, *Electric Circuits*. Reading, Mass.: Addison-Wesley, 1990.

PROBLEMS

Write PSpice statements for problems 3-1 to 3-10. Assume that the first node is the positive terminal and the second node is the negative terminal.

3-1. A resistor R_1 is connected between nodes 3 and 4 and has a nominal value of $R = 10 \ k\Omega$. The operating temperature is 55°C, and it has the form

```
R₁ = R * [1 + 0.2 * (T − TO) + 0.002 * (T − TO)²]
```

3-2. A resistor R_1 is connected between nodes 3 and 4 and has a nominal value of $R = 10 \ k\Omega$. The operating temperature is 55°C, and it has the form

```
R₁ = R * 1.01⁴·⁵*(T−TO)
```

3-3. A polynomial voltage source Y that is connected between nodes 1 and 2 is controlled by a voltage source V_1 connected between nodes 4 and 5. The source is given by

$$Y = 0.1V_1 + 0.2V_1^2 + 0.05V_1^3$$

3-4. A polynomial current source I that is connected between nodes 1 and 2 is controlled by a voltage source V_1 connected between nodes 4 and 5. The source is given by

$$Y = 0.1V_1 + 0.2V_1^2 + 0.05V_1^3$$

3-5. A voltage source V_{out} that is connected between nodes 5 and 6 is controlled by a voltage source V_1 and has a voltage gain of 25. The controlling voltage is connected between nodes 10 and 12. The source is expressed as

$$V_{out} = 25V_1$$

3-6. A current source I_{out} that is connected between nodes 5 and 6 is controlled by a current source I_1 and has a current gain of 10. The voltage through which the controlling current flows is V_C. The current source is given by

$$I_{out} = 10I_1.$$

3-7. A current source I_{out} that is connected between nodes 5 and 6 is controlled by a voltage source V_1 between nodes 8 and 9. The transconductance is 0.05 mhos. The current source is given by

$$I_{out} = 0.05V_i$$

3-8. A voltage source V_{out} that is connected between nodes 5 and 6 is controlled by a current source I_1 and has a transresistance of 150 Ω. The voltage through which the controlling current flows is V_C. The voltage source is expressed as

$$V_{out} = 150I_1$$

3-9. A nonlinear resistance R that is connected between nodes 4 and 6 is controlled by a voltage source V_1 and has a resistance of the form

$$R = V_1 + 0.2V_1^2$$

3-10. A nonlinear transconductance G_m that is connected between nodes 4 and 6 is controlled by a current source. The voltage through which the controlling current flows is V_1. The transconductance has the form

$$G_m = V_1 + 0.2V_1^2$$

3-11. For the circuit in Fig. P3-11, calculate and print **(a)** the voltage gain, $A_v = V_{out}/V_{in}$; **(b)** the input resistance, R_{in}; and **(c)** the output resistance, R_{out}.

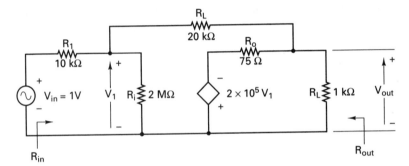

Figure P3-11

3-12. For the circuit in Fig. P3-11, calculate and plot the dc transfer characteristic V_{out} versus V_{in}. The input voltage is varied from 0 to 10 V with an increment of 0.5 V.

3-13. A dc circuit is shown in Fig. P3-13. Use PSpice to calculate and print Thévenin's equivalent circuit with respect to terminals a and b.

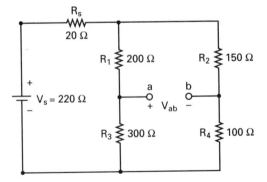

Figure P3-13

3-14. Use PSpice to calculate all node voltages and branch currents of Fig. P3-13. Assume an operating temperature of 55°C.

3-15. For Fig. P3-13, use PSpice to calculate and plot the dc transfer characteristics, V_{ab} versus V_S. The input voltage V_S is varied from 0 to 20 V with an increment of 5 V. The resistance R_1 changes by $\pm 20\%$. Print the results with the .PRINT command, and display the transfer characteristics with Probe.

3-16. Repeat Example 3-3 with a .TF command **(a)** if the input is I_s and the output is the voltage V(2,4) between nodes 2 and 4, **(b)** if the input is I_s and the output is the current through V_{in}, and **(c)** if the input is V_{in} and the output is the current through V_x.

3-17. Repeat Problem 3-13 for the circuit of Fig. P3-17.

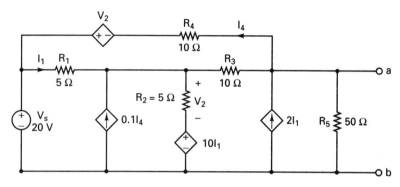

Figure P3-17

3-18. Repeat Problem 3-14 for the circuit of Fig. P3-17.

3-19. Repeat Problem 3-15 for the circuit of Fig. P3-17.

Transient Analysis

4-1 INTRODUCTION

A transient analysis deals with the behavior of an electric circuit as a function of time. If a circuit contains an energy storage element(s), a transient can also occur in a dc circuit after a sudden change due to switches opening or closing. SPICE allows simulating transient behaviors, by assigning initial conditions to circuit elements, generating sources, and the opening and closing of switches. The simulation of transients in circuits with linear elements requires modeling of

Resistors
Capacitors and inductors
Model parameters of elements
Operating temperature
Modeling of transient sources
Transient sources
Transient output variables
Output commands
Transient analysis

The SPICE simulations of transient circuits are illustrated by examples. Students are encouraged to apply the techniques for transient analysis of simple circuit laws and to verify the SPICE results by hand calculations. We have already discussed the statements for resistors in Section 3-2, assigning model parameters of elements in Section 3-3, and specifying the operating temperature in Section 3-4

The voltage and current relationships of inductors and capacitors are shown in Fig. 4-1.

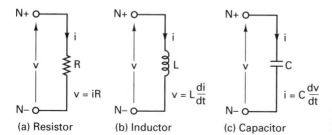

(a) Resistor   (b) Inductor   (c) Capacitor   **Figure 4-1**   Voltage and current relationships.

4-2.1 Capacitor

The symbol for a capacitor is C. The name of a capacitor must start with C, and it takes the following general form:

```
C⟨name⟩ N+ N- CNAME CVALUE IC=V0
```

$N+$ is the positive node and $N-$ is the negative node. The voltage of node $N+$ is assumed positive with respect to node $N-$, and the current flows from node $N+$ through the capacitor to node $N-$. CNAME is the model name, and CVALUE is the nominal value of the capacitor. IC defines the initial (time-zero) voltage of the capacitor, V_0.

The model parameters are shown in Table 4-1. The model command is discussed in Section 3-3. If CNAME is omitted, CVALUE is the capacitance in farads, and the CVALUE can be positive or negative but *must* not be zero. If CNAME is included, the capacitance that depends on voltage and temperature is calculated from

$$CAP = CVALUE * C * (1 + VC1 * V + VC2 * V^2)$$
$$* [1 + TC1 * (T - T0) + TC2 * (T - T0)^2]$$

TABLE 4-1   MODEL PARAMETERS FOR CAPACITORS

Name	Meaning	Units	Default
C	Capacitance multiplier		1
VC1	Linear voltage coefficient	Volt^{-1}	0
VC2	Quadratic voltage coefficient	Volt^{-2}	0
TC1	Linear temperature coefficient	°C^{-1}	0
TC2	Quadratic temperature coefficient	°C^{-2}	0

where T is the operating temperature in degrees Celsius, and T0 is the room temperature in degrees Celsius.

Some Capacitor Statements

```
C1        6     5      10UF
CLOAD     12    11     5PF     IC=2.5V
CINPUT    15    14     ACAP    10PF
C2        20    19     ACAP    20NF    IC=1.5V
.MODEL  ACAP  CAP  (C=1  VC1=0.01  VC2=0.002  TC1=0.02  TC2=0.005)
```

Notes
1. The model parameter C is a capacitance multiplier, rather than the value of the capacitance. It scales the actual capacitance value, CVALUE. Thus, $C = 1.1$ means that CVALUE is multiplied by 1.1, *not* that CVALUE is 1.1 F.
2. The initial condition (if any) applies only if the UIC (use initial condition) option is specified in the .TRAN command statement, which is described in Section 4-7.2.

4-2.2 Inductor

The symbol for an inductor is L. The name of an inductor must start with L, and it takes the following general form:

```
L⟨name⟩ N+ N- LNAME LVALUE IC=I0
```

N+ is the positive node, and N− is the negative node. The voltage of N+ is assumed positive with respect to node N−, and the current flows from node N+ through the inductor to node N−. LNAME is the model name, and LVALUE is the nominal value of the inductor. IC defines the initial (time-zero) current of the inductor, I_0.

The model parameters of an inductor are shown in Table 4-2. Refer to Section 3-3 for a discussion of the model statement. If LNAME is omitted, LVALUE is the inductance in henrys, and LVALUE can be positive or negative but *must* not be zero. If LNAME is included, the inductance that depends on the

TABLE 4-2 MODEL PARAMETERS FOR INDUCTORS

Name	Meaning	Units	Default
L	Inductance multiplier		1
IL1	Linear current coefficient	Amps^{-1}	0
IL2	Quadratic current coefficient	Amps^{-2}	0
TC1	Linear temperature coefficient	$°C^{-1}$	0
TC2	Quadratic temperature coefficient	$°C^{-2}$	0

current and temperature is calculated from

$$IND = LVALUE * L * (1 + IL1 * I + IL2 * I^2)$$
$$* [1 + TC1 * (T - T0) + TC2 * (T - T0)^2]$$

where T is the operating temperature in degrees Celsius, and T0 is the room temperature in degrees Celsius.

Notes
    1. The model parameter *L* is an inductance multiplier, rather than the value of the inductance. It scales the actual inductance value, LVALUE. L = 1.1 means that LVALUE is multiplied by 1.1, *not* that LVALUE is 1.1 F.
    2. The initial condition (if any) applies only if the UIC (use initial condition) option is specified on the .TRAN command statement, which is described in Section 4-7.2.

Some Inductor Statements

```
L1        6    5     10MH
LLOAD     12   11    5UH     IC=0.2MA
LLINE     15   14    LMOD    5MH
LCHOKE    20   19    LMOD    2UH    IC=0.5A
.MODEL    LMOD  IND  (L=1  IL1=0.1  IL2=0.002  TC1=0.02  TC2=0.005)
```

4.3 MODELING OF TRANSIENT SOURCES

PSpice allows the generation of dependent (or independent) voltage and current sources. Independent sources can be time-variant. A nonlinear source can also be simulated by a polynomial.

The independent voltage and current sources that can be modeled by PSpice follow:

    Exponential
    Pulse
    Piecewise linear
    Sinusoidal
    Single-frequency frequency modulation

Note. These sources are explained below as voltages; however, the explanations are equally applicable to currents.

4-3.1 Exponential Source

The waveform and parameters of an exponential waveform are shown in Fig. 4-2 and Table 4-3. The symbol of exponential sources is EXP, and the general form is

```
EXP (V1  V2  TRD  TRC  TFD  TFC)
```

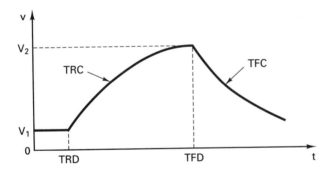

Figure 4-2 Exponential waveform.

V_1 and V_2 *must* be specified by the user and can be either voltages or currents. (TSTEP is the incrementing time during transient [.TRAN] analysis.) In an EXP waveform, the voltage remains V_1 for the first TRD seconds. Then the voltage rises exponentially from V_1 to V_2, with a rise-time constant of TRC. After a time of TFD, the voltage falls exponentially from V_2 to V_1, with a fall-time constant of TFC.

TABLE 4-3 MODEL PARAMETERS OF EXP SOURCES

Name	Meaning	Units	Default
V1	Initial voltage	Volts	None
V2	Pulsed voltage	Volts	None
TRD	Rise delay time	Seconds	0
TRC	Rise-time constant	Seconds	TSTEP
TFD	Fall delay time	Seconds	TRD + TSTEP
TFC	Fall-time constant	Seconds	TSTEP

Note. The values of an EXP waveform as well as the values of other time-dependent waveforms at intermediate time points are determined by PSpice by means of linear interpolation.

Typical model statements. For $V_1 = 0$, $V_2 = 1$ V, TRD = 2 ns, TRC = 20 ns, TFD = 60 ns, and TFC = 30 ns, the model statement is

```
EXP   (0   1   2NS   20NS   60NS   30NS)
```

With TRD = 0, the statement becomes

```
EXP   (0   1   0     20NS   60NS   30NS)
```

With $V_1 = -1$ V and $V_2 = 2$ V, it is

```
EXP   (-1   2 2NS   20NS   60NS   30NS)
```

4-3.2 Pulse Source

The waveform and parameters of a pulse waveform are shown in Fig. 4-3 and Table 4-4. The symbol of a pulse source is PULSE and the general form is

PULSE (V1   V2   TD   TR   TF   PW   PER)

V_1 and V_2 *must* be specified by the user and can be either voltages or currents. TSTEP and TSTOP are the incrementing time and stop time, respectively, during transient (.TRAN) analysis.

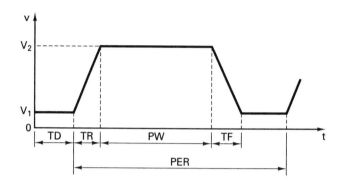

Figure 4-3   Pulse waveform.

TABLE 4-4   MODEL PARAMETERS OF
PULSE SOURCES

Name	Meaning	Units	Default
V1	Initial voltage	Volts	None
V2	Pulsed voltage	Volts	None
TD	Delay time	Seconds	0
TR	Rise time	Seconds	TSTEP
TF	Fall time	Seconds	TSTEP
PW	Pulse width	Seconds	TSTOP
PER	Period	Seconds	TSTOP

Typical statements.   For $V_1 = -1$ V, $V_2 = 1$ V, TD = 2 ns, TR = 2 ns, TF = 2 ns, PW = 50 ns, and PER = 100 ns, the model statement is

PULSE (−1   1   2NS   2NS   2NS   50NS   100NS)

With $V_1 = 0$, $V_2 = 1$ V, the model becomes

PULSE (0   1   2NS   2NS   2NS   50NS   100NS)

With $V_1 = 0$, $V_2 = -1$ V, the model becomes

PULSE (0   −1   2NS   2NS   2NS   50NS   100NS)

4-3.3 Piecewise Linear Source

A point in a waveform can be described by (T_i, V_i) or (T_i, I_i), and every pair of values (T_i, V_i) or (T_i, I_i) specifies the source value at time T_i. The voltage at times between the intermediate points is determined by PSpice by using linear interpolation. The symbol of a piecewise linear source is PWL, and the general form is

```
PWL (T1  V1  T2  V2 ... TN  VN)
```

The model parameters of a PWL waveform are given in Table 4-5.

TABLE 4-5 MODEL PARAMETERS OF
PWL SOURCES

Name	Meaning	Units	Default
T_i	Time at a point	Seconds	None
V_i	Voltage at a point	Volts	None

Typical statement. The model statement for the typical waveform in Fig. 4-4 is

```
PWL  (0  0  5  3  10US  3V  15US  6V  40US  6V  45US  2V  60US  2V  65US  0)
```

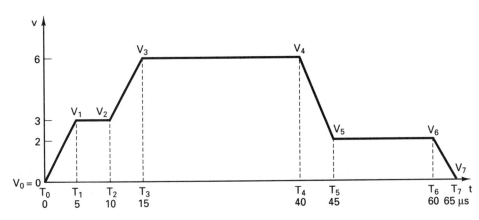

Figure 4-4 Piecewise linear waveform.

4-3.4 Single-Frequency Frequency Modulation

The symbol of a source with single-frequency frequency modulation is SFFM, and the general form is

```
SFFM (VO  VA  FC  MOD  FS)
```

The model parameters of an SFFM waveform are given in Table 4-6.

TABLE 4-6 MODEL PARAMETERS OF SFFM SOURCES

Name	Meaning	Units	Default
VO	Offset voltage	Volts	None
VA	Amplitude of voltage	Volts	None
FC	Carrier frequency	Hertz	1/TSTOP
MOD	Modulation index		0
FS	Signal frequency	Hertz	1/TSTOP

V_O and V_A *must* be specified by the user and can be either voltages or currents. TSTOP is the stop time during transient (.TRAN) analysis. The waveform is of the form

$$V = V_O + V_A \sin[(2\pi F_C t) + M \sin(2\pi F_S t)]$$

Typical statements.   For $V_O = 0$, $V_A = 1$ V, $F_C = 30$ MHz, MOD = 5, and $F_S = 5$ kHz, the model statement is

```
SFFM (0    1V    30MHZ    5    5KHZ)
```

With $V_O = 1$ mV and $V_A = 2$ V, the model becomes

```
SFFM (1MV  2V    30MHZ    5    5KHZ)
```

4-3.5 Sinusoidal Source

The symbol of a sinusoidal source is SIN, and the general form is

```
SIN (VO  VA  FREQ  TD  ALP   THETA)
```

The model parameters of the SIN waveform are given in Table 4-7.

TABLE 4-7 MODEL PARAMETERS OF SIN SOURCES

Name	Meaning	Units	Default
VO	Offset voltage	Volts	None
VA	Peak voltage	Volts	None
FREQ	Frequency	Hertz	1/TSTOP
TD	Delay time	Seconds	0
ALPHA	Damping factor	1/seconds	0
THETA	Phase delay	Degrees	0

V_O and V_A *must* be specified by the user and can be either voltages or currents. TSTOP is the stop time during transient (.TRAN) analysis. The waveform stays at 0 for a time of TD, and then the voltage becomes an exponentially damped sine wave. An exponentially damped sine wave is described by

$$V = V_O + V_A e^{-\alpha(t-t_d)} \sin[(2\pi f(t - t_d) - \theta]$$

and this is shown in Fig. 4-5.

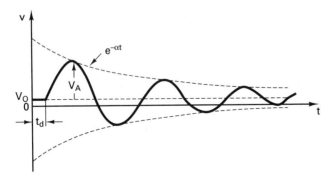

Figure 4-5 Damped sinusoidal waveform.

Typical Statements

```
SIN (0   1V   10KHZ   10US     1E5)
SIN (1   5V   10KHZ   1E5     30DEG)
SIN (0   2V   10KHZ   30DEG)
SIN (0   2V   10KHZ)
```

4-4 TRANSIENT SOURCES

The transient sources are time-variant and can be either independent or dependent. They can be voltages or currents, as shown in Fig. 4-6. The dependent sources are discussed in Section 3-6.

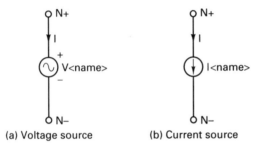

(a) Voltage source   (b) Current source

Figure 4-6 Voltage and current sources.

4-4.1 Independent Voltage Source

The symbol of an independent voltage source is V, and the general form for assigning dc and transient values is

```
V⟨name⟩ N+ N−   [DC ⟨value⟩]
+      [(transient value)
+      [PULSE] [SIN] [EXP] [PWL] [SFFM] [source arguments]]
```

N+ is the positive node, and N− is the negative node, as shown in Fig. 4-6(a). Positive current flows from node N+ through the voltage source to the negative node N−. The voltage source need not be grounded. For the dc and transient values, the default value is zero. None or all of the dc and transient values may be specified.

A source can be assigned either a dc value or a transient value. The source is set to the dc value in dc analysis. The time-dependent source (e.g., PULSE, EXP, or SIN) is assigned for transient analysis. A voltage source may be used as an **ammeter** in PSpice by inserting a zero-valued voltage source into the circuit for the purpose of measuring current.

Typical Statements

```
V1      15  0   6V
V2      15  0   DC  6V
VPULSE  10  0   PULSE (0  1  2NS  2NS  2NS  50NS  100NS)
VPULSE  12  0   PULSE (0  1  2NS  2NS  2NS  50NS  100NS)
VIN     12  3   DC  15V    SIN (0  2V  10KHZ)
```

Note. VIN assumes 15 V for dc analysis, and a sine wave of 2 V at 10 kHz for transient analysis. This allows source specifications for different analyses in the same statement.

4-4.2 Independent Current Source

The symbol of an independent current source is I, and the general form for assigning dc and transient values is

```
I⟨name⟩ N+ N−   [DC ⟨value⟩]
+      [(transient value)
+      [PULSE] [SIN] [EXP] [PWL] [SFFM] [source arguments]]
```

Note. The first column with + (*plus*) signifies continuation of the PSpice statement. After the + sign, the statement can continue in any column.

N+ is the positive node, and N− is the negative node, as shown in Fig. 4-6(b). Positive current flows from node N+ through the current source to the negative node N−. The current source need not be grounded. The source specifications are similar to those for independent voltage sources.

Typical Statements

```
I1       15  0  2.5MA
I2       15  0  DC  2.5MA
IPULSE   10  0  PULSE (0  1V  2NS  2NS  2NS  50NS  100NS)
IIN      25  22  DC  2           SIN (0  2V   10KHZ)
```

4-5 TRANSIENT OUTPUT VARIABLES

The output variables of transient analysis are similar to those of the dc sweep, discussed in Section 3-7.

Example 4-1

An *RLC* circuit with a step input is shown in Fig. 4-7. Write the various currents and voltages in forms that are allowed by PSpice.

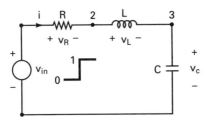

Figure 4-7 An *RLC* circuit with a step input.

Solution

SYMBOLS	PSpice VARIABLES	MEANING
i_R	I(R)	The current through resistor R
i_L	I(L)	The current through inductor L
i_C	I(C)	The current through capacitor C
i_{in}	I(VIN)	The current flowing into voltage source v_{in}
v_3	V(3)	Voltage of node 3 with respect to ground
$v_{2,3}$	V(2,3)	Voltage of node 2 with respect to node 3
$v_{1,2}$	V(1,2)	Voltage of node 1 with respect to node 2
v_R	V(R)	Voltage of resistor R, where the first node (as defined in the circuit file) is positive with respect to the second node
v_L	V(L)	Voltage of inductor L, where the first node (as defined in the circuit file) is positive with respect to the second node
v_C	V(C)	Voltage of inductor L, where the first node (as defined in the circuit file) is positive with respect to the second node

Note. SPICE and older versions of PSpice do not permit measuring voltage across a resistor, an inductor, and a capacitor, for example, V(R1), V(L1), and V(C1)]. Such statements are applicable only to outputs by .PLOT and .PRINT commands.

4-6 TRANSIENT OUTPUT COMMANDS

The output commands are similar to those for the dc sweep. The .PRINT, .PLOT, and .PROBE statements for transient outputs are

```
.PRINT TRAN ⟨output variables⟩
.PLOT TRAN ⟨output variables⟩
         +[⟨lower limit⟩ value⟩, ⟨(upper limit) value⟩]
.PROBE
```

Some Statements

```
.PRINT TRAN V(5) V(4,7) (0,10V) IB(Q1) (0,50MA) IC(Q1) (−50MA, 50MA)
.PLOT  TRAN V(5) V(4,7) (0,10V) IB(Q1) (0, 50MA) IC(Q1) (−50MA, 50MA)
```

Notes
1. Having two .PRINT statements for the same variables will not produce two tables. PSpice will ignore the first statement and produce output for the second statement.
2. The *x*-axis is the time, by default. In the last statement, the range for voltages V(5) and V(4,7) is 0 to 10V, that for current IB(Q1) is 0 to 50 MA, and that for the current IC(Q1) is −50MA to 50 MA.

4-7 TRANSIENT RESPONSE

A transient response determines the output in the time domain in response to an input signal in the time domain. Let us consider the *RC* circuit shown in Fig. 4-8(a), which is subjected to a step input voltage. Assuming that the switch S_1 is closed at time $t = 0$, the current i_c can be described by the equation

$$V_S = Ri_C + \frac{1}{C} \int i_C \, dt + V_C(t = 0) \tag{4-1}$$

where the initial condition is $V_C(t = 0) = 0$. Solving Eq. (4-1) gives the capacitor voltage $v_C(t)$ as

$$v_C(t) = V_S[1 - e^{-t/RC}] \tag{4-2}$$

The plot of $v_C(t)$ against time *t* gives the transient response of the capacitor voltage $v_C(t)$, shown in Fig. 4-8(b). The method for calculating the transient analysis bias point differs from that of the dc analysis bias point. The dc bias point is also known as the *regular bias point*. In the regular (dc) bias point, the initial values of the circuit nodes do not contribute to the operating point or to the linearized parameters. The capacitors and inductors are considered open-circuited and short-circuited, respectively, whereas in the transient bias point, the initial values of the circuit nodes are taken into account in calculating the bias

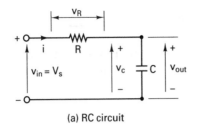

(a) RC circuit

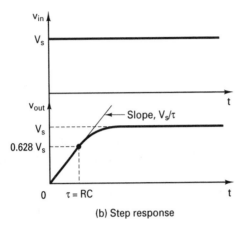

(b) Step response

Figure 4-8 Transient response of an *RC* circuit.

point and the small-signal parameters of the nonlinear elements. The capacitors and inductors, which may have initial values, therefore remain as parts of the circuit.

The determination of the transient analysis requires statements involving

.IC   Initial transient conditions
.TRAN   Transient analysis

4-7.1 .IC (Initial Transient Conditions)

The various nodes can be assigned to initial voltages during transient analysis, and the general form for assigning initial values is

```
.IC  V(1)=A1  V(1)=A2 ... V(N)=AN
```

where A1, A2, A3, . . . are the initial values for node voltages V(1), V(2), V(3), . . . , respectively. These initial values are used by PSpice to calculate the transient analysis bias point and the linearized parameters of nonlinear devices for transient analysis. After the transient analysis bias point has been calculated, the transient analysis starts, and the nodes are released. It should be noted that these initial conditions do not affect the regular bias-point calculation during dc analysis or the dc sweep.

For the .IC statement to be effective, UIC (use initial conditions) *should not* be specified in the .TRAN command.

The .IC command is not necessary in the linear circuits that are illustrated by examples in this chapter. Students are encouraged to run circuits with an .IC statement.

Statement for Initial Transient Conditions

```
.IC   V(1)=2.5   V(5)=1.7V   V(7)=0.5
```

4-7.2 .TRAN (Transient Analysis)

Transient analysis can be performed by the .TRAN command, which has one of the general forms

```
.TRAN        TSTEP   TSTOP   [TSTART TMAX]   [UIC]
.TRAN[/OP]   TSTEP   TSTOP   [TSTART TMAX]   [UIC]
```

TSTEP is the printing increment, TSTOP is the final time (or stop time), and TMAX is the maximum size of internal time step. TMAX allows the user to control the internal time step. TMAX can be smaller or larger than the printing time, TSTEP. The default value of TMAX is TSTOP/50.

The transient analysis always starts at time = 0. However, it is possible to suppress the printing of the output for a time of TSTART. TSTART is the initial time at which the transient response is printed. In fact, PSpice analyzes the circuit from $t = 0$ to TSTART, but it does not print or store the output variables. Although PSpice computes the results with an internal time step, the results are generated by interpolation for a printing step of TSTEP. Figure 4-9 shows the relationships of TSTART, TSTOP, and TSTEP.

In transient analysis, only the node voltages of the transient analysis bias point are printed. However, the .TRAN command can control the output for the transient response bias point. An .OP command with a .TRAN command, namely, .TRAN/OP, will print the small-signal parameters during transient analysis.

If UIC is not specified as an option at the end of the .TRAN statement, PSpice calculates the transient analysis bias point before the beginning of transient analysis. PSpice uses the initial values specified with the .IC command.

If UIC (use initial conditions) is specified as an option at the end of the .TRAN statement, PSpice does not calculate the transient analysis bias point before the beginning of transient analysis. However, PSpice uses the initial values specified with the IC= initial conditions for capacitors and inductors, which are discussed in Section 4-2. Therefore, if UIC is specified, the initial values of the capacitors and inductors *must* be supplied. The .TRAN statement requires a .PRINT, .PLOT, or .PROBE statement to get the results of the transient analysis.

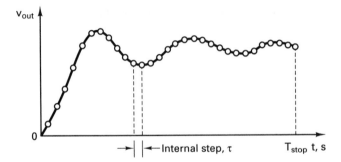

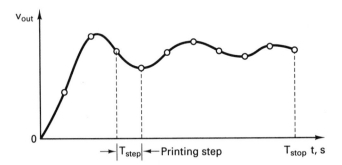

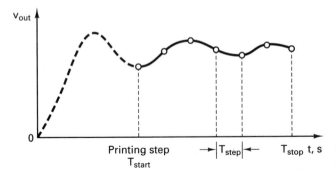

Figure 4-9 Response of transient analysis.

Statements for Transient Analysis

```
. TRAN     5US    1MS
. TRAN     5US    1MS  200US   0. 1NS
. TRAN     5US    1MS  200US   0. 1NS   UIC
. TRAN/OP  5US    1MS  200US   0. 1NS   UIC
```

Example 4-2

A pulse input, as shown in Fig. 4-10(b), is applied to the *RLC* circuit of Fig. 4-10(a). Use PSpice to calculate and plot the transient response from 0 to 400 μs with a time increment of 1 μs. The capacitor voltage V(3) and the current through R_1 I(R1) are to

---

Transient Analysis   Chap. 4

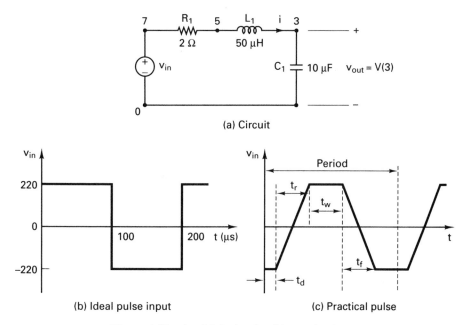

(a) Circuit

(b) Ideal pulse input   (c) Practical pulse

Figure 4-10   An *RLC* circuit with a pulse input.

be plotted. The circuit file's name is EX4-2.CIR, and the outputs are to be stored in file EX4-2.OUT. The results should also be made available for display and hard copy by the .PROBE command.

Solution   The circuit file contains the following statements:

Example 4-2   Pulse response of an *RLC* circuit

```
▲ *    PULSE  (-VS  +VS   TD   TR   TF   PW   PER)    ; Pulse input
   VIN   7    0    PULSE  (-220V   220V   0   1NS   1NS   100US   200US)
▲ ▲ R1   7    5    2
    L1    5    3    50UH
    C1    3    0    10UF
▲ ▲ ▲ * .TRAN  TSTEP   TSTOP                  ; Command for transient analysis
      .TRAN   1US      400US
      *.PRINT  TRAN   V(R1)   V(L1)   V(C1)  ; Prints to the output file
      *.PLOT   TRAN   V(3)    I(R1)          : Plots in the output file
      .PROBE                                 ; Graphical waveform analyzer
  .END                                       ; End of circuit file
```

It should be noted that with the .PROBE command, there is no need for the .PLOT command; .PLOT generates the plot in the output file, while .PROBE sends graphical output to the monitor screen that can directly be dumped into a plotter and/ or a printer. The .PRINT and .PLOT commands are made ineffective by placing an asterisk (*) in front of them. The results of the .PRINT and .PLOT statements can be obtained by printing the contents of the output file EX4-2.OUT.

The results of the transient response that are produced on the display by .PROBE command are shown in Fig. 4-11.

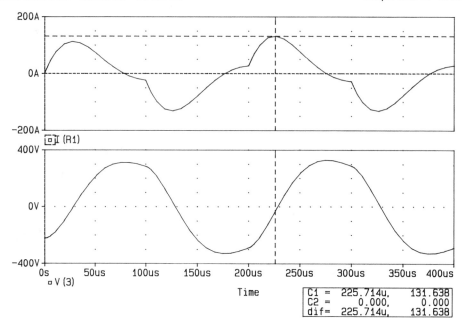

200A

0A

-200A

□I (R1)

400V

0V

-400V

0s   50us   100us   150us   200us   250us   300us   350us 400us

□ V (3)

Time

C1 =	225.714u,	131.638
C2 =	0.000,	0.000
dif=	225.714u,	131.638

Figure 4-11   Pulse response for Example 4-2.

Example 4-3

Three *RLC* circuits with $R = 2 \ \Omega$, $1 \ \Omega$, and $8 \ \Omega$ are shown in Fig. 4-12(a). The inputs are identical step voltages, as shown in Fig. 4-12(b). Use PSpice to calculate and plot the transient response from 0 to 400 μs with an increment of 1 μs. The capacitor voltages are the outputs V(3), V(6), and V(9), which are to be plotted. The circuit is to be stored in the file EX4-3.CIR, and the outputs are to be stored in the file EX4-3.OUT. The results should also be made available for display and hard copy by the .PROBE command.

Solution   The description of the circuit file is similar to that of Example 4-2, except the input is a step voltage instead of a pulse voltage. The circuit may be regarded as three *RLC* circuits having three separate inputs. The step signal can be represented by piecewise linear source, which is described in general by

```
PWL (T1 V1 T2 V2 ... TN VN)
```

where VN is the voltage at time TN.

Assuming a rise time of 1 ns, the step voltage of Fig. 4-12(b) can be described by

```
PWL (0  0  1NS  1V  1MS  1V)
```

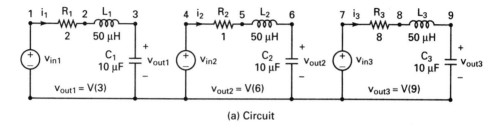

(a) Circuit

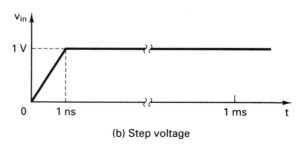

(b) Step voltage

Figure 4-12 *RLC* circuits with step-pulse input voltages.

The listing of the circuit file follows.

Example 4-3   Step response of series *RLC* circuits

```
▲ VI1  1   0    PWL (0  0  1NS  1V  1MS  1V)  ; step of 1 V
  VI2  4   0    PWL (0  0  1NS  1V  1MS  1V)  ; step of 1 V
  VI3  7   0    PWL (0  0  1NS  1V  1MS  1V)  ; step of 1 V
▲▲ R1  1   2    2
   L1  2   3    50UH
   C1  3   0    10UF
   R2  4   5    1
   L2  5   6    50UH
   C2  6   0    10UF
   R3  7   8    8
   L3  8   9    50UH
   C3  9   0    10UF
▲▲▲ .TRAN  1US   400US          ; Transient analysis
  *.PLOT   TRAN  V(3)  V(6)  V(9)  ; Plots in the output file
   .PROBE                       ; Graphical waveform analyzer
        *
  .END                          ; End of circuit file
```

With the .PROBE command, there is no need for .PLOT command. The .PLOT command is made ineffective by preceding it with an asterisk (*). The results of the .PLOT statement can be obtained by printing the contents of the output file EX4-3.OUT. The results of the transient analysis that are produced on the display by the .PROBE command are shown in Fig. 4-13.

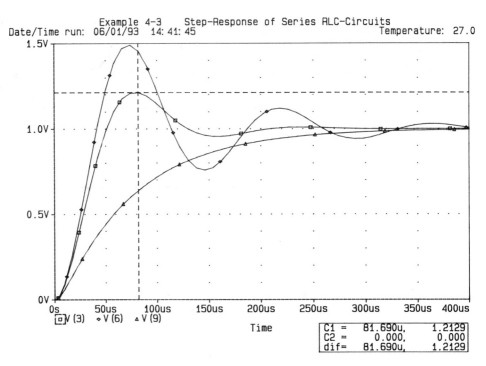

```
C1 =     81.690u,    1.2129
C2 =      0.000,     0.000
dif=     81.690u,    1.2129
```

Figure 4-13   Step response of *RLC* circuits for Example 4-3.

Example 4-4

Repeat Example 4-2 if the input voltage is a sine-wave of $v_{in} = 10 \sin(2\pi \times 5000t)$.
Solution   For a sinusoidal voltage $v_{in} = 10 \sin(2\pi \times 5000t)$, the model is

```
SIN (0   10V   5KHZ)
```

The circuit file contains the following statements:

Example 4-4   An *RLC* circuit with sinusoidal input voltage

```
▲ *  SIN (VO   VA   FREQ)          ; Simple sinusoidal source
   VIN  7  0  SIN (0  10V  5KHZ) ; Sinusoidal input voltage
▲▲ R1  7  5  2
   L1  5  3  50UH
   C1  3  0  10UF
▲▲▲ .TRAN  1US    500US             ; Transient analysis
    .PLOT  TRAN  V(3)  V(7)         ; Plots in the output file
    .PROBE                          ; Graphical waveform analyzer
  .END                              ; End of circuit file
```

The results of the transient response that are produced on the display by the
.PROBE command are shown in Fig. 4-14. The results of the .PLOT statement can
be obtained by printing the contents of the output file EX4-4.OUT.

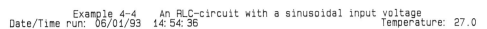

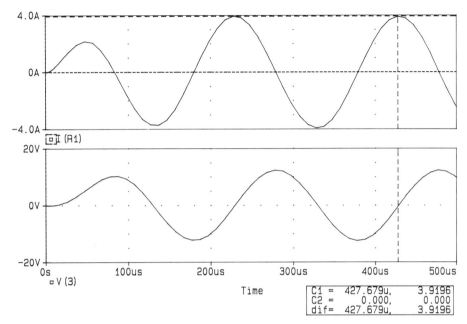

Figure 4-14 Transient response for Example 4-4.

Example 4-5

For the circuit of Fig. 4-15(a), calculate and plot the transient response from 0 to 1 ms with a time increment of 5 μs. The output voltage is taken across resistor R_2, and the input voltage is shown in Fig. 4-15(b). The results should be made available for display and hard copy by using Probe. The model parameters for the resistor are R=1, TC1=0.02, TC2=0.005; for the capacitor, C=1, VC1=0.01, VC2=0.002, TC1=0.02, TC2=0.005; and for the inductor, L=1, IL1=0.1, IL2=0.002, TC1=0.02, TC2=0.005. The operating temperature is 50°C.

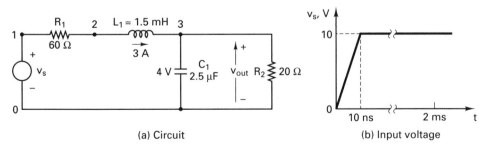

(a) Circuit (b) Input voltage

Figure 4-15 Circuit for Example 4-5.

Solution The circuit file contains the following statements:

Example 4-5 **Transient response of an *RLC* circuit**

```
▲ VS  1  0  PWL (0   0   10NS  10V  2MS 10V) ; Step voltage of PWM waveform
▲▲ R1  1  2  RMOD   60HM          ; Resistance with model RMOD
   L1  2  3  LMOD   1.5MH  IC=3A  ; Initial current of 3 A and model LMOD
   C1  3  0  CMOD   2.5UF  IC=4V  ; Initial voltage of 4 V and model name
                                    CMOD

   R2  3  0  RMOD   20HM
   .TEMP  50                      ; Operating temperature of 50 degrees C
   *   Model statements for resistor, inductor, and capacitor
.MODEL RMOD RES (R=1  TC1=0.02  TC2=0.005)
.MODEL CMOD CAP (C=1  VC2=0.01  VC2=0.002  TC1=0.02  TC2=0.005)
.MODEL LMOD IND (L=1  IL1=0.1   IL2=0.002  TC1=0.02  TC2=0.005)
▲▲▲ .TRAN  5US   1MS   UIC        ; Transient analysis with UIC
    *.PLOT   TRAN  V(3)  V(1)      ; Not needed with .PROBE
    .PROBE                         ; Graphical waveform analyzer
.END                              ; End of circuit file
```

The results of the simulation that are obtained by .PROBE are shown in Fig. 4-16. With .PROBE, the .PRINT command is redundant. The inductor has an initial current of 3 A, which is taken into consideration by UIC in the .TRAN command.

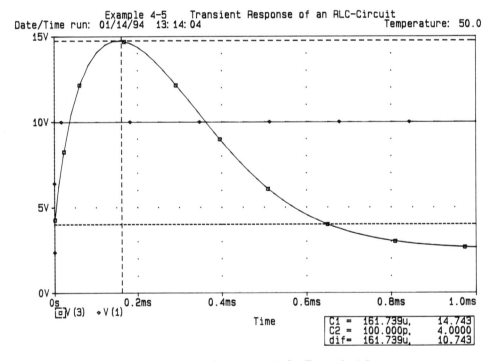

Figure 4-16 Transient response for Example 4-5.

Example 4-6

Repeat Example 4-5, assuming that the voltage across the capacitor is set by the .IC command instead of an IC condition, and UIC is *not* specified.

Solution In order to set the capacitor voltage by the .IC command, the following statement should be added:

```
.IC    V(3)=4V           ; Node 3 is set to 4 V
```

and the UIC is removed from the .TRAN statement as follows:

```
.TRAN   5US   1MS         ; Transient analysis without UIC
```

The result of the simulation is shown in Fig. 4-17. Notice that the response is completely different because the node voltage of the capacitor was set to 4 V.

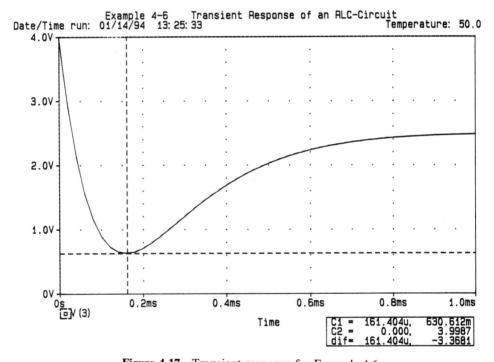

Figure 4-17 Transient response for Example 4-6.

4-8 SWITCHES

PSpice allows the simulation of a special kind of switch, shown in Fig. 4-18, whose resistance varies continuously depending on the voltage or current. When the switch is on, the resistance is R_{on}, and when it is off, the resistance becomes R_{off}.

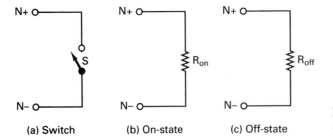

Figure 4-18 A switch with variable resistance.

(a) Switch   (b) On-state   (c) Off-state

Two types of switches are permitted in PSpice:

    Voltage-controlled switch
    Current-controlled switch

Note. The voltage- and current-controlled switches are not available in SPICE2, but they are available in SPICE3.

4-8.1 Voltage-Controlled Switch

The symbol for a voltage-controlled switch is *S*. The name of this switch must start with *S*, and it takes the general form of

S⟨name⟩   N+   N−   NC+   NC−   SNAME

N+ and N− are the two nodes of the switch. The current is assumed to flow from N+ through the switch to node N−. NC+ and NC− are the positive and negative nodes of the controlling voltage source, as shown in Fig. 4-19. SNAME is the model name. The resistance of the switch varies depending on the voltage across the switch. The type name for a voltage-controlled switch is VSWITCH, and the model parameters are shown in Table 4-8.

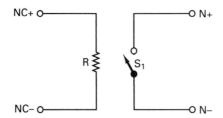

Figure 4-19 Voltage-controlled switch.

TABLE 4-8 MODEL PARAMETERS FOR VOLTAGE-CONTROLLED SWITCH

Name	Meaning	Units	Default
VON	Control voltage for on-state	Volts	1.0
VOFF	Control voltage for off-state	Volts	0
RON	On resistance	Ohms	1.0
ROFF	Off resistance	Ohms	10^6

Voltage-controlled switch statement

```
S1      6  5  4   0  SMOD
.MODEL   SMOD   VSWITCH  (RON=0.5 ROFF=10E+6 VON=0.7 VOFF=0.0)
```

Notes

1. R_{ON} and R_{OFF} must be greater than zero and less than 1/GMIN. The value of GMIN can be defined as an option, as described in the .OPTIONS command in Section 6-10. The default value of conductance, GMIN, is 1E−12 mhos.

2. The ratio of R_{OFF} to R_{ON} should be less than 1E+12.

3. The difficulty due to high gain of an ideal switch can be minimized by choosing the value of R_{OFF} as high as permissible and that of R_{ON} as low as possible as compared to other circuit elements, within the limits of allowable accuracy.

Example 4-7

A circuit with a voltage-controlled switch is shown in Fig. 4-20. If the input voltage is $V_S = 200 \sin(2000\pi t)$, plot the voltage at node 3 and the current through the load resistor R_L for a time duration of 0 to 1 ms with an increment of 5 μs. The model parameters of the switch are RON=5M, ROFF=10E+9, VON=25M, and VOFF=0.0. The results should be available for display by Probe.

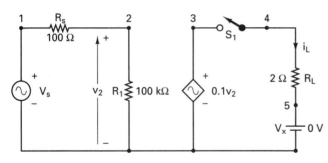

Figure 4-20 A circuit with a voltage-controlled switch.

Solution The voltage source $V_X = 0$ V is inserted to monitor the output current. The listing of the circuit file is as follows.

Example 4-7 A voltage-controlled switch

```
▲ VS    1   0   SIN (0  200V  1KHZ) ; Sinusoidal voltage of 200 V peak
▲▲ RS   1   2   100OHM
   R1   2   0   100KOHM
   E1   3   0   2   0   0.1  ; V-controlled source with a gain of 0.1
   RL   4   5   2OHM
   VX   5   0   DC   0V      ; Measures the load current
   S1   3   4   3   0   SMOD ; V-controlled switch with model SMOD
    *    Switch model descriptions
   .MODEL    SMOD   VSWITCH (RON=5M ROFF=10E+9 VON=25M VOFF=0.0)
▲▲▲ .TRAN  5US   1MS              ; Transient analysis
    *.PLOT   TRAN   V(3) I(VX)    ; Not needed with .PROBE
    .PROBE                        ; Graphical waveform analyzer
  .END                           ; End of circuit file
```

The results of the simulation that are obtained by Probe are shown in Fig. 4-21, which is the output of a diode rectifier. The switch S_1 behaves as a diode. The .PROBE statement has made .PLOT redundant.

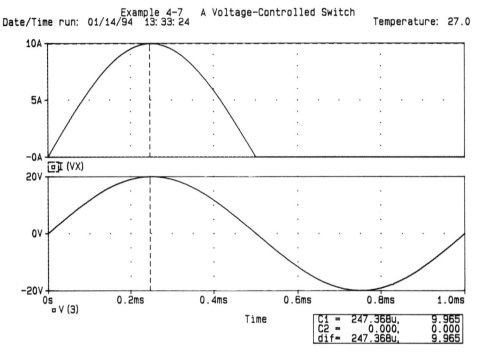

C1 =	247.368u,	9.965
C2 =	0.000,	0.000
dif=	247.368u,	9.965

Figure 4-21   Transient response for Example 4-7.

Example 4-8

The *RLC* circuit of Fig. 4-22(a) is subjected to dc transients by opening and closing switch S_1, as shown in Fig. 4-22(b). Use PSpice to calculate and plot the inductor current i_L and the capacitor voltage v_C from 0 to 20 ms with a time increment of 5 μs. The model parameters of the switch are RON=0.01, ROFF=10E+5, VON=0.1V, and VOFF=0V.

Solution   The listing of the circuit file is as follows:

Example 4-8   Dc transients

```
▲ VS    1   0   DC    200V   ; Voltage source of 100 V dc
  VG    8   0   PULSE (0V  10V    5MS    1US    1US    5MS    10.01MS)
▲▲ RG   8   0   10MEG         ; High resistance for continuity
  R1    1   2   4.7K
  R2    2   0   1.5K
  R3    2   3   2.5K
  R4    4   0   5K
  R5    4   5   1K
  R6    5   6   150
  R7    6   0   5K
```

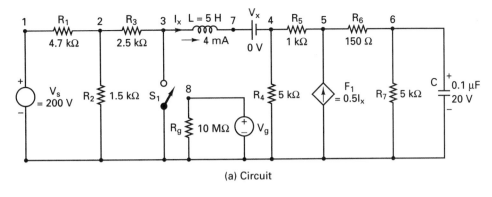

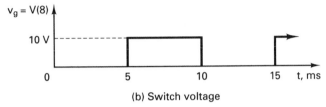

(a) Circuit

(b) Switch voltage

Figure 4-22 Dc circuit subjected to transients.

```
VX     7   4   DC    0V              ; Measures current through L₁
C1     6   0   0.1UF   IC=10V        ; Capacitor with initial voltage
L1     3   7   5H      IC=4MA        ; Inductor with initial current
F1     0   5   VX        0.5         ; Current-controlled current source
S1     3   0   8    0     SMOD       ; Voltage switch with model SMOD
.MODEL   SMOD   VSWITCH  (RON=0.01 ROFF=10E+5 VON=0.1V VOFF=0V)
▲▲▲ .TRAN    5US   20MS    UIC       ; Transient analysis with UIC
     .PROBE                          ; Graphical waveform analyzer
.END                                 ; End of circuit file
```

The inductor current I(L1) (= I(VX)) and capacitor voltage V(6), which are
obtained by the .PROBE command, are shown in Fig. 4-23. Since the circuit has two
energy-storage elements, it exhibits the characteristics of a second-order system.

4-8.2 Current-Controlled Switch

The symbol for a current-controlled switch is *W*. The name of the switch must
start with *W*, and it takes the general form of

W⟨name⟩ N+ N− VN   WNAME

N+ and N− are the two nodes of the switch. V_N is a voltage source through
which the controlling current flows, as shown in Fig. 4-24. WNAME is the model
name. The resistance of the switch depends on the current through the switch.
The type name for a current-controlled switch is ISWITCH, and the model parameters are shown in Table 4-9.

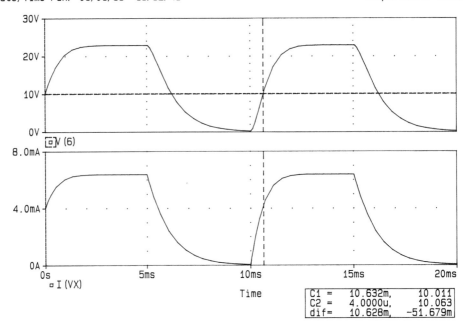

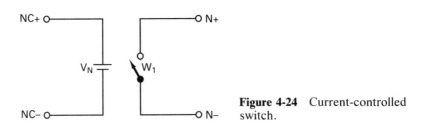

Figure 4-23 Dc transient responses for Example 4-8.

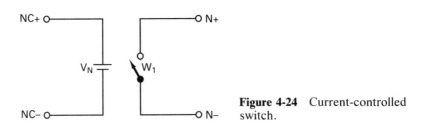

Figure 4-24 Current-controlled switch.

TABLE 4-9 MODEL PARAMETERS FOR CURRENT-CONTROLLED SWITCH

Name	Meaning	Units	Default
ION	Control current for on-state	Amps	1E-3
IOFF	Control current for off-state	Amps	0
RON	On resistance	Ohms	1.0
ROFF	Off resistance	Ohms	10^6

Current-controlled switch statement

```
W1          6  5  VN  RELAY
.MODEL    RELAY ISWITCH (RON=0.5 ROFF=10E+6 ION=0.07 IOFF=0.0)
```

Note. The current through voltage source V_N controls the switch. The voltage source V_N must be an independent source, and it can have a zero or a finite value. The limitations of the parameters are similar to those for the voltage-controlled switch.

Example 4-9

A circuit with a current-controlled switch is shown in Fig. 4-25. Plot the capacitor voltage and the current through the inductor for a time duration of 0 to 160 μs with an increment of 1 μs. The model parameters of the switch are RON=1E+6 ROFF=0.001, ION=1MA, IOFF=0. The results should be available for display by Probe.

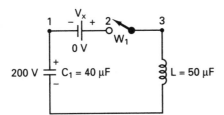

Figure 4-25 Circuit with a current-controlled switch.

Solution The voltage source $V_X = 0$ V is inserted to monitor the controlling current. The listing of the circuit file is as follows.

Example 4-9   A current-controlled switch

```
▲▲ C1   1   0   40UF  IC=200        ; With an initial voltage of 200 V
   VX    2   1   DC   0V            ; Dummy voltage source of Vₓ=0
   W1    2   3   VX   SMOD          ; I-controlled switch with model SMOD
   .MODEL   SMOD  ISWITCH (RON=1E+6 ROFF=0.001 ION=1MA IOFF=0)
   L1    3   0   50UF
▲▲▲ .TRAN   1US   160US   UIC        ; Transient analysis with UIC
   .PLOT  TRAN   V(1)   I(VX)       ; Not needed with .PROBE
   .PROBE                          ; Graphical waveform analyzer
.END                               ; End of circuit file
```

The results of the simulation that are obtained by Probe are shown in Fig. 4-26. The switch W_1 acts as diode and allows only positive current flow. The initial voltage on the capacitor is the driving source.

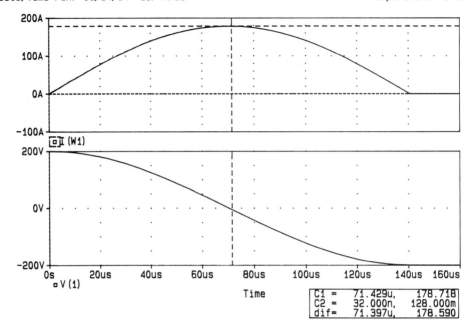

Figure 4-26 Transient response for Example 4-9.

SUMMARY

The symbols and statements for elements are

C   Capacitor

      C⟨name⟩ N+ N− CNAME CVALUE IC=VO

L   Inductor

      L⟨name⟩ N+ N− LNAME LVALUE IC=IO

R   Resistor

      R⟨name⟩ N+ N− RNAME RVALUE

S   Voltage-controlled switch

      S⟨name⟩ N+ N− NC+ NC− SNAME

W   Current-controlled switch

`W⟨name⟩ N+ N− VN WNAME`

The symbols and statements for transient sources are

EXP   Exponential source

`EXP (V1 V2 TRD TRC TFD TFC)`

POLY   Polynomial source

`POLY(n) ⟨(controlling) nodes⟩ ⟨(coefficients) values⟩`

PULSE   Pulse source

`PULSE (V1  V2  TD TR TF PW PER)`

PWL   Piecewise linear source

`PWL (T1 V1 T2 V2 ... TN VN)`

SFFM   Single-frequency frequency modulation

`SFFM (VO  VA  FC MOD FS)`

SIN   Sinusoidal source

`SIN (VO  VA  FREQ  TD ALP THETA)`

E   Voltage-controlled voltage source

`E⟨name⟩ N+ N−  NC+ NC−  ⟨(voltage gain) value⟩`

F   Current-controlled current source

`F⟨name⟩ N+ N−  VN ⟨(current gain) value⟩`

G   Voltage-controlled current source

`G⟨name⟩ N+ N−  NC+ NC− ⟨(transconductance) value⟩`

H   Current-controlled voltage source

`H⟨name⟩ N+ N−  VN ⟨(transresistance) value⟩`

I	Independent current source

```
I⟨name⟩ N+ N-  [DC ⟨value⟩]
+      [(transient value)
+      [PULSE] [SIN] [EXP] [PWL] [SFFM] [source arguments]]
```

V	Independent voltage source

```
V⟨name⟩ N+ N-  [DC ⟨value⟩]
+      [(transient value)
+      [PULSE] [SIN] [EXP] [PWL] [SFFM] [source arguments]]
```

The commands that are generally used for transient analysis follow.

.END	End of circuit
.IC	Initial transient conditions
.MODEL	Model
.OP	Operating point
.PLOT	Plot
.PRINT	Print
.PROBE	Probe
.TEMP	Temperature
.TRAN	Transient analysis
UIC	Use initial conditions
.WIDTH	Width

REFERENCES

1. M. H. Rashid, *SPICE For Power Electronics and Electric Power*. Englewood Cliffs, N.J.: Prentice Hall, 1993.
2. *PSpice Manual,* Irvine, Calif.: MicroSim Corporation, 1992.
3. W. H. Hayt and J. E. Kemmerly, *Engineering Circuit Analysis*. New York: McGraw-Hill, 1990.
4. J. David Irwin, *Basic Engineering Circuit Analysis*. New York: Macmillan, 1989.
5. D. E. Johnson, J. Hilburn, and J. R. Johnson, *Basic Electric Circuit Analysis*. Englewood Cliffs, N.J.: Prentice Hall, 1990.
6. J. W. Nilson, *Electric Circuits*. Reading, Mass.: Addison-Wesley, 1990.

PROBLEMS

Write the PSpice statements for the following problems, 4-1 to 4-6, 4-19, and 4-20.

4-1. A capacitor C_1 is connected between nodes 5 and 6 and has a value of 10 pF and an initial voltage of -20 V.

4-2. A capacitor C_1 is connected between nodes 5 and 6 and has a nominal value of $C = 10$ pF. The operating temperature is $T = 55°C$. The capacitance, which is a function of its voltage and the operating temperature, is given by

```
C₁ = C * (1 + 0.01 * V + 0.002 * V²)
       * [1 + 0.03 * (T − TO) + 0.05 * (T − TO)²]
```

4-3. An inductor L_1 is connected between nodes 5 and 6; it has a value of 0.5 mH and carries an initial current of 0.04 mA.

4-4. An inductor L_1 is connected between nodes 3 and 4 has a nominal value of $L = 1.5$ mH. The operating temperature is $T = 55°C$. The inductance, which is a function of its current and the operating temperature, is given by

```
L₁ = L * (1 + 0.01 * I + 0.002 * I²)
       * [1 + 0.03 * (T − TO) + 0.05 * (T − TO)²]
```

4-5. The various voltage or current waveforms that are connected between nodes 4 and 5 are shown in Fig. P4-5.

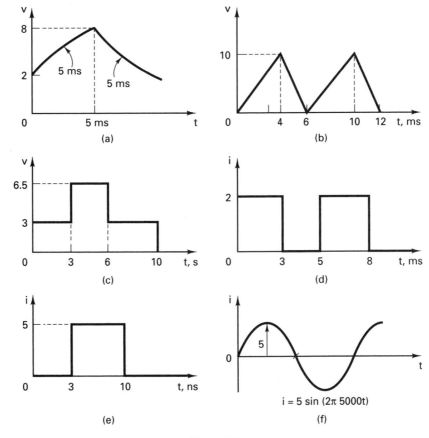

Figure P4-5

4-6. A voltage source that is connected between nodes 4 and 5 is given by

$$v = 2 \sin\{2\pi \, 50{,}000t + 5 \sin(2\pi \, 1000t\}$$

4-7. The *RLC* circuit of Fig. P4-7 is to be simulated to calculate and plot the transient response from 0 to 2 ms with an increment of 10 μs. The voltage across resistor *R* is the output. The input and output voltages are to be plotted in the output file. The results should also be made available for display and hard copy by the .PROBE command.

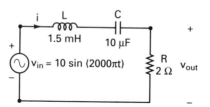

Figure P4-7

4-8. Repeat Problem 4-7 for the circuit of Fig. P4-8, where the output is taken across capacitor *C*.

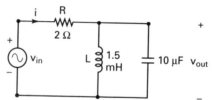

Figure P4-8

4-9. Repeat Problem 4-7 for the circuit of Fig. P4-9, where the output is the current i_s through the circuit.

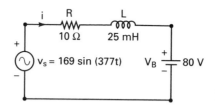

Figure P4-9

4-10. Repeat Problem 4-7 if the input is a step input, as shown in Fig. 4-10(b).

4-11. Repeat Problem 4-8 if the input is a step input, as shown in Fig. 4-10(b).

4-12. The *RLC* circuit of Fig. P4-12(a) is to be simulated to calculate and plot the transient response from 0 to 2 ms with an increment of 5 μs. The input is a step current, as shown in Fig. P4-12(b). The voltage across resistor *R* is the output. The input and output voltages are to be plotted in the output file. The results should also be made available for display and hard copy by the .PROBE command.

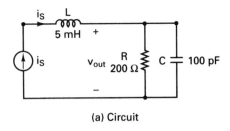

(a) Circuit

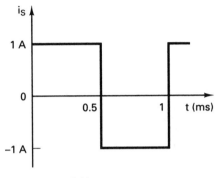

(b) Input current

Figure P4-12

4-13. Repeat Problem 4-12 for the circuit of Fig. P4-13, where the output is taken across capacitor C.

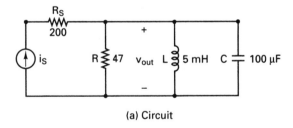

(a) Circuit

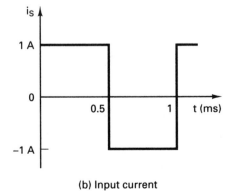

(b) Input current

Figure P4-13

4-14. Plot the transient response of the circuit in Fig. P4-14 from 0 to 5 ms with a time increment of 25 μs. The output voltage is taken across the capacitor. Use Probe for graphical output.

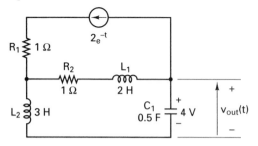

Figure P4-14

4-15. Repeat Problem 4-14 for the circuit in Fig. P4-15.

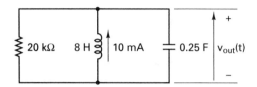

Figure P4-15

4-16. For the circuit in Fig. P4-16(a), calculate and plot the transient response of the output voltage from 0 to 2 ms with a time increment of 5 μs. The input voltage is shown in Fig. P4-16(b). The results should be available for display and hard copy by Probe.

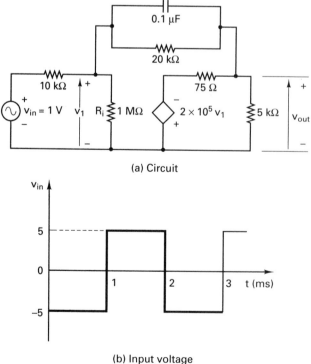

(a) Circuit

(b) Input voltage

Figure P4-16

4-17. Repeat Problem 4-16 for the input voltage shown in Fig. P4-17.

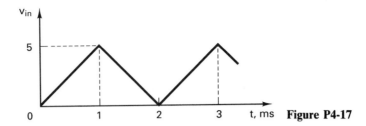

Figure P4-17

4-18. The parallel *RLC* circuit of Fig. P4-18(a) is supplied from a step input current as shown in Fig. P4-18(b). Use PSpice to calculate and plot the capacitor voltage v_C from 0 to 500 μs with a time increment of 1 μs for R = 200 Ω, 500 Ω, and 1 kΩ.

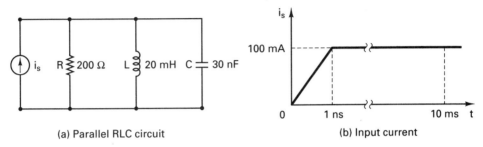

(a) Parallel RLC circuit

(b) Input current

Figure P4-18

4-19. A switch that is connected between nodes 5 and 4 is controlled by a voltage source between nodes 3 and 0. The switch will conduct if the controlling voltage is 0.5 V. The on-state resistance is 0.5 Ω, and the off-state resistance is 2E+6 Ω.

4-20. A switch that is connected between nodes 5 and 4 is controlled by a current. The voltage source V_1 through which the controlling current flows is connected between nodes 2 and 0. The switch will conduct if the controlling current is 0.55 mA. The on-state resistance is 0.5 Ω, and the off-state resistance is 2E+6 Ω.

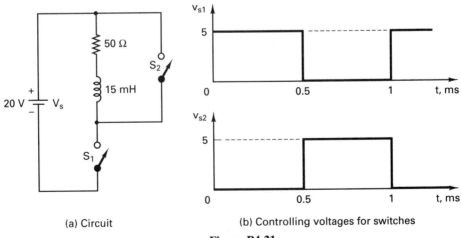

(a) Circuit

(b) Controlling voltages for switches

Figure P4-21

4-21. For the circuit in Fig. P4-21, plot the transient response of the load and source current for 5 cycles of the switching period, with a time increment of 25 μs. The model parameters of the voltage-controlled switches are RON=0.025, ROFF=1E+8, VON=0.05, and VOFF=0. The output should also be available for display and hard copy by Probe.

4-22. An *RLC* circuit is shown in Fig. P4-22. The voltage-controlled switch S_1 is closed to position *a* at $t = 0$ and then moved instantaneously to position *b* at $t = 10$ ms. Use PSpice to calculate and plot the capacitor voltage v_C from 0 to 20 ms with a time increment of 5 μs. Assume that the model parameters of the switch are RON=0.01, ROFF=10E+5, VON=0.1V, and VOFF=0V.

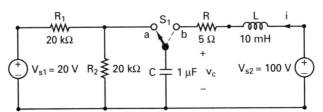

Figure P4-22

AC Circuit Analysis

5-1 INTRODUCTION

Sources in a circuit are time-variant. They are alternating current, or ac, and have both magnitude and phase displacement. Practical sources are generally sinusoidal or near sinusoidal. The behavior of a circuit is normally evaluated with a sinusoidal source. The steady-state voltages and currents that are normally used to evaluate the performance of a circuit can be calculated by applying the circuit laws that are applicable to dc circuits. The transient behavior is also important under some conditions, but the analysis becomes more complex compared to steady-state analysis.

To simplify the steady-state analysis of ac circuits with sinusoidal inputs, the circuit elements (R, L, and C) are represented in complex numbers, and all voltages and currents are expressed in phasor qualities. For example, a sinusoidal voltage of V_m with a phase delay of ϕ is represented as

$$V = V_m\underline{/\phi}$$

Inductance (L) and capacitor (C) are expressed as impedances:

$$Z_L = j2\pi L = j\omega$$

$$Z_C = -j/(2\pi C) = -j/(\omega C)$$

SPICE is an ideal software tool for simulating a circuit and for studying the behavior of voltages, currents, and power flow under steady-state and transient conditions. We have discussed the SPICE representations of circuit elements and sources in Chapters 3 and 4. In this chapter, we will simulate the steady-state analysis of a circuit and include coupled and nonlinear magnetic elements. Students are encouraged to apply the basic circuit laws and to verify the SPICE results by hand calculations.

In ac analysis, the output variables are sinusoidal quantities and are represented by complex numbers. An output variable can have magnitude in decibels, phase, group delay, real part, and imaginary part. The output variables listed in Sections 5-2.1 and 5-2.2 are augmented by adding a suffix as follows:

SUFFIX	MEANING
(none)	Peak magnitude
M	Peak magnitude
DB	Peak magnitude in decibels
P	Phase in radians
G	Group delay (ΔPHASE/ΔFREQUENCY)
R	Real part
I	Imaginary part

5-2.1 Voltage Output

The statements for ac analysis are similar to those for the dc sweep and the transient analysis, provided the suffixes are added as follows:

VARIABLES	
VM(5)	Magnitude of voltage at node 5 with respect to ground
VM(4,2)	Magnitude of voltage at node 4 with respect to node 2
VDB(R1)	DB magnitude of voltage across resistor R_1, where the first node (as defined in the circuit file) is assumed to be positive with respect to second node
VP(D1)	Phase of anode voltage of diode D_1 with respect to cathode
VCM(Q3)	Magnitude of the collector voltage of transistor Q_3 with respect to ground
VDSP(M6)	Phase of the drain-source voltage of MOSFET M_6
VBP(T1)	Phase of voltage at port B of transmission line T_1
VR(2,3)	Real part of voltage at node 2 with respect to node 3
VI(2,3)	Imaginary part of voltage at node 2 with respect to node 3

5-2.2 Current Output

The statements for ac analysis are similar to those for the dc sweep and the transient responses. However, only the currents through the elements in Table 5-1 are available. For all other elements, a zero-valued voltage source must be placed in series with the device (or device terminal) of interest. Then a print or plot statement should be used for the current through this voltage source.

TABLE 5-1 CURRENT THROUGH
ELEMENTS FOR AC ANALYSIS

First letter	Element
C	Capacitor
I	Independent current source
L	Inductor
R	Resistor
T	Transmission line
V	Independent voltage source

VARIABLES	MEANING
IM(R5)	Magnitude of current through resistor R_5
IR(R5)	Real part of current through resistor R_5
II(R5)	Imaginary part of current through resistor R_5
IM(VIN)	Magnitude of current through source v_{in}
IR(VIN)	Real part of current through source v_{in}
II(VIN)	Imaginary part of current through source v_{in}
IAG(T1)	Group delay of current at port A of transmission line T_1

Example 5-1

For the circuit in Fig. 5-1, write the various voltages and currents in forms that are allowed by PSpice. The dummy voltage source of 0 V is introduced to measure current, I_L.

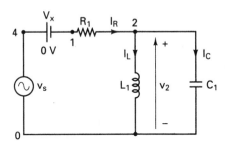

Figure 5-1 Circuit for Example 5-1.

Solution

PSpice VARIABLES		MEANING
V_2	VM(2)	The peak magnitude of voltage at node 2
$\underline{/V_2}$	VP(2)	The phase angle of voltage at node 2
$V_{1,2}$	VM(1,2)	The peak magnitude of voltage between nodes 1 and 2
$\underline{/V_{1,2}}$	VP(1,2)	The phase angle of voltage between nodes 1 and 2
I_R	IM(VX)	The magnitude of current through voltage source v_X
$\underline{/I_R}$	IP(VX)	The phase angle of current through voltage source v_X

PSpice VARIABLES		MEANING
I_L	IM(L1)	The magnitude of current through inductor L_1
$\underline{/I_L}$	IP(L1)	The phase angle of current through inductor L_1
I_C	IM(C1)	The magnitude of current through capacitor C_1
$\underline{/I_C}$	IP(C1)	The phase angle of current through capacitor C_1

5-3 INDEPENDENT AC SOURCES

The statements for a voltage and current source have the following general forms, respectively:

```
V⟨name⟩  N+  N-  [AC⟨(magnitude) value⟩ ⟨(phase) value⟩]
I⟨name⟩  N+  N-  [AC⟨(magnitude) value⟩ ⟨(phase) value⟩]
```

The ⟨(magnitude) value⟩ is the peak value of sinusoidal voltage. The ⟨(phase) value⟩ is in degrees.

Some Typical Statements

```
VAC   5  6  AC  1V          ; Ac specification of 1V with 0 delay
VACP  5  6  AC  1V  45DEG   ; Ac specification of 1V with 45 delay
IAC   5  6  AC  1A          ; Ac specification of 1A with 0 delay
IACP  5  6  AC  1A  45DEG   ; Ac specification of 1V with 45 delay
```

5-4 AC ANALYSIS

The ac analysis calculates the frequency response of a circuit over a range of frequencies. Let us consider an RC circuit, as shown in Fig. 5-2(a). The output voltage is taken across the capacitor C. Using the voltage-divider rule, the voltage gain $G(j\omega)$, which is the ratio of the output voltage $V_{out}(j\omega)$ to the input voltage $V_{in}(j\omega)$ can be expressed as

$$G(j\omega) = \frac{V_{out}(j\omega)}{V_{in}(j\omega)} = \frac{-j/(\omega C)}{R - j/(\omega C)} = \frac{1}{1 + j\omega RC}$$

$$= \frac{1}{1 + j\omega\tau} \tag{5-1}$$

where $\tau = RC$

Thus, the magnitude $|G(j\omega)|$ can be found from

$$|G(j\omega)| = \frac{1}{[1 + (\omega\tau)^2]^{1/2}} = \frac{1}{[1 + (\omega/\omega_o)^2]^{1/2}} \tag{5-2}$$

and the phase angle ϕ of $G(j\omega)$ is given by

$$\phi = -\tan^{-1}(\omega\tau) = -\tan^{-1}(\omega/\omega_o) \tag{5-3}$$

where $\omega_o = 1/RC = 1/\tau$

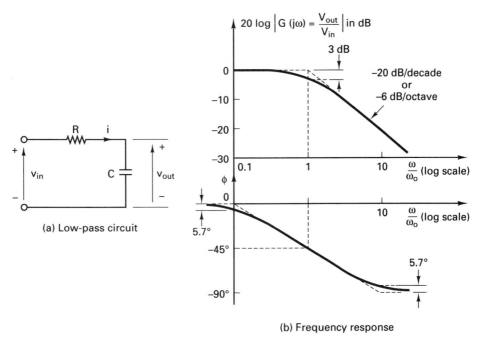

Figure 5-2 Frequency response of an *RC* circuit.

The frequency response is shown in Fig. 5-2(b). If the frequency is doubled, it is called an *octave* in frequency axis. If the frequency is increased by a factor of 10, it is called a *decade*. Thus for a decade increase in frequency, the magnitude changes by -20 dB, and the magnitude plot is a straight line with a slope of -20dB/decade or -6dB/octave. The magnitude curve is therefore defined by two straight-line asymptotes, which meet at the *corner frequency* (or *break frequency*), ω_0. The difference between the actual magnitude curve and the asymptotic curve is the largest at the break frequency. The error can be found by substituting $\omega = \omega_0$, $|G(j\omega)| = 1/\sqrt{2}$, and $20\log_{10}(1/\sqrt{2}) = -3$ dB. This error is symmetrical with respect to the break frequency. The break frequency is also known as the *3-db frequency*.

If the circuit contains nonlinear devices or elements, it is necessary to obtain the small-signal parameters of the elements before calculating the frequency response. Prior to the frequency response (or ac analysis), PSpice determines the small-signal parameters of the elements. The method for calculating the bias point for ac analysis is identical to that for dc analysis. The details of the bias points can be printed by the .OP command.

The command for performing frequency response takes one of the following general forms:

```
.AC   LIN   NP   FSTART   FSTOP
.AC   OCT   NP   FSTART   FSTOP
.AC   DEC   NP   FSTART   FSTOP
```

NP is the number of points in a frequency sweep. FSTART is the starting frequency, and FSTOP is the ending frequency. Only one of LIN, OCT, or DEC must be specified in the statement. LIN, OCT, or DEC specify the type of sweep, as follows:

LIN *Linear sweep:* The frequency is swept linearly from the starting frequency to the ending frequency, and NP becomes the total number of points in the sweep. The next frequency is generated by adding a constant to the present value. LIN is used if the frequency range is narrow.

OCT *Sweep by octave:* The frequency is swept logarithmically by octave, and NP becomes the number of points per octave. The next frequency is generated by multiplying the present value by a constant larger than unity. OCT is used if the frequency range is wide.

DEC *Sweep by decade:* The frequency is swept logarithmically by decade, and NP becomes the number of points per decade. DEC is used if the frequency range is the widest.

PSpice does not print or plot any output by itself for ac analyses. The results of ac sweep are obtained by .PRINT, .PLOT, or .PROBE statements.

Some Statements for Ac Analysis

```
.AC  LIN  201  100HZ  300HZ
.AC  LIN  1    60HZ   120HZ
.AC  OCT  10   100HZ  10KHZ
.AC  DEC  100  1KHZ   1MEGHZ
```

Notes

1. FSTART must be less than FSTOP and cannot be zero.
2. NP = 1 is permissible, and the second statement calculates the frequency response at 60 Hz only.
3. Before performing the frequency-response analysis, PSpice automatically calculates the bias point to determine the linearized circuit parameters around the bias point.
4. All independent voltage and current sources that have ac values are inputs to the circuit. At least one source must have an ac value; otherwise, the analysis would not be meaningful.
5. If a group delay output is required by a suffix of G, as mentioned in Section 3-3, the frequency steps should be small, so that the output changes smoothly.

Example 5-2

For the *RLC* circuit shown in Fig. 4-12(a), with R = 2 Ω, 1 Ω, and 8 Ω, calculate and print the frequency response over the frequency range from 100 Hz to 100 kHz with a decade increment and 100 points per decade. The peak magnitude and phase angle of the voltage across the capacitors are to be plotted on the output file. The results should also be available for display and hard copy by the .PROBE command.

Solution The circuit file is similar to that of Example 4-3, except the statements for the type of analysis and output are different. The input voltage is a type, and the frequency is variable. We can consider a voltage source with a peak magnitude of 1V.

The frequency-response analysis is invoked by the .AC command. For NP = 100, FSTART = 100 Hz, and FSTOP = 100 kHz, the statement is

```
.AC   DEC   100   100 100KHZ
```

The magnitude and phase of voltage V(3) are specified as VM(3) and VP(3). The statement to plot is

```
.PLOT   AC   VM(3)   VP(3)
```

The circuit file contains the following statements.

Example 5-2   Frequency response of *RLC* circuit

```
▲ VI1   1   0   AC   1V      ; Ac voltage of 1 V
  VI2   4   0   AC   1V      ; Ac voltage of 1 V
  VI3   7   0   AC   1V      ; Ac voltage of 1 V
▲▲ R1   1   2   2
   L1   2   3   50UH
   C1   3   0   10UF
   R2   4   5   1
   L2   5   6   50UH
   C2   6   0   10UF
   R3   7   8   8
   L3   8   9   50UH
   C3   9   0   10UF
▲▲▲ *    .AC   DEC   NP   FSTART   FSTOP        ; Format for .AC command
     .AC   DEC   100      100HZ   100KHZ        ; Ac sweep
     .PLOT      AC   VM(3)      VP(3)           ; Plots on the output file
     .PROBE                                     ; Graphical waveform analyzer
.END                                            ; End of circuit file
```

The results of the frequency response that are obtained on the display by the .PROBE command are shown in Fig. 5-3. The results of the .PLOT statement can be obtained by printing the contents of the output file EX5-2.OUT. A circuit file can contain both .AC and .TRAN commands in order to perform two analyses.

Example 5-3

An ac circuit is shown in Fig. 5-4. Use PSpice to calculate the *rms* input current I_1, Thévenin's *rms* voltage $V_{Th} = V_{ab}$ and impedance Z_{Th}.

Solution V_s = 120 V (*rms*), and peak voltage $V_m = \sqrt{2} \times 120 = 169.7$ V. Thévenin's voltage V_{Th} can be found by measuring open-circuit voltage V_{ab} between terminals *a* and *b*. The listing of the circuit file for determining V_{Th} and transient response follows.

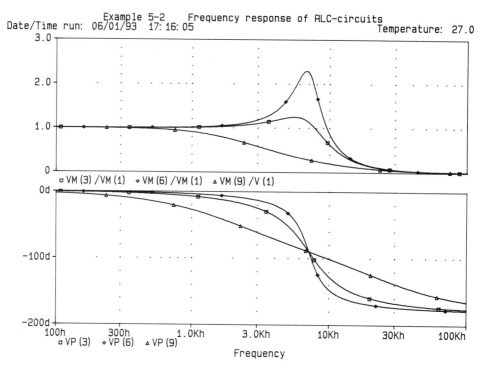

□ VM (3) /VM (1)   ◇ VM (6) /VM (1)   ▵ VM (9) /V (1)

□ VP (3)   ◇ VP (6)   ▵ VP (9)

Frequency

Figure 5-3 Frequency responses of the *RLC* circuit for Example 5-2.

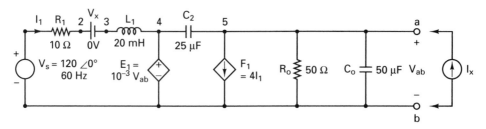

Figure 5-4 Ac circuit with controlled sources.

Example 5-3   Ac circuit analysis

```
▲ VS  1  0  AC  120V   ; Ac voltage of 120 V (peak)
  IX  5  0  AC  0A     ; Only for finding Thévenin's impedance
▲▲ R1  1  2  10
  RO  5  0  50
  L1  3  4  20MH
  C1  4  5  25UF
  CO  5  0  50UF
  VX  2  3  DC      0V  ; Measures current through R₁
  F1  5  0  VX      4   ; Current-controlled current source
  E1  4  0  5  0  1M    ; Voltage-controlled voltage source
```

```
▲▲▲ .AC  LIN  1  60Hz  120HZ  ; Only one ac sweep
    .PRINT  AC  IM(R1)  IP(R1)          ; Prints the input current
    .PRINT  AC  VR(5)  VI(5)  VM(5)  VP(5) ; Prints the output voltage
.END                                   ; End of circuit file
```

The results of the ac sweep, which are obtained from the output file EX5-3.OUT are shown below:

```
****      AC ANALYSIS                    TEMPERATURE =   27.000 DEG C
      FREQ        IM(R1)        IP(R1)        VM(5)        VP(5)
    6.000E+01    9.579E+00   -3.754E+01    1.107E+03    8.774E+01
```

$$I_1 = 9.579 \, \underline{/-37.54°} \text{ A}, \quad \text{and} \quad V_{Th} = V_{ab} = 1107 \, \underline{/87.74°} \text{ V}$$

Since the .TF command can be applied to ac circuits, we need to find a way to measure the output impedance Z_{Th}. The simplest way is to apply a test current I_x between terminals a and b with voltage source V_s short-circuited. The circuit file EX5-3.CIR can be modified by adding a test current I_x of 1 A between nodes 5 and 0, and by setting V_S to 0 V for ac analysis. These statements become

```
VS  1  0  AC  0V  ; Ac voltage of 0 V
IX  5  0  AC  1A  ; Only for finding Thévenin's resistance
```

The results of simulation for Thévenin's output impedance are shown below:

```
   VR(5)        VI(5)        VM(5)        VP(5)
-1.646E+01    2.372E+01    2.887E+01    1.248E+02
```

$$Z_{Th} = -16.46 + j\, 23.72 = 28.87 \, \underline{/124.8°} \, \Omega$$

Example 5-4

The ac circuit shown in Fig. 5-5 is supplied from a three-phase balanced supply. Use PSpice (a) to calculate and plot the instantaneous currents i_a, i_b, i_c, and i_n, and total (average) input power P_{in} from 0 to 50 ms with a time increment of 5 μs, and (b) to

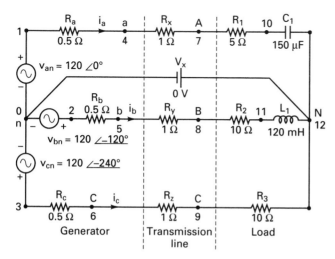

Figure 5-5 Three-phase circuit.

calculate the *rms* magnitudes and phase angles of currents: I_a, I_b, I_c, and I_n.
Solution V_{an} = 120 V (*rms*), and peak phase voltage $V_m = \sqrt{2} \times 120 = 169.7$ V.
The listing of the circuit file follows.

Example 5-4   Three-phase circuit

```
▲ Van   1   0   AC  120V    0   SIN(0  169.7V  60HZ)
  Vbn   2   0   AC  120V  120   SIN(0  169.7V  60HZ  0  0  120DEG)
  Vcn   3   0   AC  120V  240   SIN(0  169.7V  60HZ  0  0  240DEG)
▲▲ RA   1   4   0.5
   RB   2   5   0.5
   RC   3   6   0.5
   RX   4   7   1
   RY   5   8   1
   RZ   6   9   1
   R1   7  10   5
   R2   8  11   10
   R3   9  12   10
   C1  10  12   150UF
   L2  11  12   120MH
   VX  12   0   DC   0V    ; Measures the neutral line current
▲▲▲ .TRAN  5US  50MS              ; Transient analysis
    .AC   LIN   1   60HZ 120HZ  ; Only one ac sweep
    *   Prints magnitude and phase of input currents and output voltages
    .PRINT  AC  IM(RA)  IP(RA)  VM(7,12)  VP(7,12)
    .PRINT  AC  IM(RB)  IP(RB)  VM(8,12)  VP(8,12)
    .PRINT  AC  IM(RC)  IP(RC)  VM(9,12)  VP(9,12)
    .PRINT  AC  IM(VX)  IP(VX)  ; Neutral line current
    .PROBE                      ; Graphical waveform analyzer
.END                            ; End of circuit file
```

(a) The currents I(RA), I(RB), I(RC) and I(VX), which are obtained by the .PROBE command, are shown in Fig. 5-6. The total (average) input power P_{in} is shown in Fig. 5-7. Due to unbalanced loads, the line currents are unbalanced, and a current flows through the neutral line. For a balanced supply with balanced loads, the neutral current will be zero.

(b) The results of the ac sweep, which are obtained from the output file EX5-4.OUT, are shown below.

```
****     AC ANALYSIS                   TEMPERATURE =  27.000 DEG C
    IM(R1)        IP(R1)       VM(5)       VP(5)
  9.579E+00   -3.754E+01   1.107E+03   8.774E+01
```
$I_1 = 9.579 \; \underline{/-37.54°}$ A, and $V_{Th} = V_{ab} = 1107 \; \underline{/87.74°}$ V
```
    IM(RA)        IP(RA)       VM(7,12)    VP(7,12)
  6.369E+00    6.982E+01   1.170E+02   -4.394E+00
```
$I_a = 6.369 \; \underline{/69.82°}$ A, and $V_{AN} = 117 \; \underline{/-4.394°}$ V
```
    IM(RB)        IP(RB)       VM(8,12)    VP(8,12)
  2.571E+00    4.426E+01   1.191E+02    1.218E+02
```
$I_b = 2.571 \; \underline{/44.26°}$ A, and $V_{BN} = 119.1 \; \underline{/121.8°}$ V
```
    IM(RC)        IP(RC)       VM(9,12)    VP(9,12)
  1.043E+01   -1.200E+02   1.043E+02   -1.200E+02
```

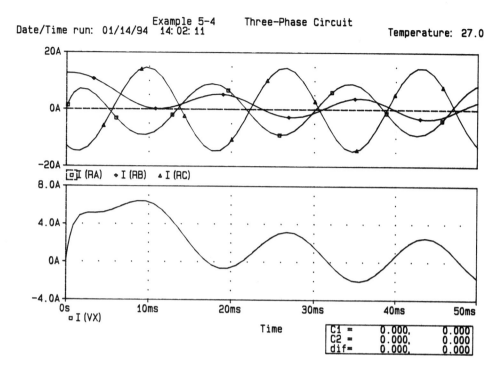

Figure 5-6 Instantaneous current for Example 5-4.

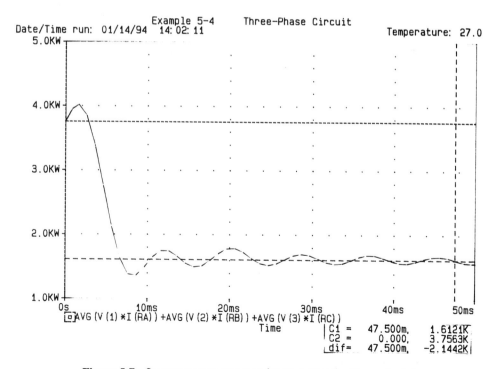

Figure 5-7 Instantaneous average input power for Example 5-4.

$$I_c = 10.43 \; \underline{/-120°} \; \text{A}, \quad \text{and} \quad V_{CN} = \underline{/-120°} \; \text{V}$$

$$\begin{array}{cc} \text{IM (VX)} & \text{IP (VX)} \\ 1.729\text{E}+00 & -1.330\text{E}+02 \end{array}$$

$$I_n = 1.729 \; \underline{/-133°} \; \text{A}$$

5-5 MAGNETIC ELEMENTS

The magnetic elements are mutual inductors (transformers). The symbol for mutual coupling is K. The general form of coupled inductors is

```
K⟨name⟩  L⟨(1st inductor) name⟩  L⟨(2nd inductor) name⟩
+        ⟨(coupling) value⟩
```

For couple inductors, K⟨name⟩ couples two or more inductors, and ⟨(coupling) value⟩ is the coefficient of coupling k. The value of the coefficient of coupling must be greater than 0 and less than or equal to 1, $0 < k \leq 1$.

The inductors can be coupled in either order. In terms of the *dot* convention as shown in Fig. 5-8(a), PSpice assumes a dot on the first node of each inductor.

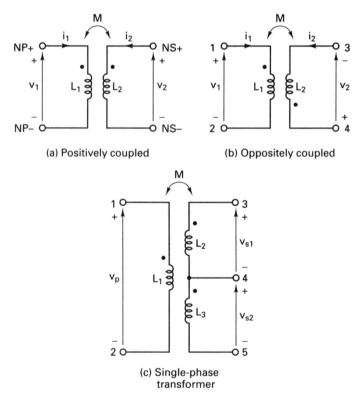

(a) Positively coupled   (b) Oppositely coupled

(c) Single-phase
transformer

Figure 5-8   Coupled inductors.

The mutual inductance is determined from

$$M = k\sqrt{L_1 L_2}$$

In the time domain, the voltages of coupled inductors are expressed as

$$v_1 = L_1 \frac{di_1}{dt} + M \frac{di_2}{dt}$$

$$v_2 = M \frac{di_1}{dt} + L_2 \frac{di_2}{dt}$$

In frequency domain, the voltages are expressed as

$$V_1 = j\omega L_1 I_1 + j\omega M I_2$$

$$V_2 = j\omega M I_1 + j\omega L_2 I_2$$

where ω is the frequency in radians per second.

Some Coupled Inductor Statements

```
KTR    LA  LB   0.9
KIND   L1  L2   0.98
```

The coupled inductors in Fig. 5-8(a) can be written as a single-phase transformer (with $k = 0.9999$):

```
*  PRIMARY
L1         1   2    0.5MH
*  SECONDARY
L2         3   4    0.5MH
*  MAGNETIC COUPLING
KXFRMER  L1  L2  0.9999
```

If the dot in the second coil is changed as shown in Fig. 5-8(b), the coupled inductors are written as

```
L1         1   2  0.5MH
L2         4   3  0.5MH
KXFRMER  L1  L2  0.9999
```

A transformer with a single primary coil and center-tapped secondary, as shown in Fig. 5-8(c), can be written as

```
*  PRIMARY
L1    1   2    0.5MH
*  SECONDARY
L2    3   4    0.5MH
L3    4   5    0.5MH
```

```
*  MAGNETIC COUPLING
K12  L1   L2   0.9999
K13  L1   L3   0.9999
K23  L2   L3   0.9999
```

These three statements can be written in PSpice as KALL L1 L2 L3 0.9999.

Notes

1. The name K*xx* need not be related to the names of the inductors it is coupling. However, it is a good practice, because it is convenient to identify the inductors involved in the coupling.

2. The polarity (or dot) is determined by the order of the nodes in the L . . . statements and not by the order of the inductors in the K . . . statement— e.g., (K12 L1 L2 0.9999) has the same result as (K12 L2 L1 0.9999).

For a nonlinear inductor, the general form is

```
K⟨name⟩  L⟨(inductor) name⟩  ⟨(coupling) value⟩
+        ⟨(model) name⟩  [(size) value]
```

For an iron-core transformer, k is very high and is greater than 0.9999. The model type name for a nonlinear magnetic inductor is CORE; the model parameters are shown in Table 5-2. The [(size) value] scales the magnetic cross section and defaults to 1. It represents the number of lamination layers, so that only one model statement can be used for a particular lamination type of core.

TABLE 5-2 MODEL PARAMETERS FOR NONLINEAR MAGNETIC ELEMENTS

Name	Meaning	Units	Default
AREA	Mean magnetic cross section	centimeters^2	0.1
PATH	Mean magnetic path length	centimeters	1.0
GAP	Effective air-gap length	centimeters	0
PACK	Pack (stacking)		1.0
MS	Magnetic saturation	A/meters	1E+6
ALPHA	Mean field parameter		1E−3
A	Shape parameter		1E+3
C	Domain wall-flexing constant		0.2
K	Domain wall-pinning constant		500

If the ⟨(model) name⟩ is specified, then the mutual coupling inductor becomes a nonlinear magnetic core, and the inductors specify the number of turns instead of inductance. The list of the coupled inductors may be just one inductor. The magnetic core's B-H characteristics are analyzed using the Jiles-Atherton model [2]. The procedures to adjust the model parameters to specified B-H characteristics are described in Appendix C.

The statements for the coupled inductors in Fig. 5-8(a) are similar to the following:

```
*    Inductor L1 of 100 turns:
L1        1    2    100
*    Inductor L2 of 10 turns:
L2        3    4    10
*    Nonlinear coupled inductors with model CMOD
K12   L1  L2  0.9999   CMOD
*    Model for the nonlinear inductors
.MODEL CMOD CORE  (AREA=2.0 PATH=62.8 GAP=0.1 PACK=0.98)
```

Note. A nonlinear magnetic model is not available in SPICE2.

Example 5-5

A circuit with two coupled inductors is shown in Fig. 5-9. If the input voltage is 120 V peak, calculate the magnitude and phase of the output current for frequencies from 60 to 120 Hz with a linear increment. The total number of points in the sweep is 2. The coefficient of coupling for the transformer is 0.999.

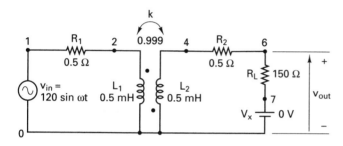

Figure 5-9 Circuit with two coupled inductors.

Solution It is important to note that the primary and the secondary windings have a common node. Without this, PSpice will give an error message, because there is no dc path from the nodes of the secondary to the ground. The voltage source, $V_X = 0$ V, is connected to measure the output current I_L. The circuit file contains the following statements.

Example 5-5 Coupled inductors

```
▲ VIN  1   0   AC   120    ; Ac voltage is 120 V peak and 0 degree phase.
▲▲ R1  1   2   0.5
   *   The dot convention is followed in inductors L1 and L2.
   L1  2   0   0.5MH
   L2  0   4   0.5MH
   *   Magnetic coupling coefficient is 0.999.
   K12  L1   L2   0.999 ; Order of L1 and L2 is not significant.
   R2  4   6   0.5
   RL  6   7   150
   VX  7   0   DC  0V   ; Measure the load current.
▲▲▲ *  AC linear sweep from 60 Hz to 120 Hz with 2 points
   .AC  LIN  2  60HZ  120HZ
   *    Print the magnitude and phase of output current. Some versions of
   *    PSpice and SPICE do not permit reference to currents through
   resistors
```

```
*    e.g., IM(RL)  IP(RL).
.PRINT   AC  IM(VX)  IP(VX)  IM(RL)  IP(RL)
.END                ; End of circuit file
```

The results of the simulation that are obtained from the output file EX5-5.OUT follow.

```
****      AC ANALYSIS                    TEMPERATURE =  27.000 DEG C
  FREQ         IM(VX)      IP(VX)       IM(RL)       IP(RL)
  6.000E+01    2.809E-01  -1.107E+02   2.809E-01   -1.107E+02
  1.200E+02    4.790E-01  -1.271E+02   4.790E-01   -1.271E+02
              JOB CONCLUDED
              TOTAL JOB TIME           8.18
```

5-6 TRANSMISSION LINES

The symbol for a lossless transmission line is T. A transmission line has two ports—input and output [7,8]. The general form of a transmission line is

```
T⟨name⟩  NA+  NA-  NB+  NB-  Z0=⟨value⟩  [TD=⟨value⟩]
+           [F=⟨value⟩  NL=⟨value⟩]
```

T⟨name⟩ is the name of the transmission line. NA+ and NA− are the nodes at the input port. NB+ and NB− are the nodes at the output port. NA+ and NB+ are defined as the positive nodes. NA− and NB− are defined as the negative nodes. The positive current flows from NA+ to NA− and from NB+ to NB−. Z_0 is the characteristic impedance.

The length of the line can be expressed in either of two forms: (1) the transmission delay TD may be specified, or (2) the frequency F may be specified together with NL, which is the normalized electrical length of the transmission line with respect to wavelength in the line at frequency F. If the frequency F is specified but NL is not, then the default value of NL is 0.25—that is, F has quarter-wave frequency. It should be noted that one of the options for expressing the length of the line must be specified—that is, TD or at least F must be specified. The block diagram of transmission line is shown in Fig. 5-10(a).

A coaxial line, as shown in Fig. 5-10(b), can be represented by two propagating lines, where the first line (T_1) models the inner conductor with respect to the shield, and the second line (T_2) models the shield with respect to the outside. This is shown as follows:

```
T1  1  2  3  4  Z0=50    TD=1.5NS
T2  2  0  4  0  Z0=150   TD=1NS
```

A lossy transmission line can be represented by "lumped line segments," as shown in Fig. 5-11. R, L, G, and C are the per-unit-length values of resistance, inductance, conductance, and capacitance, respectively.

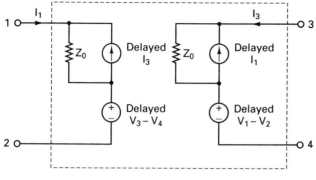

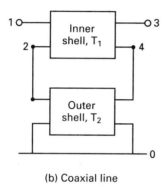

(a) Transmission line

(b) Coaxial line

Figure 5-10 Transmission line.

The general form of a lossy transmission line is

```
T⟨name⟩   NA+   NA−   NB+   NB−
+      LEN=⟨value⟩   R=⟨value⟩   L=⟨value⟩   G=⟨value⟩   C=⟨value⟩
```

where LEN is the per-unit electrical length.

It should be noted that line parameters can be expressions; R and G can be Laplace's expressions, thereby permitting frequency-dependent effects, such as skin effect and dielectric loss. Expressions are discussed in Section 6-2.

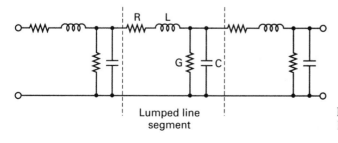

Lumped line segment

Figure 5-11 Lossy transmission line.

Some Transmission Statements

```
T1     1   2    3    4    Z0=50   TD=10NS
T2     4   5    6    7    Z0=50   F=2MHZ
TTRM   9   10   11   12   Z0=50   F=2MHZ   NL=0.4
TR     2   3    7    9    LEN=1.1  R=0.45  L=450U  G=3.5U  C=105P
```

Note. During the transient (.TRAN) analysis, the internal time step of PSpice is limited to be no more than one-half of the smallest transmission delay. Thus short transmission lines will cause long run times.

5-7 MULTIPLE ANALYSES

A source may be assigned the specifications of dc, ac, and transient analysis for different analyses. The statements for voltage and current sources have the following general forms, respectively:

```
V⟨name⟩   N+      N−
+         [DC ⟨value⟩]
+         [AC ⟨(magnitude) value⟩ ⟨(phase) value⟩]
+         [⟨(transient specifications)⟩]

I⟨name⟩   N+      N−
+         [DC ⟨value⟩]
+         [AC ⟨(magnitude) value⟩ ⟨(phase) value⟩]
+         [⟨(transient specifications)⟩]
```

The source is set to the dc value in dc analysis. It is set to an ac value in ac analysis. The time-dependent source (e.g., PULSE, EXP, SIN, etc.) is assigned for transient analysis. This allows source specifications for different analyses in the same statement.

Typical Statements

```
VIN   25   22   DC   2V   AC   1V   30   SIN (0   2V   10KHZ)
IIN   25   22   DC   2A   AC   1A   30DEG  SIN (0   2A 10KHZ)
```

Notes
1. VIN assumes 2 V for dc analysis, 1 V with a delay angle of 30° for ac analysis, and a sine wave of 2 V at 10 kHz for transient analysis.
2. IIN assumes 2 A for dc analysis, 1 A with a delay angle of 30° for ac analysis, and a sine wave of 2 A at 10 kHz for transient analysis.

Example 5-6

A series-parallel *RLC* circuit is shown in Fig. 5-12. The output voltage is taken across the capacitor. Write the circuit descriptions for the following:

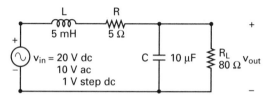

Figure 5-12 Series-parallel *RLC* circuit.

(a) to calculate and plot the instantaneous output voltage v_{out} from 0 to 5 ms with a time increment of 10 μs

(b) to calculate and plot the dc transfer characteristics, V_{out} versus V_S, from 0 to 20 V with a 10 V increment

(c) to calculate and plot the output voltage V_{out} over the frequency range from 100 Hz to 100 kHz with a decade increment and 100 points per decade

The peak magnitude and phase angle of the voltage across the capacitors are to be plotted on the output file. The results should also be available for display and hard copy by the .PROBE command.

Solution   The step input voltage is described by a piecewise linear waveform

```
PWL (0   0   1NS   1V   10MS   1V)
```

The listing of the circuit file follows.

Example 5-5   Dc/ac/transient analysis

```
▲ VS   1   0   DC   20V   AC   10V   PWL (0   0   1NS   1V   10MS   1V)  ; Input voltages
▲▲ L    1   2   5MH
   R    2   3   5
   C    3   0   10UF
   RL   3   0   80
▲▲▲ .DC    VS    0V    20V    10V        ; Dc sweep
    .TRAN   10US    5MS               ; Transient analysis
    .AC     DEC   100    100HZ    10KHZ  ; Ac sweep
    .PROBE                          ; Graphical waveform analyzer
.END                               ; End of circuit file
```

During the dc analysis, the capacitors are open-circuited, and inductors are short-circuited. Students are encouraged to plot the results of .TRAN, .DC, and .AC analysis, and to verify the results with hand calculations. The values of R, L, or C can be varied by the .PARAM and .STEP commands discussed in Sections 6-11 and 6-15, respectively.

SUMMARY

```
V⟨name⟩  N+       N−
+        [DC ⟨value⟩]
+        [AC ⟨(magnitude) value⟩ ⟨(phase) value⟩]
+        [(transient specifications)]
```

```
I⟨name⟩  N+      N–
+        [DC ⟨value⟩]
+        [AC ⟨(magnitude) value⟩ ⟨(phase) value⟩]
+        [(transient specifications)]


K⟨name⟩  L⟨(1st inductor) name⟩  L⟨(2nd inductor) name⟩
+        ⟨(coupling) value⟩


K⟨name⟩  L⟨(inductor) name⟩  ⟨(coupling) value⟩
+        ⟨(model) name⟩ [(size) value]


T⟨name⟩  NA+  NA–  NB+  NB–  Z0=⟨value⟩   [TD=⟨value⟩]
+        [F=⟨value⟩  NL=⟨value⟩]


T⟨name⟩  NA+  NA–  NB+  NB–
+     LEN=⟨value⟩  R=⟨value⟩ L=⟨value⟩  G=⟨value⟩ C=⟨value⟩


T⟨name⟩  NA+  NA–  NB+  NB–  Z0=⟨value⟩   [TD=⟨value⟩]
+        [F=⟨value⟩  NL=⟨value⟩]
```

REFERENCES

1. M. H. Rashid, *SPICE for Power Electronics and Electric Power.* Englewood Cliffs, N.J.: Prentice Hall, 1993.
2. *PSpice Manual,* Irvine, Calif.: MicroSim Corporation, 1992.
3. J. D. Irwin, *Basic Engineering Circuit Analysis.* New York: Macmillan, 1989.
4. C. R. Paul, S. A. Nassar, and L. E. Unnewehr, *Introduction to Electrical Engineering.* New York: McGraw-Hill, 1992.
5. H. W. Jackson, *Introduction to Electric Circuits.* Englewood Cliffs, N.J.: Prentice Hall, 1986.
6. R. C. Dorf, *Introduction to Electric Circuits.* New York: Wiley, 1989.
7. Mohamed E. El-Hawary, *Electrical Power Systems—Design and Analysis.* Reston, Va.: Reston, 1983, Chapter 4.
8. Charles A. Gross, *Power System Analysis.* New York: Wiley, 1986, Chapter 4.

PROBLEMS

5-1. Use PSpice to calculate and print the frequency response of the circuit shown in Fig. P4-7. The results should be available for display and hard copy by the .PROBE command.

5-2. Use PSpice to calculate and print the frequency response of the circuit shown in Fig. P4-8. The results should be available for display and hard copy by the .PROBE command.

5-3. Use PSpice to calculate and print the frequency response of the circuit shown in Fig. P4-9. The results should be available for display and hard copy by the .PROBE command.

5-4. Use PSpice to calculate and print the frequency response of the circuit shown in Fig. P4-12. The results should be available for display and hard copy by the .PROBE command.

5-5. Use PSpice to calculate and print the frequency response of the circuit shown in Fig. P4-13. The results should be available for display and hard copy by the .PROBE command.

5-6. Plot the frequency response of the circuit in Fig. P5-6 from 10 Hz to 100 kHz with a decade increment and 10 points per decade. The output voltage is taken across the capacitor. Print and plot the magnitude and phase angle of the output voltage. Assume a source voltage of 1 V peak.

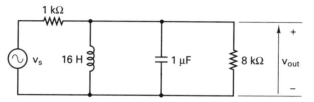

Figure P5-6

5-7. For the circuit in Fig. P5-7, the frequency response is to be calculated and printed over the frequency range from 1 Hz to 100 kHz with a decade increment and 10 points per decade. The peak magnitude and phase angle of the output voltage are to be plotted on the output file. The results should also be available for display and hard copy by the .PROBE command.

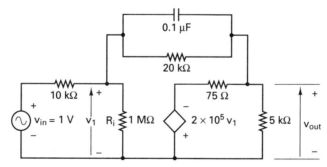

Figure P5-7

5-8. Repeat Problem 5-7 for the circuit in Fig. P5-8.

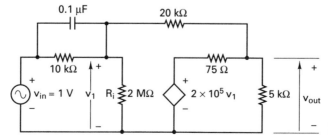

Figure P5-8

5-9. An ac circuit is shown in Fig. P5-9. Use PSpice **(a)** to calculate and plot the instantaneous output voltage v_{ab} from 0 to 33 ms with a time increment of 5 μs, and **(b)** to calculate the *rms* input current I_1, Thévenin *rms* voltage $V_{Th} = V_{ab}$ and Thévenin impedance Z_{Th}.

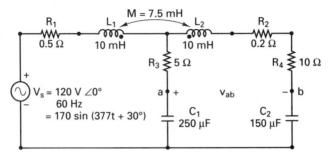

Figure P5-9

5-10. A feedback circuit is shown in Fig. P5-10. Use PSpice **(a)** to calculate and plot the voltage gain $|V_{out}/V_s|$ against the normalized frequency $u = f/f_o = 0$ to 5, where f_o is the resonant frequency, and **(b)** to calculate Thévenin *rms* voltage $V_{Th} = V_{ab}$ and Thévenin impedance Z_{Th}.

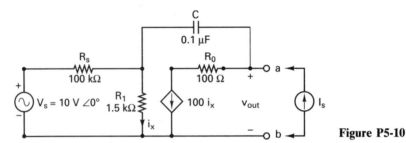

Figure P5-10

5-11. The ac circuit shown in Fig. P5-11 is supplied from a three-phase balanced supply. Use PSpice **(a)** to calculate and plot the instantaneous currents i_a, i_b, i_c, and i_n, and total (average) input power P_{in} from 0 to 50 ms with a time increment of 5 μs, and **(b)** to calculate the *rms* magnitudes and phase angles of currents: I_a, I_b, I_c, and I_n. Compare the results with those of Example 5-4.

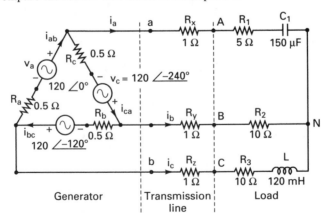

Figure P5-11

5-12. A single-phase transformer, as shown in Fig. P5-12, has a center-tapped primary, where $L_p = 1.5$ mH, $L_s = 1.3$ mH, and $K_{ps} = K_{sp} = 0.999$. Write PSpice statements.

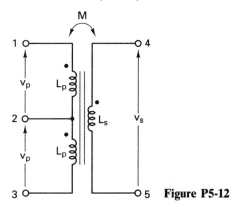

Figure P5-12

5-13. A three-phase transformer, which is shown in Fig. P5-13, has $L_1 = L_2 = L_3 = 1.2$ mH, and $L_4 = L_5 = L_6 = 0.5$ mH. The coupling coefficients between the primary and secondary windings of each phase are $K_{14} = K_{41} = K_{25} = K_{52} = K_{36} = K_{63} = 0.9999$. There is no cross-coupling with other phases. Write PSpice statements.

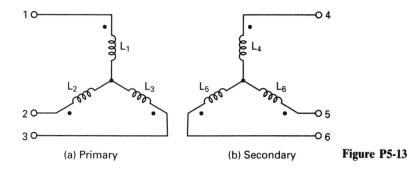

(a) Primary (b) Secondary **Figure P5-13**

Advanced SPICE Commands and Analysis

6-1 INTRODUCTION

In addition to the commands described in Chapters 3, 4, and 5, PSpice has other commands that can enhance and simplify the analysis of electrical and electronics circuits. PSpice allows one (1) to model an element based on its parameters, (2) to model a small circuit that is repeated a number of times in the main circuit, (3) to use a model that is defined in another file, (4) to create a user-defined function, (5) to use parameters instead of number values, and (6) to use parameter variations. The following commands and features are available:

1. Behavioral modeling
2. .SUBCKT Subcircuit
3. .ENDS End of subcircuit
4. .FUNC Function
5. .GLOBAL Global
6. .INC Include file
7. .LIB Library file
8. .NODESET Nodeset
9. .OPTIONS Options
10. .PARAM Parameter
11. .FOUR Fourier analysis
12. .NOISE Noise analysis
13. .SENS Sensitivity analysis
14. .STEP Parametric analysis
15. .DC Dc parametric sweep

16. .MC Monte Carlo analysis

17. DEV/LOT Device tolerances

18. .WCASE Worst-case analysis

6-2 BEHAVIORAL MODELING

PSpice allows modeling of a device in terms of the relation between its input and output. The relationship is specified as an extension to a voltage-controlled voltage source (E) or a voltage-controlled current-source (G), and takes one of the following forms:

VALUE

TABLE

LAPLACE

FREQ

This type of modeling is known as *behavioral modeling* and is only available with the Analog Behavioral Modeling Option of PSpice. However, the student's version of PSpice allows it.

6-2.1 VALUE

The VALUE extension allows a relation to be described by a mathematical expression, and can be specified by the following general forms:

```
E⟨name⟩   N+   N-   VALUE = { ⟨expression⟩ }
G⟨name⟩   N+   N-   VALUE = { ⟨expression⟩ }
```

The keyword VALUE indicates that the relation is described in an ⟨expression⟩. The ⟨expression⟩ itself must be enclosed by braces { }. It can have the functions shown in Table 6-1, and can contain the arithmetical operators (+, −, *, and /) along with parentheses.

Notes

1. VALUE *should* be followed by a space.

2. The ⟨expression⟩ *must* fit on one line.

Typical Statements

```
ESQROOT 2  3   VALUE = {2V*SQRT(V(5)*V(2))}   ; square roots
EPWR    1  2   VALUE = {V(4,3)*I(VX)}         ; product of v and i
ELOG    3  0   VALUE = {10V*LOG(I(VS)/10mA)}  ; log of current ratio
GVCO    4  5   VALUE = {15MA*SIN(6.28*10kHz*TIME*(10V*V(7,3)))}
GRATIO  3  6   VALUE = {V(8,1)/V(7,3)}        ; voltage ratio
```

TABLE 6-1 FUNCTIONS

Function	Meaning
ABS(x)	$\lvert x \rvert$ (absolute value)
SQRT(x)	$\sqrt{x}$
EXP(x)	e^x
LOG(x)	$\ln(x)$ (log of base e)
LOG10(x)	$\log(x)$ (log of base 10)
PWR(x,y)	$\lvert x \rvert^y$
PWRS(x,y)	$+\lvert x \rvert^y$ (if $x > 0$), $-\lvert x \rvert^y$ (if $x < 0$)
SIN(x)	$\sin(x)$ (x in radians)
COS(x)	$\cos(x)$ (x in radians)
TAN(x)	$\tan(x)$ (x in radians)
ARCTAN(x)	$\tan^{-1}(x)$ (result in radians)

The value can be used to simulate linear and nonlinear resistances (or conductances). A resistance can be regarded as a current-controlled voltage source. For example, the statement

```
ERES   2   3   VALUE = {I(VX)*5K}
```

is a linear resistance with a value of 5 kΩ. VX is connected in series with ERES, and it measures the current through ERES.

Similarly, a conductance can be regarded as a voltage-controlled current source. For example, the statement

```
GCOND   2   3   VALUE = {V(2,3)*0.01}
```

is a linear conductance with a value of 0.01 mho. It should be noted that the controlling nodes are the same as the output nodes.

6-2.2 TABLE

The TABLE extension allows a relation to be described by a table, and can be specified by the following general forms:

```
E⟨name⟩ N+  N-  TABLE { ⟨expression⟩ } =
+               ⟨ ⟨(input) value⟩, ⟨(output) value⟩ ⟩*
G⟨name⟩ N+  N-  TABLE { ⟨expression⟩ } =
+               ⟨ ⟨(input) value⟩, ⟨(output) value⟩ ⟩*
```

The keyword TABLE indicates that the relation is described by a table of data. The table consists of pairs of values. The first value in a pair is the input, and the second value is the corresponding output. The ⟨expression⟩ is the input value and is used to find the corresponding output from the look-up table. If an input value falls between two entries, the output is found by linear interpolation. If the input falls outside the table's range, the output remains a constant that corresponds to the smallest (or largest) input.

Notes
   1. TABLE *must* be followed by a space.
   2. The input to the table is ⟨expression⟩, which *must* fit in one line.
   3. The TABLE's input *must* be in order from lowest to highest.

Typical Statements

The TABLE can represent the current-voltage characteristics of a diode,

```
EDIODE   5    6    TABLE { I(VDIODE) } =
+      (0.0,   0.5)    (10E-3, 0.870)  (20E-3, 0.98)  (30E-3, 1.058)
+      (40E-3, 1.115)  (50E-3, 1.173)  (60E-3, 1.212) (70E-3, 1.250)
```

where I(VDIODE) is the diode current and produces an output voltage EDIODE. The TABLE can be used to represent a constant power load P = 2 kW, with a voltage-controlled current source,

```
GCONST  2   3   TABLE  {2K/V(2,3)} = (-2K, -2K) (2K, 2K)
```

where GCONST will try to dissipate 2 kW of power regardless of the voltage across it. If the voltage V(2,3) is very small, the formula 2K/V(2,3) can give a high value of current. The TABLE limits the current between −2 kA and 2 kA.

6-2.3 LAPLACE

The LAPLACE extension allows a relation to be described by a Laplace transform function and can be specified by the following general forms:

```
E⟨name⟩ N+  N-  LAPLACE { ⟨expression⟩ } = { ⟨transform⟩ }
G⟨name⟩ N+  N-  LAPLACE { ⟨expression⟩ } = { ⟨transform⟩ }
```

The keyword LAPLACE indicates that the relation is described in Laplace's domain of *s*. The ⟨expression⟩ is the input to the Laplace's transform, and it follows the same rules as in Section 6-2.1. The output, which is defined by the ⟨transform⟩ relation, gives the value of E⟨name⟩ or G⟨name⟩.

Typical Statements

The output voltage of a lossy integrator with a time constant of 1 ms and an input voltage V(2,3) can be described by

```
EINTR  4   0   LAPLACE {V(2, 3)} = {1/(1 + 0.001*s)}
```

The output voltage of a lossless integrator with an input voltage V(2,3) can be described by

```
EINTR  4   0   LAPLACE {V(2, 3)} = {1/s}
```

A frequency-dependent impedance (with *R* and *L*) can be written as

```
GRL    2   3   LAPLACE {V(2, 3)} = {1/(1 + 0.001*s)}
```

The frequency-dependent impedance of a capacitor can be written as

```
GCAP   5   4   LAPLACE {V(5,4)} = {s}
```

Notes
  1. LAPLACE *must* be followed by a space.
  2. ⟨expression⟩ and ⟨transform⟩ *must* each fit on one line.
  3. Voltages, currents, and TIME must not appear in a Laplace transform.
  4. The LAPLACE device uses much more computer memory than does the built-in capacitor (*C*) device and should be avoided, if possible.

6-2.4 FREQ

The FREQ extension allows a relation to be described by a frequency-response table, and can be specified by the following general forms:

```
E⟨name⟩ N+ N− FREQ { ⟨expression⟩ } =
+ ⟨ ⟨(frequency) value⟩, ⟨(magnitude in dB) value⟩, ⟨(phase) value⟩ ⟩*
G⟨name⟩ N+ N− FREQ { ⟨expression⟩ } =
+ ⟨ ⟨(frequency) value⟩, ⟨(magnitude in dB) value⟩, ⟨(phase) value⟩ ⟩*
```

The keyword FREQ indicates that the relation is described by a table of frequency responses. The table contains the magnitude (in dB) and the phase (in degrees) of the response for each frequency. The ⟨expression⟩ is the input to the table at a specified frequency, and follows the same rules as in Section 6-2.1. The output, which is defined by the tabular relation at the input frequency, gives the value of E⟨name⟩ or G⟨name⟩.

If the frequency falls between two entries, the magnitude and phase of the output are found by linear interpolation. Phase is interpolated linearly, and the magnitude is interpolated logarithmically with frequencies. If the input frequency falls outside the table's range, the magnitude and the phase remain constants corresponding to the smallest (or largest) frequency.

Typical Statements

The output voltage of a low-pass filter with input voltage V(2) can be expressed by

```
ELOWPASS 2  0  FREQ {V(2)} = (0,    0,    0)
+              (5 kHz,    0,   −57.6)  (6kHz0,   40,   −69.2)
```

Notes

1. FREQ *should* be followed by a space.

2. ⟨expression⟩ *must* fit on one line.

3. The FREQ's frequencies *must* be in order from lowest to highest.

6-3 .SUBCKT (Subcircuit)

PSpice allows one to define a small circuit as a *subcircuit,* which then can be called upon in several places in the main circuit. The general form for subcircuit definition (or description) is

```
.SUBCKT  SUBNAME  [(⟨two or more⟩ nodes)]
```

Whenever there is a subcircuit definition, there is normally a call statement in order to use the subcircuit during the simulation. The symbol for a subcircuit call is X. The general form of a call statement is

```
X⟨name⟩  [(⟨two or more⟩ nodes)]  SUBNAME
```

SUBNAME is the name of the subcircuit definition, and ⟨(two or more) nodes⟩ are the nodes of the subcircuit. X⟨name⟩ causes the referenced subcircuit to be inserted into the circuit, so that the given nodes in the subcircuit call replace the argument nodes in the definition. The subcircuit name SUBNAME may be considered as equivalent to a subroutine name in FORTRAN programming, where X⟨name⟩ is the call statement and ⟨(two or more) nodes⟩ are the variables or arguments of the subroutine.

Subcircuits can be nested; subcircuit *A* can call other subcircuits, but the nesting cannot be circular. That is, if subcircuit *A* contains a call to subcircuit *B*, then subcircuit *B* must not contain a call to subcircuit *A*.

The number of nodes in the subcircuit calling statement must be the same as that of its definition. The subcircuit definition can have only element statements and .MODEL statements, but must not contain other dot statements.

6-4 .ENDS (End of Subcircuit)

A subcircuit must end with an .ENDS statement. The end of a subcircuit definition has the general form

```
.ENDS  SUBNAME
```

SUBNAME is the name of the subcircuit, and it indicates which subcircuit description is to be terminated. If the .ENDS statement is missing, all subcircuit descriptions are terminated.

End of Subcircuit Statements

```
.ENDS OPAMP
.ENDS
```

Note. The name of the subcircuit can be omitted; however, it is advisable to identify the name of the subcircuit to be terminated, especially if there is more than one subcircuit.

Example 6-1

The equivalent circuit of a bipolar junction transistor (BJT) is shown in Fig. 6-1. Write the subcircuit call and subcircuit description.

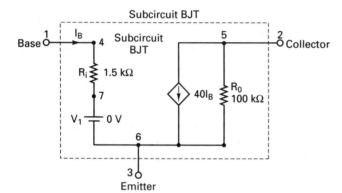

Figure 6-1 Bipolar junction transistor (BJT) subcircuit.

Solution   The list of statements for the subcircuit call and description follows.

```
*   The call statement X1 to be connected to input nodes 1, 2, and 3. The subcircuit
*   name is BJT.  Nodes 1, 2, and 3 are referred to the main circuit file;
*   they do not interact with the nodes of the subcircuit.
X1    1        2          3          BJT
*   base   emitter   collector    model name
*   The subcircuit definition for BJT.  Nodes 4, 5, 6, and 7 are referred
*   to the subcircuit; they do not interact with the nodes of the
*   main circuit.
.SUBCKT    BJT      4        6          5
*       model name base    emitter   collector
RI    4    7    1.5K
RO    5    6    100K
*   A dummy voltage source to measure the controlling current IB
VX    7    6    DC    0V
F1    5    6    VX    40        ; Current-controlled current source
.ENDS   BJT                     ; End of subcircuit definition BJT
```

PSpice allows one to generate user-defined functions. If there are several similar expressions in a single circuit file, a user-defined function can simplify the circuit description and can be accessed by an .INC statement (Section 6-7) near the beginning of the circuit file. The general form of a function statement is

```
.FUNC    FNAME (arg)    ⟨function⟩
```

The ⟨function⟩ can have the functions shown in Table 6-1, and may refer to other previously defined functions. FNAME (arg) is the name of the function with argument (arg). FNAME must not be the same as the built-in functions in Table 6-1, for example, sin or cos. The number of arguments (arg) must be the same as the number of variables in the ⟨function⟩. FNAME can be without arguments, but the parentheses are still required; that is, FNAME ( ) is acceptable.

Some Function Statements

```
.FUNC    E(x)        exp(x)
.FUNC    Sinh(x)     (E(x)+E(-x))/2
.FUNC    MIN(C,D)    (C+D-ABS(C-D))/2
.FUNC    MAX(C,D)    (C+D+ABS(C+D))/2
.FUNC    IND(I(VX))  (A0+A1*I(VX)+A2*I(VX)*I(VX))  ; Polynomial form
```

Notes

1. The definition of the ⟨function⟩ *must* fit on one line.
2. In-line comments *must not* be used after the ⟨function⟩ definition.
3. The last statement illustrates a current-dependent nonlinear inductor of polynomial form.

6-6 .GLOBAL (Global)

PSpice allows one to define global nodes, which are accessible by all subcircuits. The general statement is

```
.GLOBAL   N
```

where N is the node number. For example,

```
.GLOBAL   4
```

makes node 4 global to the circuit file and subcircuit(s).

The statement for including the contents of another file in the circuit file has the following form:

```
.INC  NFILE
```

NFILE is the name of the file to be included and can be any character string that is a legal file name for computer systems. NFILE may contain any statements except a title line, and can have a comment line(s). It can contain an .END statement, which simply indicates the end of the NFILE. Up to four levels of "including" are allowed in an .INC statement. It should be noted that the .INC statement simply brings everything from the included file into the circuit file and takes up space in main memory (RAM).

Some Include File Statements

```
.INC  OPAMP.CIR
.INC  a:INVERTER.CIR
.INC  c:\LIB\NOR.CIR
```

6-8 .LIB (Library File)

A library file may be referenced in the circuit file by using the following statement:

```
.LIB  FNAME
```

FNAME is the name of the library file to be called. A library file may contain comments, .MODEL statements, subcircuit definitions, .LIB statements, and .END statements. No other statements are permitted. If FNAME is omitted, PSpice looks for the default file, EVAL.LIB, that comes with PSpice programs. The library file FNAME may call for another library file.

When a .LIB command calls for a file, it does not bring the whole text of the library file into the circuit file. It simply reads those models or subcircuits that are called by the main circuit file. As a result, only those models or subcircuit descriptions that are needed by the main circuit file take up the main memory (RAM) space.

Note. Check the PSpice files for the default library file name.

Some Library File Statements

```
.LIB                 ; (Default file is EVAL.LIB.)
.LIB  EVAL.LIB       ; (Library file EVAL.LIB is on the default drive.)
.LIB  C:\LIB\EVAL.LIB ; (Library file EVAL.LIB is in directory file LIB
                       on drive C.)
.LIB  D:\LIB\EVAL.LIB ; (Library file EVAL.LIB is in directory file LIB
                       on drive D.)
```

In calculating the operating bias point, some or all of the nodes of the circuit may be assigned initial guesses to help dc convergence by the statement as,

```
.NODESET  V(1)=A1 V(2)=A2 ... V(N)=AN
```

V(1), V(2), . . . are the node voltages, and A1, A2, . . . are the respective values of the initial guesses. Once the operating point is found, the .NODESET command has no effect during the dc sweep or transient analysis. This command can be used to avoid a convergence problem, for example, on flip-flop circuits to break the tie-in condition. This command is not normally used, and it should not be confused with the .IC command, which sets the initial conditions of the circuits during the operating-point calculations for transient analysis.

Statement for Nodeset

```
.NODESET  V(4)=1.5V V(6)=0 V(25)=1.5V
```

6-10 OPTIONS

PSpice allows various options to control and to limit parameters for the various analysis. The general form is

```
.OPTIONS [⟨(options) name⟩] [⟨(options) name⟩=⟨value⟩]
```

The options can be listed in any order. There are two types of options: (1) those without values, and (2) those with values. The options without values are used as flags of various kinds, and only the option name is mentioned. Table 6-2 shows the options without values.

The options with values are used to specify certain optional parameters. The option names and their values are specified. Table 6-3 shows the options with values. The commonly used options are NOPAGE, NOECHO, NOMOD, TNOM, CPTIME, NUMDGT, GMIN, and LIMPTS.

TABLE 6-2 LIST OF OPTIONS WITHOUT VALUES

Option	Effects
NOPAGE	Suppresses paging and printing of a banner for each major section of output
NOECHO	Suppresses listing of the input file
NODE	Causes output of net list (node table)
NOMOD	Suppresses listing of model parameters
LIST	Causes summary of all circuit elements (or devices) to be output
OPTS	Causes values for all options to be output
ACCT	Summary and accounting information is output at the end of all the analysis.
WIDTH	Same as .WIDTH OUT= statement

TABLE 6-3 LIST OF OPTIONS WITH VALUES

Option	Effects	Units	Default
DEFL	MOSFET channel length (L)	Meter	100u
DEFW	MOSFET channel width (W)	Meter	100u
DEFAD	MOSFET drain diffusion area (AD)	Meter^{-2}	0
DEFAS	MOSFET source diffusion area (AS)	Meter^{-2}	0
TNOM	Default temperature (also the temperature at which model parameters are assumed to have been measured)	Degrees Celsius	27
NUMDGT	Number of digits output in print tables		4
CPTIME	CPU time allowed for a run	Second	1E6
LIMPTS	Maximum points allowed for any print table or plot		201
ITL1	Dc and bias-point "blind" iteration limit		40
ITL2	Dc and bias-point "educated guess" iteration limit		20
ITL4	Iteration limit at any point in transient analysis		10
ITL5	Total iteration limit for all points in transient analysis (ITL5=0 means ITL5=infinite)		5000
RELTOL	Relative accuracy of voltages and currents		0.001
TRTOL	Transient analysis accuracy adjustment		7.0
ABSTOL	Best accuracy of currents	Amp	1pA
CHGTOL	Best accuracy of charges	Coulomb	0.01pC
VNTOL	Best accuracy of voltages	Volt	1uV
PIVREL	Relative magnitude required for pivot in matrix solution		1E-13
GMIN	Minimum conductance used for any branch	Ohm^{-1}	1E-12

Options Statements

```
.OPTIONS  NOPAGE NOECHO NOMOD DEFL=20U DEFW=15U DEFAD=50P DEFAS=50P
.OPTIONS  ACCT LIST RELTOL=.005
```

Job statistics summary. If the option ACCT is specified on the .OPTIONS statement, PSpice will print various statistics about the run at the end. This option is not normally required for most circuit simulations. This list follows the format of the output.

ITEM	MEANING
NUNODS	Number of distinct circuit nodes before subcircuit expansion.
NCNODS	Number of distinct circuit nodes after subcircuit expansion. If there are no subcircuits, NCNODS = NUNODS.
NUMNOD	Total number of distinct nodes in circuit. This is NCNODS plus the internal nodes generated by parasitic resistances. If no device has parasitic resistances, NUMNOD = NCNODS.
NUMEL	Total number of devices (or elements) in circuit

ITEM	MEANING
	after subcircuit expansion. This includes all statements that do not begin with . or X.
DIODES	Number of diodes after subcircuit expansion.
BJTS	Number of bipolar transistors after subcircuit expansion.
JFETS	Number of junction FETs after subcircuit expansion.
MFETS	Number of MOSFETs after subcircuit expansion.
GASFETS	Number of GaAs MESFETs after subcircuit expansion.
NUMTEM	Number of different temperatures.
ICVFLG	Number of steps of dc sweep.
JTRFLG	Number of print steps of transient analysis.
JACFLG	Number of steps of ac analysis.
INOISE	1 or 0: noise analysis was/was not done.
NOGO	1 or 0: run did/did not have an error.
NSTOP	The circuit matrix is conceptually (not physically) of dimension NSTOP × NSTOP.
NTTAR	Actual number of entries in circuit matrix at beginning of run.
NTTBR	Actual number of entries in circuit matrix at end of run.
NTTOV	Number of terms in circuit matrix that come from more than one device.
IFILL	Difference between NTTAR and NTTBR.
IOPS	Number of floating-point operations needed to do one solution of circuit matrix.
PERSPA	Percent sparsity of circuit matrix.
NUMTTP	Number of internal time steps in transient analysis.
NUMRTP	Number of times in transient analysis that a time step was too large and had to be cut back.
NUMNIT	Total number of iterations for transient analysis.
MEMUSE/MAXMEM	Amount of circuit memory used/available in bytes. There are two memory pools. Exceeding either one will abort the run.
COPYKNT	Number of bytes that were copied in the course of doing memory management for this run.

ITEM	MEANING
READIN	Time spent reading and error checking the input file.
SETUP	Time spent setting up the circuit-matrix pointer structure.
DCSWEEP	Time spent and iteration count for calculating dc sweep.
BIASPNT	Time spent and iteration count for calculating bias point and bias point for transient analysis.
MATSOL	Time spent solving circuit matrix (this time is also included in the total for each analysis). The iteration count is the number of times the rows or columns were swapped in the course of solving it.
ACAN	Time spent and iteration count for ac analysis.
TRANAN	Time spent and iteration count for transient analysis.
OUTPUT	Time spent preparing .PRINT tables and .PLOT plots.
LOAD	Time spent evaluating device equations (this time is also included in the total for each analysis).
OVERHEAD	Other time spent during run.
TOTAL JOB TIME	Total run time, excluding the time to load the program files PSPICE1.EXE and PSPICE2.EXE into memory.

6-11 .PARAM (Parameter)

PSpice allows one to use a parameter instead of a numerical value. This parameter can be integrated into arithmetic expressions. The parameter definition is one of the following forms:

```
.PARAM  ⟨ PNAME = ⟨value⟩ or { ⟨ expression ⟩ } ⟩*
```

The keyword .PARAM is followed by a list of names with values or expressions. The ⟨value⟩ must be a constant and does not need to be enclosed in braces. However, the ⟨expression⟩ does need braces, { }, and must use only previously defined parameters.

PNAME is the parameter name, and it cannot be a predefined parameter, such as TIME (time), TEMP (temperature), VT (thermal voltage), or GMIN (shunt conductance for semiconductor p-n junctions).

Some .PARAM Statements

```
.PARAM VSUPPLY = 22V
.PARAM VCC = 15 V, VEE = -15V
.PARAM BANDWIDTH = {50kHz/20}
.PARAM PI = 3.14159, TWO_PI ={2*3.14159}
```

Once a parameter is defined, it can be used in place of a numerical value; for example, both of the following statements,

```
VCC    1    0    {SUPPLY}
VEE    0    5    {SUPPLY}
```

will change the value of SUPPLY. As an additional example, each of the following statements,

```
.FUNC    IND(I(VX))    (A0+A1*I(VX)+A2*I(VX)*I(VX))    ; Polynomial form
.PARAM   INDUCTOR = IND(I(VX))
L1    1    3    {INDUCTOR}
```

will change the value of INDUCTOR in a polynomial form.

It should be noted that the parameters defined by the .PARAM statement are global and can be used anywhere in the circuit, including inside of subcircuits. However, parameters can be made local to subcircuits by having parameters as arguments to subcircuits. For example,

```
.SUBCKT FILTER 1  2 PARAMS: CENTER=100kHz, BANDWIDTH=10kHZ
```

is a subcircuit definition for a bandpass filter with nodes 1 and 2, and with parameters CENTER (center frequency), and BANDWIDTH (bandwidth). The keyword PARAMS separates the nodes list from the parameter list. The parameters can be given new values by changing the values of the arguments while calling the subcircuit FILTER. For example,

```
X1    4    6   FILTER  PARAMS: CENTER=200kHz
```

will override the default value of 100 kHz with a CENTER value of 200 kHz.

A defined parameter can be used in the following cases:

1. All model parameters
2. Device parameters, such as AREA, L, NRD, Z0, and IC-values on capacitors and inductors, and TC1 and TC2 for resistors
3. All parameters of independent voltage and current sources (V and I devices).
4. Values in a .NODESET statement (Section 6-9) and .IC statement (Section 4-7.1)

A defined parameter *cannot* be used in the following cases:

1. For the transmission-line parameters NL and F
2. For the in-line temperature coefficients for resistors
3. For the E, F, G, and H device polynomial coefficient values
4. For replacing node numbers
5. For values in analysis statements (.TRAN, .AC, .DC, etc.)

6-12 FOURIER ANALYSIS

The results of a transient analysis are in discrete forms. These sampled data can be used to calculate the coefficients of a Fourier series. A periodic waveform can be expressed in a Fourier series as

$$v(\theta) = C_0 + \sum_{n=1}^{\infty} C_n \sin(n\theta + \phi_n)$$

where $\theta = 2\pi ft$
f = frequency in hertz
C_0 = dc component
C_n = nth harmonic component

PSpice uses the results of the transient analysis to perform the Fourier analysis up to nth harmonics. The statement takes one of these general forms:

```
.FOUR   FREQ N V1   V2   V3 ... VN
.FOUR   FREQ N I1   I2   I3 ... IN
```

FREQ is the fundamental frequency. V1, V2, . . . (or I1, I2, . . .) are the output voltages (or currents) for which the Fourier analysis is desired. A .FOUR statement must be used with a .TRAN statement. The output voltages (or currents) must have the same forms as in the .TRAN statement for transient analysis. If the number of harmonics N is not specified, the dc component, fundamental, and second through ninth harmonics (or 10 coefficients) are calculated by default.
   Fourier analysis is performed over the interval (TSTOP-PERIOD) to TSTOP, where TSTOP is the final (or stop) time for the transient analysis, and PERIOD is one period of the fundamental frequency. Therefore, the duration of the transient analysis must be at least one period long, PERIOD. At the end of the analysis, PSpice determines the dc component and the amplitudes up to the nth harmonic.
   PSpice does print a table, automatically, showing the results of Fourier analysis and does not require the .PRINT, .PLOT, or .PROBE statement.

Statement for Fourier Analysis

```
.FOUR   100KHZ   V(2,3),  V(3),  I(R1),  I(VIN)
```

Example 6-2

For Example 4-7, calculate the coefficients of the Fourier series for the load current if the fundamental frequency is 1 kHz (up to the eleventh harmonic is desired).
Solution   The listing of the circuit file follows.

Example 6-2   Fourier analysis

```
▲ VS  1   0   SIN (0   200V 1KHZ)  ; Sinusoidal voltage of 200 V peak
▲▲ RS  1   2   100OHM
    R1  2   0   100KOHM
    E1  3   0   2   0   0.1  ; V-controlled source with a gain of 0.1
    RL  4   5   2OHM
    VX  5   0   DC   0V     ; Measures the load current
    S1  3   4   3   0   SMOD  ; V-controlled switch with model SMOD
    * Switch model descriptions
    .MODEL SMOD   VSWITCH (RON=5M ROFF=10E+9 VON=25M VOFF=0.0)
▲▲▲ .TRAN  5US  1MS              ; Transient analysis
    *   Fourier analysis of load current at a fundamental frequency of 1 kHz
    .FOUR   1KHZ   11   I(VX)   ; Fourier analysis
    .PLOT   TRAN   V(3) I(VX)   ; Not needed with .PROBE
    .PROBE                      ; Graphical waveform analyzer
.END                           ; End of circuit file
```

The results of the Fourier analysis, which are obtained from the output file EX6-2.OUT, follow.

```
****   FOURIER ANALYSIS                    TEMPERATURE =   27.000 DEG C
FOURIER COMPONENTS OF TRANSIENT RESPONSE I(VX)
DC COMPONENT = 3.171399E+00
```

HARMONIC NO	FREQUENCY (HZ)	FOURIER COMPONENT	NORMALIZED COMPONENT	PHASE (DEG)	NORMALIZED PHASE (DEG)
1	1.000E+03	4.982E+00	1.000E+00	3.647E-05	0.000E+00
2	2.000E+03	2.115E+00	4.245E-01	-9.000E+01	-9.000E+01
3	3.000E+03	3.896E-07	7.819E-08	-8.424E+01	-8.424E+01
4	4.000E+03	4.234E-01	8.499E-02	-9.000E+01	-9.000E+01
5	5.000E+03	1.318E-07	2.646E-08	-7.547E+01	-7.547E+01
6	6.000E+03	1.818E-01	3.648E-02	-9.000E+01	-9.000E+01
7	7.000E+03	4.353E-08	8.737E-09	2.692E+01	2.692E+01
8	8.000E+03	1.012E-01	2.032E-02	-9.000E+01	-9.000E+01
9	9.000E+03	4.223E-08	8.476E-09	2.707E+01	2.707E+01
10	1.000E+04	6.460E-02	1.297E-02	-9.000E+01	-9.000E+01
11	1.100E+04	7.012E-08	1.407E-08	1.778E+01	1.778E+01

```
TOTAL HARMONIC DISTORTION = 4.351438E+01 PERCENT
```

Notes
1. If the order of highest harmonics N is not specified in the .FOUR statement, PSpice calculates up to the ninth harmonic, or 10 coefficients. For example,

the statement

```
.FOUR 1kHZ I(VX)
```

will print up to the ninth harmonic in the output file.

2. One could add the statement .FOUR 1KHZ I(VX) in the circuit file of Example 4-7.

6-13 NOISE ANALYSIS

Resistors and semiconductor devices generate noise. The level of the noise depends on the frequency. The various types of noise that are generated by resistors and semiconductor devices are discussed in Appendix B. Noise analysis is done in conjunction with ac analysis and requires an .AC command. For each frequency of the ac analysis, the noise level of each generator in a circuit (e.g., resistors and transistors) is calculated, and their contributions to the output nodes are computed by summing the rms noise values. The gain from the input source to the output voltage is calculated. From this gain, the equivalent input noise level at the specified source is calculated by PSpice.

The statement for performing noise analysis has the following form:

```
.NOISE V(N+, N-)  SOURCE  M
```

where V(N+, N−) is the output voltage across nodes N+ and N−. The output could be at a node N, such as V(N).

SOURCE is the name of an independent voltage or current source at which the equivalent input noise will be generated. It should be noted that SOURCE is not a noise generator; rather it is a place at which to compute the equivalent noise input. For a voltage source, the equivalent input is in $V/\sqrt{Hz}$; and for a current source, it is in $A/\sqrt{Hz}$.

M is the print interval that permits PSpice to print a table for the individual contributions of all generators to the output nodes for every mth frequency. There is no need for a .PRINT or .PLOT command for printing a table of all contributions. If the value of M is not specified, then PSpice does not print a table of individual contributions. The output noise and equivalent noise can also be printed by the .PRINT or .PLOT command.

Statements for Noise Analysis

```
.NOISE  V(4,5)   VIN
.NOISE  V(6)     IIN
.NOISE  V(10)    V1    10
```

Note. The .PROBE command cannot be used for noise analysis.

Example 6-3

For the circuit in Fig. 6-2(a), calculate and print the equivalent input and output noise if the frequency of the source is varied from 1 Hz to 100 kHz. The frequency should be increased by a decade with 1 point per decade. The amplifier is represented by the subcircuit of Fig. 6-2(b). The operating temperature is 40°C.

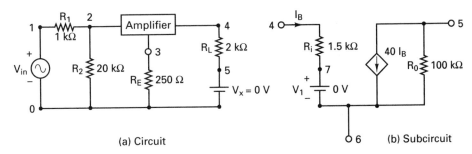

(a) Circuit   (b) Subcircuit

Figure 6-2 Circuit for Example 6-3.

Solution The input source is of ac type. The listing of the circuit file follows.

Example 6-3 Noise analysis

```
▲ VIN  1  0  AC  1V   ; Ac input voltage of 1 V
▲▲ R1  1  2  1K
   R2   2  0  20K
   RE   3  0  250
   RL   4  5  2K
   VX   5  0  DC  0V        ; Measures the load current
   X1   2  3  4   BJT       ; Subcircuit call for subcircuit BJT
   *    The subcircuit definition for BJT.  Nodes 4, 5, 6, and 7 are
   *    referred to the subcircuit; they do not interact with
   *    the nodes of the main circuit.
   .SUBCKT    BJT      4        6         5
   *          model name    base     emitter    collector
   RI   4  7  1.5K
   RO   5  6  100K
   VX   7  6  DC  0V        ; Measures the controlling current IB
   F1   5  6  VX  40        ; Current-controlled current source
   .ENDS   BJT             ; End of subcircuit definition BJT
   .TEMP  40               ; Operating temperature is 40 degrees
   .OPTIONS  NOPAGE  NOECHO    ; Options
▲▲▲ *  Ac sweep from 1 Hz to 100 kHz with a decade increment and 10 points
    *  per decade
    .AC    DEC   1   1HZ   100KHz
    *  Noise analysis without printing details of individual contributions
    .NOISE   V(4)    VIN
    *  PSpice prints the details of equivalent input and output noise
    .PRINT   NOISE   ONOISE   INOISE
    .PROBE                   ; Graphical waveform analyzer
 .END                       ; End of circuit file
```

The results of noise analysis, which are stored in the output file EX6-3.OUT due to the .PRINT statement, follow. The node voltages are printed automatically by PSpice.

```
****      AC ANALYSIS                              TEMPERATURE =  40.000 DEG C
           FREQ           ONOISE          INOISE
        1.000E+00       4.325E-08       7.245E-09
        1.000E+01       4.325E-08       7.245E-09
        1.000E+02       4.325E-08       7.245E-09
        1.000E+03       4.325E-08       7.245E-09
        1.000E+04       4.325E-08       7.245E-09
        1.000E+05       4.325E-08       7.245E-09
             JOB CONCLUDED
             TOTAL JOB TIME                   1.53
```

Notes. The individual contributions of each element can be obtained by adding the parameter M in the NOISE statement. For instance, to print the individual contributions at every second frequency, the NOISE statement will read

```
.NOISE  V(4)    VIN    2
```

Students are encouraged to run the circuit file EX6-3.CIR with this statement and to look at the output file EX6-3.OUT. The effects of including M (= 2) will be obvious.

6-14 .SENS (Sensitivity Analysis)

The sensitivity of output voltages or currents with respect to every circuit and device parameter can be calculated by the .SENS statement, which has the following general form:

```
.SENS  ((one or more output) variables))
```

The .SENS statement calculates the bias point and the linearized parameters around the bias point. In this analysis the inductors are assumed to be short circuits, and capacitors are assumed to be open circuits. If the output variable is a current, then that current must be through a voltage source. The sensitivity of each output variable with respect to all the device values and model parameters is calculated, and the .SENS statement prints the results automatically. Therefore, it should be noted that a .SENS statement may generate a huge amount of data if many output variables are specified.

Statement for Sensitivity Analysis

```
.SENS  V(5)   V(2.3)   I(V2)   I(V5)
```

Example 6-4

For the circuit in Fig. 6-2(a), calculate and print the sensitivity of output voltage V(4) with respect to each circuit element. The amplifier is represented by the subcircuit in Fig. 6-2(b). The operating temperature is 40°C.

Solution The listing of the circuit file follows.

Example 6-4 Dc sensitivity analysis

```
▲ VIN   1    0    DC    5V      ; Dc input voltage of 5 V
▲▲ R1   1    2    1K
   R2   2    0    20K
   RE   3    0    250
   RL   4    5    2K
   VX   5    0    DC    0V      ; Measures the load current
   X1   2    3    4    BJT      ; Subcircuit call for subcircuit BJT
   *   The subcircuit definition for BJT.  Nodes 4, 5, 6, and 7 are
   *   referred to the subcircuit; they do not interact with
   *   the nodes of the main circuit.
   .SUBCKT    BJT     4      6         5
   *      model name   base   emitter   collector
   RI   4    7    1.5K
   RO   5    6    100K
   VX   7    6    DC    0V      ; Measures the controlling current IB
   F1   5    6    VX    40      ; Current-controlled current source
   .ENDS   BJT                  ; End of subcircuit definition BJT
   .TEMP  40                    ; Operating temperature is 40 degrees.
   .OPTIONS  NOPAGE  NOECHO     ; Options
▲▲▲ *  Sensitivity analysis calculates the dc bias point and prints
   *   the current through the input source, I(VIN), before computing
   *   the sensitivity.
   *   It calculates and prints the sensitivity analysis of output
   *   voltage V(4) with respect to all elements in the circuit.
   .SENS   V(4)                 ; Sensitivity analysis
.END                           ; End of circuit file
```

The results of the sensitivity analysis are shown next. The node voltages are also printed automatically.

```
****      SMALL SIGNAL BIAS SOLUTION          TEMPERATURE =  27.000 DEG C
NODE     VOLTAGE    NODE     VOLTAGE    NODE    VOLTAGE    NODE    VOLTAGE
(    1)   5.0000   (    2)   4.3986   (    3)   3.8263   (    4)  -29.8470
(    5)   0.0000   ( X1.7)   3.8263
VOLTAGE SOURCE CURRENTS
NAME         CURRENT
VIN         -6.014E-04
VX          -1.492E-02
X1.V1        3.815E-04
TOTAL POWER DISSIPATION   3.01E-03   WATTS

****      DC SENSITIVITY ANALYSIS             TEMPERATURE =   40.000 DEG C
DC SENSITIVITIES OF OUTPUT V(4)
```

ELEMENT NAME	ELEMENT VALUE	ELEMENT SENSITIVITY (VOLTS/UNIT)	NORMALIZED SENSITIVITY (VOLTS/PERCENT)
R1	1.000E+03	3.590E−03	3.590E−02
R2	2.000E+04	−6.564E−05	−1.313E−02
RE	2.500E+02	9.600E−02	2.400E−01
RL	2.000E+03	−1.486E−02	−2.972E−01
X1.RI	1.500E+03	2.391E−03	3.587E−02
X1.R5	1.000E+05	−1.426E−06	−1.426E−03
VIN	5.000E+00	−5.969E+00	−2.985E−01
VX	0.000E+00	9.958E−01	0.000E+00
X1.V1	0.000E+00	6.268E+00	0.000E+00

JOB CONCLUDED

TOTAL JOB TIME   1.2

Notes

1. $V(4)$ is most sensitive to changes in R_E and R_L. An increase in R_E causes $V(4)$ to increase, whereas an increase in R_L causes $V(4)$ to decrease.

2. One could combine the .AC, .NOISE, and .SEN V(4) commands in the circuit file of Example 6-4 by modifying VIN 1 0 AC 1V DC 5V.

6-15 .STEP (Parametric Analysis)

The .STEP command can be used to evaluate the effects of parameter variations. It has one of the following general forms:

```
.STEP   LIN     SWNAME    SSTART    SEND    SINC
.STEP   OCT     SWNAME    SSTART    SEND    NP
.STEP   DEC     SWNAME    SSTART    SEND    NP
.STEP   SWNAME  LIST ⟨value⟩*
```

SWNAME is the sweep variable name. SSTART, SEND, and SINC are the start value, the end value, and the increment value of the sweep variables, respectively. NP is the number of steps. LIN, OCT, or DEC specifies the type of sweep, as follows:

LIN   *Linear sweep:* SWNAME is swept linearly from SSTART to SEND. SINC is the step size. LIN is used if the variable range is narrow.

OCT   *Sweep by octave:* SWNAME is swept logarithmically by octave, and NP becomes the number of steps per octave. The next variable is generated by multiplying the present value by a constant larger than unity. OCT is used if the variable range is wide.

DEC   *Sweep by decade:* SWNAME is swept logarithmically by decade, and NP becomes the number of steps per decade. The next variable is generated by multiplying the present value by a constant larger than unity. DEC is used if the variable range is very wide.

LIST   *List of values:* There are no start and end values. The values of the sweep variables are listed after the keyword LIST.

The SWNAME can be one of the following types:

Source. A name of an independent voltage or current source. During the sweep, the source's voltage or current is set to the sweep value.

Model Parameter. A model type name and then model name followed by a model parameter name in parenthesis. The parameter in the model is set to the sweep value. The model parameters L and W for MOS devices and any temperature parameters, such as TC1 and TC2 for the resistor, *cannot* be swept.

Temperature. The keyword TEMP followed by the keyword LIST. The temperature is set to the sweep value. For each value of sweep, the model parameters of all circuit components are updated to that temperature.

Global Parameter. The keyword PARAM followed by a parameter name. The parameter is set to sweep. During the sweep, the global parameter's value is set to the sweep value, and all expressions are affected by the sweep parameter value.

Some Step Statements

```
.STEP  VCE  −5V  10V  5V
```

sweeps the voltage VCE linearly.

```
.STEP  LIN  IIN  −10mA  5mA  0.1mA
```

sweeps the current IIN linearly.

```
.STEP  RES  RMOD(R)  0.9  1.1  0.001
```

sweeps linearly the model parameter R of the resistor model RMOD.

```
.STEP  DEC  NPN  QM(IS)  1E−18  1E−14  10
```

sweeps with a decade increment the parameter IS of the NPN transistor.

```
.STEP  TEMP  LIST  0  50  80  100  150
```

sweeps the temperature TEMP as listed.

```
.STEP  PARAM  Frequency  8.5kHz  10.5kHZ  50Hz
```

sweeps linearly the parameter PARAM Frequency.

Notes

1. The sweep start value SSTART may be greater than or less than the sweep end value SEND.

2. The sweep increment SINC must be greater than zero.

3. The number of points NP must be greater than zero.

4. If the .STEP command is included in a circuit file, all specified analyses (.DC, .AC, .TRAN, etc.) are done for each step.

Example 6-5

The *RLC* circuit of Fig. 6-3(a) is supplied from a step input voltage, as shown in Fig. 6-3(b). Use PSpice to calculate and plot the capacitor voltage v_C from 0 to 400 μs with a time increment of 1 μs for $R_1 = 1\ \Omega$, 2 Ω, and 8 Ω. The capacitor voltages V(3) is the output, and it is to be plotted. The results should also be available for display and hard copy by the .PROBE command.

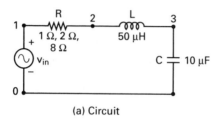

(a) Circuit

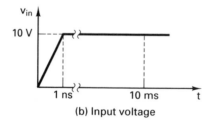

(b) Input voltage

Figure 6-3 An *RLC* circuit with a step input voltage.

Solution In Example 4-3, we found the transient responses for three values of R_1 by considering three separate *RLC* circuits. This is inefficient, because PSpice allows one to vary the values of a parameter. Thus, we will define R as a parameter and assign its values. The listing of the circuit file follows.

Example 6-5 Step response of an *RLC* circuit by parametric analysis

```
▲ VIN   1  0  PWL (0  0  1NS  1V  1MS  1V)  ; Step input of 1 V
▲▲ .PARAM   VAL = 1              ; Defining parameter VAL
    R    1   2    {VAL}          ; Resistance with variable values
    L    2   3    50UH
    C    3   0    10UF
▲▲▲ .STEP  PARAM  VAL  LIST  1  2  8  ; Assigning STEP values
    .TRAN   1US   400US          ; Transient analysis
    .PROBE                       ; Graphical waveform analyzer
.END                            ; End of circuit file
```

The capacitor voltage V(3), which is obtained by the .PROBE command, is shown in Fig. 6-4 for three values of R. The responses depict the effects of damping on a second-order system.

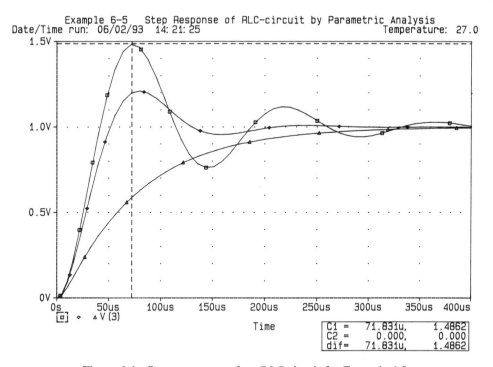

Figure 6-4   Step responses of an *RLC* circuit for Example 6-5.

Example 6-6

An *RLC* circuit is shown in Fig. 6-5. Plot the frequency response of the current through the circuit and the magnitude of the input impedance. The frequency of the source is varied from 100 Hz to 100 kHz with a decade increment and 10 points per decade. The values of the inductor L are 5 mH, 15 mH, and 25 mH.

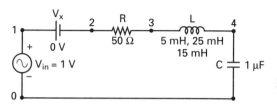

Figure 6-5   An *RLC* circuit for Example 6-6.

Solution   The listing of the circuit file follows.

Example 6-6   Input impedance characteristics by parametric analysis

```
▲ VIN    1    0    AC    1V            ; Ac voltage of 1 V
```

```
▲▲ R      2   3    50
   .PARAM   VAL = 1MH                    ; Defining parameter VAL
   L       3   4    {VAL}                ; Inductance with variable values
   C       4   0    1UF
   VX      1   2    DC   0V              ; Measures the current
▲▲▲ .STEP PARAM VAL 5MH  25MH  10MH      ; Step variations for VAL
   .AC    DEC  100  100HZ 100KHZ         ; Ac sweep from 10 Hz to 100 kHz
   .PLOT  AC   I(VX)                     ; Plots in the output file
   .PROBE                                ; Graphical waveform analyzer
.END                                     ; End of circuit file
```

The frequency response of the current through the circuit and the magnitude of the input impedance are shown in Fig. 6-6. As the inductance is increased, the resonant frequency decreases.

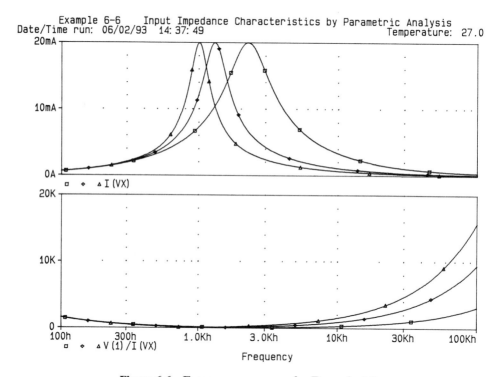

Figure 6-6   Frequency response for Example 6-6.

6-16 .DC (Dc Parametric Sweep)

PSpice allows one to evaluate the effects of parameter variations in dc analysis. These parameters could be elements and their model parameters. The general statement for the dc parametric sweep is

```
.DC  SWNAME   LIST⟨value⟩*
```

where SWNAME is the sweep variable name and can be one of the following types.

Source. A name of an independent voltage or current source. During the sweep, the source's voltage or current is set to the sweep value.

Model Parameter. A model type name and then a model name followed by a model parameter name in parenthesis. The parameter in the model is set to the sweep value. The model parameters L and W for MOS devices and any temperature parameters, such as TC1 and TC2 for the resistor, *cannot* be swept.

Temperature. The keyword TEMP followed by the keyword LIST. The temperature is set to the sweep value. For each value of sweep, the model parameters of all circuit components are updated to that temperature.

Global Parameter. The keyword PARAM followed by a parameter name. The parameter is set to sweep. During the sweep, the global parameter's value is set to the sweep value, and all expressions are evaluated.

Statements for dc Parametric Sweep

```
.DC  RES  RMOD(R)  0.9  1.1  0.001
```

sweeps linearly the model parameter R of the resistor model RMOD.

```
.DC  DEC  NPN  QM(IS)  1E-18  1E-14  10
```

sweeps with a decade increment the parameter IS of the NPN transistor.

```
.DC  TEMP  LIST  0  50  80  100  150
```

sweeps the temperature TEMP as listed values.

```
.DC  PARAM  Vsupply  -15V  15V  0.5V
```

sweeps linearly the parameter PARAM Vsupply.

Example 6-7

For the amplifier circuit of Fig. 6-2, calculate and plot the dc transfer characteristic, V_{out} versus V_{in}. The input voltage V_{in} is varied from 0 to 100 mV with an increment of 2 mV. The load resistance is varied from 10 kΩ to 30 kΩ with a 10 kΩ increment.
Solution The listing of the circuit file follows.

Example 6-7 Dc parametric sweep

```
▲ VIN  1  0  DC  100MV        ; Dc input voltage of 100 mV
▲▲ R1  1  2  1K
   R2   2  0  20K
   RE   3  0  250
   .PARAM  VAL = 10K          ; Defining parameter VAL
   RL   4  5  {VAL}           ; Load resistance with variable values
```

```
VX      5   0   DC   0V          ; Measures the load current
X1      2   3   4    BJT         ; Subcircuit call for subcircuit BJT
*    The subcircuit definition for BJT.
.SUBCKT      BJT      4       6          5
*        model name  base   emitter   collector
RI      4   7   1.5K
RO      5   6   100K
VX      7   6   DC   0V          ; Measures the controlling current IB
F1      5   6   VX   40          ; Current-controlled current source
.ENDS   BJT                      ; End of subcircuit definition BJT
▲▲▲ .STEP  PARAM  VAL 10K 30K 10K ; Step variations for VAL
.DC     VIN   0  100MV    2MV     ; Dc sweep from 0 to 100 mV
.PLOT   DC    V(4)                ; Plots in the output file
.PROBE                           ; Graphical waveform analyzer
.END                             ; End of circuit file
```

The transfer characteristic is shown in Fig. 6-7.

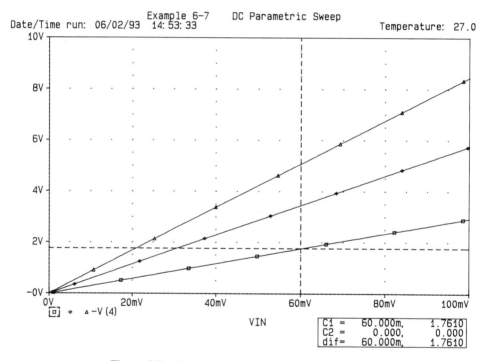

Figure 6-7 Transfer characteristic for Example 6-7.

6-17 MONTE CARLO ANALYSIS

PSpice allows one to perform the Monte Carlo (statistical) analysis of a circuit for variations of model parameters. The variations are specified by DEV and LOT tolerances in .MODEL statements. The general form of the statement for Monte Carlo analysis is

```
.MC   ⟨(# runs) value⟩   ⟨(analysis)⟩   ⟨(output variable)⟩
+     ⟨(function)⟩   [(option)]*
```

⟨(# runs) value⟩ is the number of runs to be evaluated and has an upper limit of 2,000. If the results are to be viewed with Probe, then the upper limit is 100. The first run is done with the nominal values of all components, but the subsequent runs are done with values generated randomly by PSpice. Thus, ⟨(# runs) value⟩ should be greater than 1.

⟨(analysis)⟩ specifies type of analysis and must be dc, ac, or transient.

⟨(output variable)⟩ is the output variable for which the Monte Carlo (statistical) analysis is to be done and has the same format as the .PRINT output variable.

⟨(function)⟩ specifies the desired information of the ⟨(output variable)⟩ by one of the following keywords:

YMAX	Finds the greatest difference in each waveform from the nominal run.
MAX	Finds the maximum value of each waveform.
MIN	Finds the minimum value of each waveform.
RISE_EDGE (⟨value⟩)	Finds the first occurrence of the waveform crossing above the threshold ⟨value⟩.
FALL_EDGE (⟨value⟩)	Finds the first occurrence of the waveform crossing below the threshold ⟨value⟩.

[(option)]* allows one to obtain information about the details of the parameters used in each run and to control the amount of output by one of the following keywords:

LIST	Prints out the model parameter values used for each component during each run.
OUTPUT	(output type) specifies the output after the nominal (first) run. If OUTPUT is omitted, the output is only from the nominal run. The (output type) is indicated by one of the following:

ALL	Generates output from all runs.
FIRST ⟨value⟩	Generates output only during first n runs.
EVERY ⟨value⟩	Generates output every nth run.
RUNS ⟨value⟩*	Generates output only for the listed runs, up to 25.

RANGE $(\langle(\text{low}) \text{ value}, \langle(\text{high}) \text{ value}\rangle)$ limits the range for the $\langle(\text{func-}$ tion$)\rangle$. An asterisk (*) can be used to specify all $\langle\text{values}\rangle$. For example,

```
YMAX RANGE (*, 0.5)
MAX RANGE (-1, *)
```

Some Statements for Monte Carlo Analysis

```
.MC   5   TRAN   V(3)    YMAX
.MC   6   DC     V(5)    YMAX   LIST
.MC  10   DC     V(9)    YMAX   LIST   OUTPUT   ALL
.MC  15   AC     V(2,7)  YMAX   YMAX   RANGE (*, 1.5)
```

6-18 DEV/LOT DEVICE AND LOT TOLERANCES

The device tolerances are used for variations in component values. Each variation is independent; that is, the variation of one component does not affect that of another component. A device tolerance can be specified by the keyword DEV as a percentage value

```
DEV = 15%
```

or as an absolute value

```
DEV = 0.1
```

The lot tolerances assume the same variations for all components having the same .MODEL statement; that is, devices that have the same .MODEL statement will vary together by the same amount. Thus, the variation is not independent. A lot tolerance can be specified by the keyword LOT as a percentage value

```
LOT = 5%
```

or as an absolute value

```
LOT = 0.01
```

The distribution of the deviation can be specified as uniformly distributed by including the keyword UNIFORM,

```
DEV/UNIFORM = 15%
```

Similarly, the distribution of the deviation can be specified as Gaussian distribution by the keyword GAUSS

```
DEV/GAUSS = 15%
```

It should be noted that if the distribution is not specified, it is, by default, UNI-FORM.

Some Statements for Tolerances

```
R     2   7   RMOD   2K                          ; Resistance of 2 kΩ
.MODEL RMOD  RES (R=1  DEV = 15%)                ; Uniform by default
.MODEL RMOD  RES (R=1  DEV/UNIFORM = 15%)        ; Uniform distribution
.MODEL RMOD  RES (R=1  DEV/GAUSS = 15%)          ; Gaussian distribution
```

Example 6-8

The *RLC* circuit of Fig. 6-3(a) is supplied from a step input voltage as shown in Fig. 6-3(b). The circuit parameters are

$R = 2\ \Omega \pm 15\%$ (uniform deviation)
$L = 50\ \mu H \pm 20\%$ (uniform deviation)
$C = 10\ \mu H \pm 10\%$ (Gaussian deviation)

The model parameters are as follows: for the resistor, $R = 1$; for the capacitor, $C = 1$; and for the inductor, $L = 1$.

Use PSpice to perform Monte Carlo analysis of the capacitor voltage v_C from 0 to 400 μs with a time increment of 1 μs.

(a) The greatest difference from the nominal run is to be printed for five runs.
(b) The first occurrence of the capacitor voltage crossing below 1 V is to be printed.

Solution The listing of the circuit file follows.

Example 6-8   Monte Carlo analysis for transient response
```
▲  VIN  1  0  PWL (0  0  1NS  1V  10MS  1V)  ; Step input of 1 V
▲▲  R    1  2  RMOD   2           ; Resistance with model RMOD
    L    2  3  LMOD   50UH        ; Inductance with model RMOD
    C    3  0  CMOD   10UF        ; Capacitance with model CMOD
    .MODEL  RMOD  RES (R=1 DEV=15%)
    .MODEL  LMOD  IND (L=1 DEV/UNIFORM=20%)
    .MODEL  CMOD  CAP (C=1 DEV/GAUSS=10%)
▲▲▲ .TRAN  1US  400US              ; Transient analysis
    .MC    5  TRAN  V(3)  YMAX            ; Monte Carlo Analysis
    *.MC   5  TRAN  V(3)  FALL_EDGE (1V)  ; Monte Carlo Analysis
    .PROBE                         ; Graphical waveform analyzer
.END                               ; End of circuit file
```

(a) The results of Monte Carlo analysis, which are obtained from the output file EX6-8.OUT, are shown below.

```
****   SORTED DEVIATIONS OF V(3)                      TEMPERATURE =   27.000 DEG C
                          MONTE CARLO SUMMARY
       Mean Deviation =      -.0229
       Sigma          =       .0567
              RUN                       MAX DEVIATION FROM NOMINAL

       Pass   2     .0732 (1.29 sigma) higher at T = 50.9930E-06
                    (108.62% of Nominal)
       Pass   4     .0653 (1.15 sigma) lower at T = 90.9940E-06
                    (94.611% of Nominal)
       Pass   3     .0635 (1.12 sigma) lower at T = 98.9940E-06
                    (94.656% of Nominal)
       Pass   5     .0362 (.64 sigma) lower at T = 50.9930E-06
                    (95.74% of Nominal)
```

(b) To find the first occurrence of the capacitor voltage crossing below 1 V, the .MC statement is changed as follows:

```
.MC   5   TRAN   V(3)   FALL_EDGE (1V)      ; Monte Carlo analysis
```

The results of the Monte Carlo analysis are shown below:

```
                     MONTE CARLO SUMMARY
            RUN      FIRST FALLING EDGE VALUE THRU   1

       Pass   5     135.0600E-06
                    ( 104.03% of Nominal)
       NOMINAL      129.8300E-06
       Pass   4     129.6100E-06
                    (  99.827% of Nominal)
       Pass   3     124.6000E-06
                    (  95.973% of Nominal)
       Pass   2     119.4800E-06
                    (  92.026% of Nominal)
```

6-19 SENSITIVITY/WORST-CASE ANALYSIS

PSpice allows one to perform the sensitivity and worst-case analysis of a circuit for variations of model parameters. The variations are specified by DEV and LOT tolerances on .MODEL statements. The general form of the sensitivity/worst-case analysis is

```
.WCASE  ⟨(analysis)⟩  ⟨(output variable)⟩  ⟨(function)⟩  [(option)]*
```

This command is similar to the .MC command, except that it does not include ⟨(# runs) value⟩. With a .WCASE command, PSpice first calculates the sensitivity of

---

the ⟨(output variable)⟩ to each model parameter. Once all the sensitivities are evaluated, one final run is done with all model parameters, and PSpice produces the worst-case ⟨(output variable)⟩.

Note. Either the .MC or .WCASE command can be used, but *not* both in the same circuit.

Some Statements for Sensitivity/Worst-Case Analysis

```
.WCASE   TRAN   V(3)    YMAX
.WCASE   DC     V(5)    YMAX    LIST
.WCASE   DC     V(9)    YMAX    LIST    OUTPUT    ALL
.WCASE   AC     V(2,7)  YMAX    YMAX    RANGE  (*,  1.5)
```

Example 6-9

Run the sensitivity/worst-case analysis for Example 6-8.

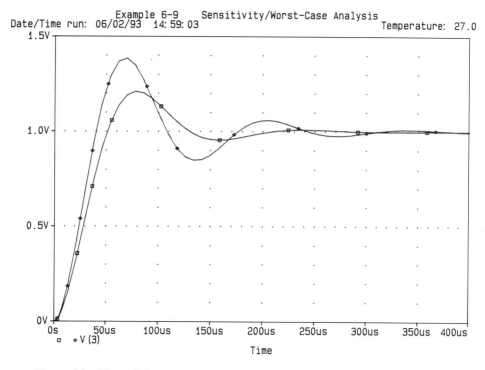

Figure 6-8 Plots of the greatest difference and the nominal run for Example 6-9.

Solution

(a) To run the sensitivity/worst-case analysis for finding the greatest difference from the nominal run, the .MC statement in Example 6-8 is changed to a .WCASE statement, as follows:

```
.WCASE   TRAN   V(3)   YMAX    ; Sensitivity/worst-case analysis
```

The plots of the greatest difference and the nominal run are shown in Fig. 6-8. The results of the worst-case analysis, which are obtained from the output file EX6-9.OUT, are shown below:

```
****     SORTED DEVIATIONS OF V(3)                TEMPERATURE =  27.000 DEG C
                        WORST-CASE SUMMARY
      RUN                      MAX DEVIATION FROM NOMINAL
      ALL DEVICES                .246  higher  at T =  58.9930E-06
                              ( 124.59% of Nominal)
```

(b) To find the first occurrence of the capacitor voltage crossing below 1 V, the .WCASE statement is changed as follows:

```
.WCASE  TRAN  V(3)  FALL_EDGE (1V)   ; Sensitivity/worst-case analysis
```

The plots of the first crossing of the capacitor voltage below 1 V and the nominal run are shown in Fig. 6-9. The results of the worst-case analysis are shown below:

WORST-CASE SUMMARY

RUN	FIRST FALLING EDGE VALUE THRU 1
NOMINAL	129.8300E−06
ALL DEVICES	91.9470E−06
	(70.82% of Nominal)

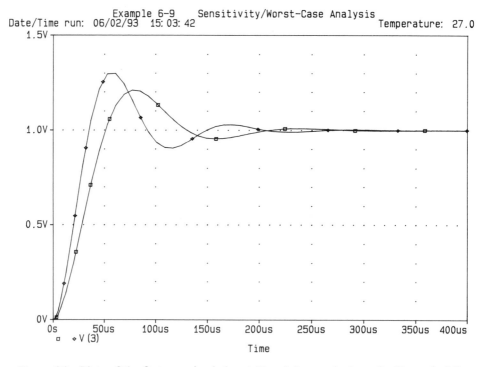

Figure 6-9 Plots of the first crossing below 1 V and the nominal run for Example 6-9.

SUMMARY

The statements for behavioral modeling are

```
E⟨name⟩   N+  N-   VALUE = { ⟨expression⟩ }
E⟨name⟩   N+  N-   TABLE { ⟨expression⟩ } =
+                  ( ⟨(input) value⟩, ⟨(output) value⟩ )*
E⟨name⟩   N+  N-   LAPLACE { ⟨expression⟩ } = { ⟨transform⟩ }
E⟨name⟩   N+  N-   FREQ { ⟨expression⟩ } =
+ ( ⟨(frequency) value⟩, ⟨(magnitude in dB) value⟩, ⟨(phase) value⟩ )*

G⟨name⟩   N+  N-   VALUE = { ⟨expression⟩ }
G⟨name⟩   N+  N-   TABLE { ⟨expression⟩ } =
+                  ( ⟨(input) value⟩, ⟨(output) value⟩ )*
G⟨name⟩   N+  N-   LAPLACE { ⟨expression⟩ } = { ⟨transform⟩ }
G⟨name⟩   N+  N-   FREQ { ⟨expression⟩ } =
+ ( ⟨(frequency) value⟩, ⟨(magnitude in dB) value⟩, ⟨(phase) value⟩ )*
```

The advanced commands and features are

.SUBCKT	Subcircuit
.ENDS	End of subcircuit
.FUNC	Function
.GLOBAL	Global
.INC	Include file
.LIB	Library file
.NODESET	Nodeset
.OPTIONS	Options
.PARAM	Parameter
.FOUR	Fourier analysis
.NOISE	Noise analysis
.SENS	Sensitivity analysis
.STEP	Parametric analysis
.DC	Dc parametric sweep
.MC	Monte Carlo analysis
DEV/LOT	Device tolerances
.WCASE	Worst-case analysis

REFERENCES

1. *PSpice Manual,* Irvine, Calif.: MicroSim, 1992.
2. P. Antognetti and G. Massobri, *Semiconductor Device Modeling with SPICE.* New York: McGraw-Hill, 1988.

3. M. H. Rashid, *SPICE For Power Electronics and Electronic Power.* Englewood Cliffs, N.J.: Prentice Hall, 1993.

PROBLEMS

6-1. For the circuit in Fig. P6-1, calculate and print the sensitivity of output voltage V_{out} with respect to each circuit element. The operating temperature is 50°C.

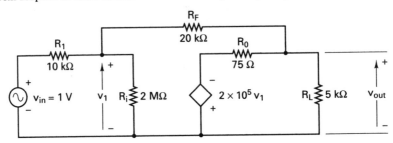

Figure P6-1

6-2. For the circuit in Fig. P6-1, calculate and print **(a)** the voltage gain, $A_v = V_{out}/V_{in}$; **(b)** the input resistance, R_{in}; and **(c)** the output resistance, R_{out}.

6-3. For the circuit in Fig. P6-1, calculate and plot the dc transfer characteristic, V_{out} versus V_{in}. The input voltage is varied from 0 to 10 V with an increment of 0.5 V for $R_F = 1$ kΩ, 20 kΩ, and 50 kΩ.

6-4. For the circuit in Fig. P6-1, calculate and print the equivalent input and output noise if the frequency of the source is varied from 10 Hz to 1 MHz. The frequency should be increased by a decade with 2 points per decade.

6-5. For the circuit in Fig. P6-5, the frequency response is to be calculated and printed over the frequency range from 1 Hz to 100 kHz with a decade increment and 10 points per decade. The peak magnitude and phase angle of the output voltage are to be plotted from the output file. The results should also be available for display and hard copy by the .PROBE command for $R_F = 1$ kΩ, 20 kΩ, and 50 kΩ.

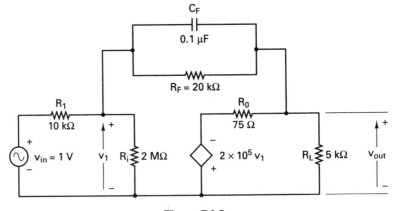

Figure P6-5

6-6. Repeat Problem 6-5 for the circuit in Fig. P6-6, for $R_F = 1$ kΩ, 20 kΩ, 50 kΩ.

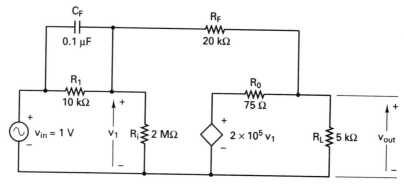

Figure P6-6

6-7. For the circuit in Fig. P6-5, calculate and plot the transient response of the output voltage from 0 to 2 ms with a time increment of 5 μs. The input voltage is shown in Fig. P6-7. The results should be available for display and hard copy by Probe for $R_F = 10$ kΩ, 20 kΩ, and 30 kΩ.

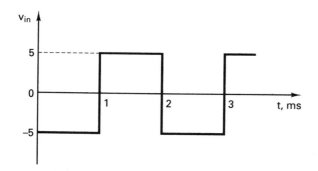

Figure P6-7

6-8. Repeat Problem 6-7 for the input voltage shown in Fig. P6-8.

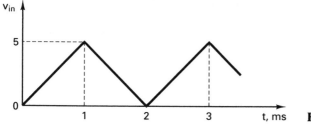

Figure P6-8

6-9. For the circuit in Fig. P6-6, calculate and plot the transient response of the output voltage from 0 to 2 ms with a time increment of 5 μs. The input voltage is shown in Fig. P6-7. The results should be available for display and hard copy by Probe for $R_F = 10$ kΩ, 20 kΩ, and 30 kΩ.

6-10. Repeat Problem 6-9 for the input voltage shown in Fig. P6-8.

6-11. Use PSpice to perform a Monte Carlo analysis for seven runs and for the dc response of Problem 6-3. The model parameter for resistors is $R = 1$. The circuit parameters having uniform deviations are the following:

$R_F = 20 \text{ k}\Omega \pm 20\%$

$R_i = 2 \text{ M}\Omega \pm 5\%$

$R_L = 5 \text{ k}\Omega \pm 10\%$

$R_o = 75 \ \Omega \pm 5\%$

$R_1 = 10 \text{ k}\Omega \pm 15\%$

(a) The greatest difference from the nominal run is to be printed.
(b) The maximum value of the output voltage is to be printed.
(c) The minimum value of the output voltage is to be printed.
(d) The first occurrence of the output voltage crossing below 1 V is to be printed.

6-12. Use PSpice to perform the worst-case analysis for Problem 6-11.

6-13. Use PSpice to perform a Monte Carlo analysis for six runs and for the frequency response of Problem 6-5. The model parameters are $R = 1$ for resistors and $C = 1$ for the capacitor. The circuit parameters having uniform deviations are the following:

$R_F = 20 \text{ k}\Omega \pm 20\%$

$R_i = 2 \text{ M}\Omega \pm 5\%$

$R_L = 5 \text{ k}\Omega \pm 10\%$

$R_o = 75 \ \Omega \pm 5\%$

$R_1 = 10 \text{ k}\Omega \pm 15\%$

$C_F = 0.1 \ \mu\text{F} \pm 10\%$

(a) The greatest difference from the nominal run is to be printed.
(b) The maximum value of the output voltage is to be printed.
(c) The minimum value of the output voltage is to be printed.
(d) The first occurrence of the output voltage crossing below 1 V is to be printed.

6-14. Repeat Problem 6-13 for the frequency response of Problem 6-6.

6-15. Use PSpice to perform the worst-case analysis for Problem 6-13.

6-16. Use PSpice to perform the worst-case analysis for Problem 6-14.

6-17. Use PSpice to perform a Monte Carlo analysis for 10 runs and for the transient response of Problem 6-7. The model parameters are $R = 1$ for resistors and $C = 1$ for the capacitor. The circuit parameters having uniform deviations are the following:

$R_F = 20 \text{ k}\Omega \pm 20\%$

$R_i = 2 \text{ M}\Omega \pm 5\%$

$R_L = 5 \text{ k}\Omega \pm 10\%$

$R_o = 75 \ \Omega \pm 5\%$

$R_1 = 10 \text{ k}\Omega \pm 15\%$

$C_F = 0.1 \ \mu\text{F} \pm 10\%$

(a) The greatest difference from the nominal run is to be printed.

(b) The maximum value of the output voltage is to be printed.

(c) The minimum value of the output voltage is to be printed.

(d) The first occurrence of the output voltage crossing below 1 V is to be printed.

6-18. Repeat Problem 6-17 for the transient response of Problem 6-8.

6-19. Repeat Problem 6-17 for the transient response of Problem 6-9.

6-20. Use PSpice to perform the worst-case analysis for Problem 6-17.

6-21. Use PSpice to perform the worst-case analysis for Problem 6-18.

6-22. Use PSpice to perform the worst-case analysis for Problem 6-19.

Semiconductor Diodes

7-1 INTRODUCTION

A semiconductor diode may be specified in PSpice by a diode statement in conjunction with a model statement. The diode statement specifies the diode name, the nodes to which the diode is connected, and its model name. The model incorporates an extensive range of diode characteristics (e.g., dc and small-signal behavior, temperature dependency, and noise generation). The model parameters take into account temperature effects, various capacitances, and physical properties of semiconductors.

7-2 DIODE CHARACTERISTICS

The typical v-i characteristic of a diode is shown in Fig. 7-1. The characteristic can be expressed by an equation known as the *Shockley diode equation*, and it is given by [4]

$$I_D = I_S(e^{V_D/\eta V_T} - 1) \tag{7-1}$$

where I_D = forward current through the diode, A

V_D = forward voltage with anode positive with respect to cathode, V

I_S = leakage (or reverse saturation) current, typically in the range of 10^{-6} A to 10^{-16} A

η = an empirical constant known as the *emission coefficient* (or *ideality factor*), whose value varies from 1 to 2.

The emission constant η depends upon the material and the physical construction of diodes. Its value ranges from 1 to 2, but it should, ideally, be 1.

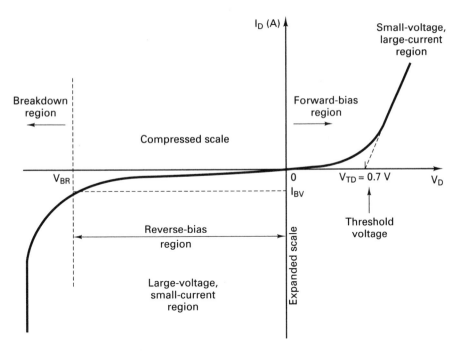

Figure 7-1 Diode characteristics.

The value V_T in Eq. (7-1) is a constant called *thermal voltage*, and it is given by

$$V_T = \frac{kT}{q} \qquad (7\text{-}2)$$

where q = electron charge: 1.6022×10^{-19} coulombs (C).

T = absolute temperature in degrees Kelvin (°K = 273 + °C).

k = Boltzmann's constant: 1.3806×10^{-23} J per °K

At a junction temperature of 25°C (77°F, 298°K), Eq. (7-2) gives

$$V_T = \frac{kT}{q} = \frac{1.3806 \times 10^{-23} \times (273 + 25)}{1.6022 \times 10^{-19}} \approx 25.8 \text{ mV}$$

At a specified temperature, the leakage current I_S remains constant for a given diode. For small-signal diodes, the typical value of I_S is 10^{-14} A.

The diode characteristic of Fig. 7-1 can be divided into three regions:

Forward-bias region, where the forward voltage $V_D > 0$

Reverse-bias region, where the reverse voltage $V_R < 0$

Breakdown region, where the breakdown voltage $V_R < -V_{BR}$

Forward-bias region. In the forward-bias region, $V_D > 0$. The diode current I_D is very small if the diode voltage V_D is less than a specific value V_{TD}

(typically 0.7 V). The diode conducts fully if V_D is higher than this value V_{TR}, which is referred to as the *threshold voltage,* or the *cut-in voltage,* or the *turn-on voltage.* Thus, the threshold voltage is a voltage at which the diode conducts.

For $V_D > 0.1$ V, which is usually the case, $I_D \gg I_S$, and Eq. (7-1) can be approximated to

$$I_D = I_S[e^{V_D/\eta V_T} - 1] \approx I_S e^{V_D/\eta V_T} \tag{7-3}$$

Taking natural (base e) logarithm on both sides, we get

$$\ln I_D = \ln I_S + \frac{V_D}{\eta V_T}$$

which, after simplification, gives the diode voltage V_D as

$$V_D = \eta V_T \ln \left(\frac{I_D}{I_S}\right) \tag{7-4}$$

Converting the natural log of base e to the logarithm of base 10, Eq. (7-4) becomes

$$V_D = 2.3 \, \eta \, V_T \log \left(\frac{I_D}{I_S}\right) \tag{7-5}$$

If I_{D1} is the diode current corresponding to diode voltage V_{D1}, Eq. (7-4) gives

$$V_{D1} = 2.3 \, \eta \, V_T \log \left(\frac{I_{D1}}{I_S}\right) \tag{7-6}$$

Similarly, if V_{D2} is the diode voltage corresponding to the diode current I_{D2}, we get

$$V_{D2} = 2.3 \, \eta \, V_T \log \left(\frac{I_{D2}}{I_S}\right) \tag{7-7}$$

Therefore, the difference in diode voltages can be expressed by

$$V_{D2} - V_{D1} = 2.3 \, \eta \, V_T \log \left(\frac{I_{D2}}{I_S}\right) - 2.3 \, \eta \, V_T \log \left(\frac{I_{D1}}{I_S}\right)$$

$$= 2.3 \, \eta \, V_T \log \left(\frac{I_{D2}}{I_{D1}}\right) \tag{7-8}$$

Thus, for a decade (factor of 10) change in diode current, that is, $I_{D2} = 10 \, I_{D1}$, the diode voltage would change by $2.3 \, \eta \, V_T$.

Reverse-bias region. In the reverse-bias region, the reverse voltage $V_R < 0$. If V_R is negative, and $|V_R| \gg V_T$, which occurs for $V_R < -0.1$, the exponential term in Eq. (7-1) becomes negligibly small compared to unity, and the reverse diode current I_R becomes,

$$I_R = I_S[e^{-|V_R|/\eta V_T} - 1] \approx -I_S \tag{7-9}$$

which indicates that the reverse diode current I_R is almost constant and equal to $-I_S$. In practical diodes, however, the reverse current I_R increases with the reverse voltage V_R.

Breakdown region. In the breakdown region, the reverse voltage V_R has a high magnitude, usually greater than 100 V. If the magnitude of the reverse voltage exceeds a specified voltage known as the *reverse breakdown voltage* V_{BR}, the magnitude of the reverse current I_R increases rapidly with a small change in reverse voltage. The reverse current corresponding to V_{BR} is denoted as the reverse current I_{BV}.

7-3 ANALYSIS OF DIODE CIRCUITS

Diodes are used in electronic circuits for signal processing. Consider the diode circuit of Fig. 7-2, in which a small-amplitude sinusoidal voltage v_s is superimposed on a dc source V_{DD}. The source V_{DD} sets the operating point (Q-point) by coordinates V_D and I_D. The operating point will vary with time as the ac signal v_s varies. If this variation is small, then the superposition theorem can be applied to find the instantaneous diode current i_D from

$$i_D = I_D + i_d$$

where I_D is the dc current due to V_{DD} only, and $v_s = 0$.

i_d is the small-signal current due to v_s only, and $V_{DD} = 0$.

Thus, the analysis of a diode circuit will consist of

Dc analysis
Small-signal ac analysis

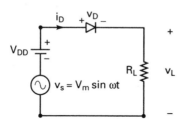

Figure 7-2 Diode circuit with a small signal superimposed on a dc source.

7-3.1 Dc Analysis

Assuming $v_s = 0$, Fig. 7.2 is reduced to Fig. 7-3(a). Using Kirchhoff's voltage law (KVL), the diode current I_D can be expressed as

$$V_{DD} = V_D + R_L I_D$$

or

$$I_D = -\frac{V_D}{R_L} + \frac{V_{DD}}{R_L} \qquad (7\text{-}10)$$

Since the diode is forward-biased, I_D is related to the diode voltage V_D by

$$I_D = I_S[e^{V_D/\eta V_T} - 1] \qquad (7\text{-}11)$$

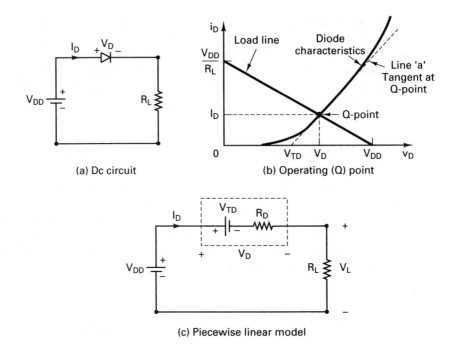

(a) Dc circuit

(b) Operating (Q) point

(c) Piecewise linear model

Figure 7-3 Piecewise linear-model.

I_D depends on V_D, which in turn depends on I_D. Equations (7-10) and (7-11) can be solved for V_D and I_D at the operating point (Q-point) either graphically or iteratively.

The diode characteristic can be represented, approximately, by a fixed voltage drop V_{TD} and a straight line, as shown in Fig. 7-3(b). The straight line a takes into account the current-dependency of the voltage drop, and it represents a fixed resistance R_D. The line a is tangent to the diode characteristic at the estimated Q-point. This representation is known as a *piecewise linear-model* consisting of two parts: a fixed part V_{TD} and a current-dependent part R_D. The resistance R_D is given, approximately, by

$$R_D \approx \frac{\eta V_T}{I_D} = \frac{0.0258\, \eta}{I_D}\bigg|_{\text{at } Q\text{-point}} \tag{7-12}$$

If the diode is replaced by its piecewise linear-model, Fig. 7-3(a) can be represented by a linear circuit, as shown in Fig. 7-3(c). The diode current I_D can be calculated from

$$I_D = \frac{V_{DD} - V_{TD}}{R_L + R_D} \tag{7-13}$$

The diode voltage V_D can be found from

$$V_D = V_{TD} + R_D I_D \tag{7-14}$$

7-3.2 Small-Signal Ac Analysis

Replacing the diode by its piecewise linear-model, we can find the instantaneous diode current i_D from Fig. 7-4(a). The diode will exhibit a junction capacitance C_D due to the time-variant signal v_s. Since the dc sources V_{DD} and V_{TD} in Fig. 7-4(a) will offer zero impedances to v_s, we can find the small-signal diode current i_d from the small-signal circuit of Fig. 7-4(b).

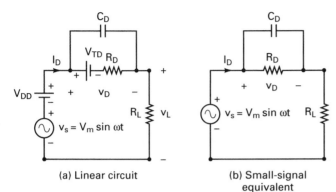

(a) Linear circuit   (b) Small-signal   **Figure 7-4** Small-signal equivalent
equivalent   circuit.

The value of diode resistance R_D will be low if the diode is forward-biased and high if it is reverse-biased. The junction capacitance will also vary depending upon its basing condition. The small-signal ac model of a diode is shown in Fig. 7-5. R_s is included to represent the contact and bulk resistance of the diode.

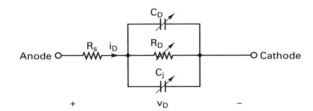

Figure 7-5 Small-signal ac diode model.

C_D is the diffusion capacitance, and its value depends on the forward diode current I_D. C_j is the depletion capacitance, and its value depends on the reverse voltage according to the following relation:

$$C_j = \frac{C_{jo}}{(1 - V_D/V_j)^m} \qquad \text{for } V_D < 0 \qquad (7\text{-}15)$$

where m is a junction-gradient coefficient, ranging from 0.33 to 0.5.

V_j is the *junction potential*. For a silicon diode, $V_j \approx 0.5$ V to 0.9 V, and for a germanium diode, $V_j \approx 0.2$ V to 0.6 V.

V_D is negative in the reverse direction and positive in the forward direction.

C_{jo} is the zero-bias junction capacitance.

We can conclude that the analysis of a diode circuit having a dc source and an ac source requires the following steps:

1. Find the dc operating point (V_D and I_D) of a diode.
2. Determine the parameters of the small-signal ac model for the diode.
3. Find the small-signal ac voltage and current of the diode.
4. Find the instantaneous diode voltage v_D and diode current i_D by superimposing the small-signal ac diode voltage v_d and diode current i_d on the voltage V_D and the dc current I_D, respectively.

7-4 DIODE MODEL

The PSpice model for a reverse-biased diode is shown in Fig. 7-6 [1, 2]. The small-signal model and the static model, which are generated by PSpice, are shown in Figs. 7-7 and 7-8, respectively. In the static model, the diode current that depends on its voltage is represented by a current source. The small-signal parameters are generated by PSpice from the operating point.

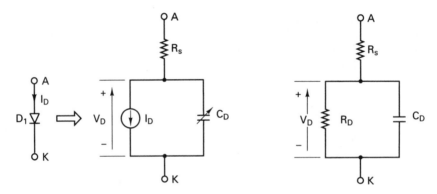

Figure 7-6 PSpice diode model with reverse-biased diode.

Figure 7-7 Small-signal diode model.

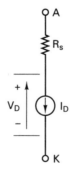

Figure 7-8 Static diode model with reverse-biased diode.

PSpice generates a complex model for diodes. The model equations that are used by PSpice are described in the manual [1] and in Antognetti [2]. In many cases, especially the level at which this book is aimed, such complex models are not necessary. Many model parameters can be ignored by the users, and PSpice assigns default values for the parameters.

The model statement of a diode has the following general form:

```
.MODEL DNAME D (P1=A1 P2=A2 P3=A3 ... PN=AN)
```

where DNAME is the model name. DNAME can begin with any character, but its word size is normally limited to eight characters. D is the type symbol for diodes. P_1, P_2, ... and A_1, A_2, ... are the model parameters and their values, respectively. The model parameters are listed in Table 7-1.

An *area factor* is used to determine the number of equivalent parallel diodes of a specified model. The model parameters that are affected by the area factor are marked by an asterisk (*) in the descriptions of the model parameters.

The diode is modeled as an ohmic resistance (value = RS/area) in series with an intrinsic diode. The resistance is attached between node NA and an internal anode node. The [(area) value] scales IS, RS, CJO, and IBV, and defaults to 1. IBV and BV are both specified as positive values.

The dc characteristic of a diode is determined by the reverse saturation current IS, the emission coefficient N, and the ohmic resistance RS. The charge storage effects are modeled by the transit time TT, a nonlinear depletion layer capacitance, which depends on the zero-bias junction capacitance CJO, the junction potential VJ, and grading coefficient M. The temperature of the reverse saturation current is defined by the gap-activation energy (or gap energy) EG and saturation-temperature exponent XTI.

TABLE 7-1 PARAMETERS OF DIODE MODEL

Name	Area	Model parameter	Units	Default	Typical
IS	*	Saturation current	Amps	1E−14	1E−14
RS	*	Parasitic resistance	Ohms	0	10
N		Emission coefficient		1	
TT		Transit time	seconds	0	0.1NS
CJO	*	Zero-bias *p-n* capacitance	Farads	0	2PF
VJ		Junction potential	Volts	1	0.6
M		Junction-gradient coefficient		0.5	0.5
EG		Activation energy	Electron volts	1.11	11.1
XTI		IS temperature exponent		3	3
KF		Flicker noise coefficient		0	
AF		Flicker noise exponent		1	
FC		Forward-bias depletion capacitance coefficient		0.5	
BV		Reverse breakdown voltage	Volts	∞	50
IBV	*	Reverse breakdown current	Amps	1E−10	

In order to simulate a Zener diode, the model in Fig. 7-9 can be used. Diode D_1 and the threshold voltage source V_{Th} represent the normal forward voltage and reverse behavior of a Zener diode. Diode D_2, the voltage source BV, and resistance R_B define the breakdown region. Diode D_2 does not conduct until $V_D = -BV$, and if the reverse voltage is increased, then diode D_2 becomes forward-biased, and the reverse current flows through R_B.

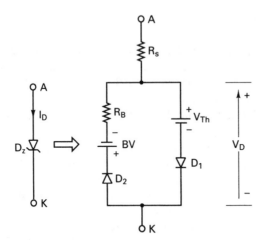

Figure 7-9 Static Zener diode model.

Reverse breakdown is modeled by an exponential increase in the reverse diode current and is determined by the reverse breakdown voltage, BV, and the current at breakdown voltage, I_{BV}.

7-5 DIODE STATEMENT

The symbol for a diode is D. The name of a diode must start with D, and it takes the general form

```
D⟨name⟩ NA  NK  DNAME  [⟨area⟩ value]
```

where NA and NK are the node and cathode nodes, respectively. The current flows from anode node NA through the diode to cathode node NK. DNAME is the model name.

Some Statements for Diodes

```
DM    5    6    DNAME
.MODEL DNAME  D (IS=0.5UA RS=6 BV=5.20 IBV=0.5UA)
D15  33    35   SWITCH  1.5
.MODEL SWITCH D (IS=100E-15 RS=16 CJO=2PF TT=12NS BV=100 IBV=100E-15)
DCLAMP  0  8    DMOD
.MODEL DMOD   D (IS=100E-15 RS=16 CJO=2PF TT=12NS BV=100 IBV=100E-15)
```

Note. Diode DM, having model name DNAME, is a Zener diode with a zener breakdown voltage of 5.2 V; the current at the zener break is 0.5 μA.

Some versions of SPICE (e.g., PSpice) support device library files that give the model parameters. The library file EVAL.LIB contains the list of device library files and the device model statements that are available with the student's version. The software PARTS of PSpice can generate SPICE models from the data-sheet parameters of diodes. The SPICE model parameters are also supplied by some manufacturers. However, some model parameters of a diode can be determined from the data sheet. The data sheet of diode 1N914, which is shown in Fig. 7-10, gives the following:

Reverse recovery time t_r = 4 ns   (From diode data)

Reverse breakdown voltage BV = 100V

Forward voltage V_D = 0.4 V at I_D = 25 μA   (From the curve of I_F versus V_F)

$$V_D = 0.6 \text{ V at } I_D = 2 \text{ mA}$$

Reverse current I_R = 25 nA at V_R = 20 V   (From diode data)

$$I_R = 5 \ \mu\text{A at } V_R = 75 \text{ V}$$

Capacitance with zero reverse voltage (from the curve of capacitance versus reverse voltage) C_{jo} = 1.7 pF.

Assuming V_T = 25.8 mV = 0.0258, V_{D1} = 0.4 V at I_{D1} = 25 μA = 0.025 mA, and V_{D2} = 0.6 V at I_{D2} = 2 mA, we can apply Eq. (7-8) to find the *emission coefficient* η as follows:

$$0.6 - 0.4 = 2.3 \ \eta \times 0.0258 \log\left(\frac{2}{0.025}\right)$$

which gives η = 1.77

For n = 1.77, and V_D = 0.6 V at I_{D2} = 2 mA, we can apply Eq. (7-1) to find the saturation current I_S

$$2 \text{ mA} = I_S[e^{0.6/(1.77 \times 0.0258)} - 1] \qquad \text{which gives } I_S = 3.93\text{E} - 9 \text{ A}$$

Assuming that I_R = 5 μA at V_R = 75 V is equal to the reverse breakdown current, then I_{BV} = 5 μA.

The transit time τ_T can be calculated approximately from

$$\tau_T = \frac{Q_{RR}}{I_{FM}}$$

where Q_{RR} = reverse recovery charge
I_{FM} = reverse recovery current

Since Q_{RR} is not specified on the data sheet, it can be related approximately to t_r and I_{FM} by

$$Q_{RR} = \frac{1}{2} t_r I_{FM} \qquad\qquad (7\text{-}16)$$

Diode Data

Computer Diodes (Glass Package)

Device No.	Package No.	V_{RRM} V Min	I_R nA @ Max	V_R V	V_F V Min	Max	@	I_F mA	C pF Max	t_{rr} ns Max	Test Cond.	Proc. No.
1N625	DO-35	30	1000	20		1.5		4		1000	(Note 1)	D4
1N914	DO-35	100	25 5000	20 75		1.0		10		4	(Note 2)	D4
1N914A	DO-35	100	25 5000	20 75		1.0		20		4	(Note 2)	D4
1N914B	DO-35	100	25 5000	20 75		0.72 1.0		5 100		4	(Note 2)	D4
1N916	DO-35	100	25 5000	20 75		1.0		10		4	(Note 2)	D4
1N916A	DO-35	100	25 5000	20 75		1.0		20		4	(Note 2)	D4
1N916B	DO-35	100	25 5000	20 75		0.73 1.0		5 30		4	(Note 2)	D4
1N3064	DO-35	75	100	50		0.575 0.650 0.710 1.0		0.250 1.0 2.0 10.0	2	4	(Note 3)	D4
1N3600	DO-35	75	100	50	0.54 0.66 0.76 0.82 0.87	0.62 0.74 0.86 0.92 1.0		1.0 10.0 50.0 100.0 200.0	2.5	4	(Note 4)	D4
1N4009	DO-35	35	100	25		1.0		30	4	2	(Note 2)	D4
1N4146	DO-35	See Data for 1N914A/914B										
1N4147	DO-35	See Data for 1N914A/914B										
1N4148	DO-35	See Data for 1N914										
1N4149	DO-35	See Data for 1N916										
1N4150	DO-35	See Data for 1N3600										
1N4151	DO-35	75	50	50		1.0		50	4	2	(Note 2)	D4
1N4152	DO-35	40	50	30	0.49 0.53 0.59 0.62 0.70 0.74	0.55 0.59 0.67 0.70 0.81 0.88		0.1 0.25 1.0 2.0 10.0 20.0	4	2	(Note 2)	D4
1N4153	DO-35	75	50	50		See 1N4152			4	2	(Note 2)	D4
1N4154	DO-35	35	100	25		1.0		30	4	2	(Note 2)	D4

Figure 7-10 Data sheet for diode 1N914 (Courtesy of National Semiconductor, Inc.).

Curve Set Number D4

Typical Electrical Characteristic Curves 25°C Ambient Temperature unless otherwise noted

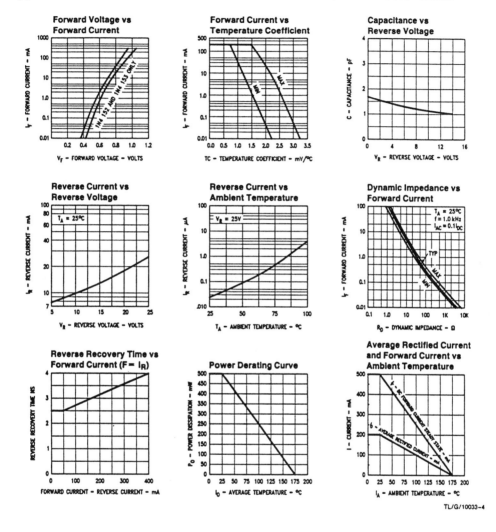

TL/G/10033-4

Figure 7-10

Thus, the transit time $TT = \tau_T = t_r/2 = 4/2 = 2$ ns.

Assuming RS = 1 Ω, the model statement for diode 1N914 is

.MODEL DMOD D(IS=3.93E-9 RS=1 BV=100V IBV=5E-6 CJO=1.7PF TT=2NS)

Note. The model parameters that are calculated above from the data sheet are approximate and will differ from those supplied by the diode manufacturer and SPICE/PSpice library.

7-7 EXAMPLES OF DC ANALYSIS

In dc circuits, the sources are dc. The diode is specified by its model parameters. If the model parameters are not specified, PSpice assumes the default values as indicated in Table 7-1. After calculating the steady-state dc voltage and current of a diode, PSpice determines the small-signal model parameters, which can be printed by the .OP command. The following examples illustrate the PSpice simulation of dc diode circuits.

Example 7-1

A diode circuit is shown in Fig. 7-11. Plot the v-i characteristic of the diode for forward voltage from 0 to 2 V and for temperature of 50°, 100°, and 150°. The diode is of type D1N914, and the model parameters are IS=3.93E−9, RS=1, BV=100V, IBV=5E−6, CJO=1.7PF, and TT=2NS.

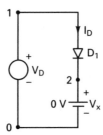

Figure 7-11 Diode circuit.

Solution The listing of the circuit file follows.

Example 7-1 Diode characteristic

```
▲ VD   1   0   DC   0V  ; Dc input voltage is overridden during dc sweep
▲▲ D1   1    2     D1N914    ; Diode with model D1N914
   *  anode cathode   model
   VX    2   0    DC   0V      ; Measures the diode current ID
   *   Diode model defines the model parameters
   .MODEL  D1N914  D (IS=3.93E-9 RS=1 BV=100V  IBV=5E-6  CJO=1.7PF TT=2NS)
   .TEMP  50  100   150    ; Operating temperatures: 50, 100, and 150 degrees
   .OPTIONS NOPAGE  NOECHO       ; Options
▲▲▲ .DC  VD  0  1V  0.01V  ; Dc sweep from 0 to 2 V with 0.01 V increment
   .PLOT   DC   I(VX)      ; Plots the diode current in the output file
   .PROBE              ; Graphical waveform analyzer
.END                    ; End of circuit file
```

The v-i characteristic that is obtained by varying the input voltage is shown in Fig. 7-12.

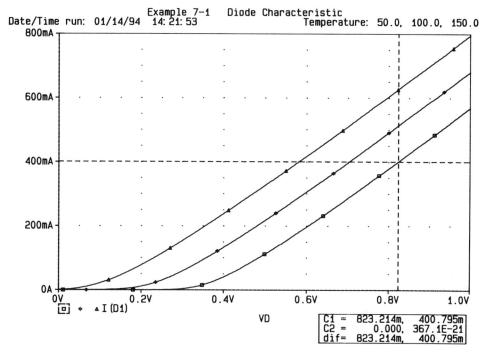

C1 =	823.214m,	400.795m
C2 =	0.000,	367.1E-21
dif=	823.214m,	400.795m

Figure 7-12   Diode forward characteristics for Example 7-1.

Example 7-2

The diode circuit of Fig. 7-13(a) has $V_{DD} = 15$ V, and $R_L = 500$ Ω. Use PSpice to calculate the operating point (V_D, I_D). The diode characteristic is represented by the following table:

I_D (mA)	0	10	20	30	40	50	60	70
V_D (V)	0.5	0.87	0.98	1.058	1.115	1.173	1.212	1.25

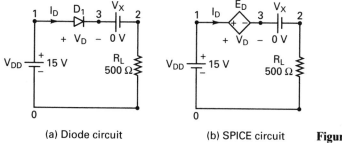

(a) Diode circuit   (b) SPICE circuit   **Figure 7-13**   Diode circuit.

Solution The diode is replaced by a voltage-controlled voltage source, shown in Fig. 7-13(b), to represent the table of the current-voltage characteristic. The listing of the circuit file follows.

Example 7-2 Diode circuit

```
▲   VD   1   0   DC   15V     ; Dc voltage of 15 V
▲▲  VX   3   2   DC   0V      ; Measures the diode current ID
    RL   2   0   500
    ED   1   3   TABLE {I(VX) } =
    +    (0.0,   0.5)    (10E-3, 0.87)   (20E-3, 0.98)   (30E-3, 1.058)
    +    (40E-3, 1.115)  (50E-3, 1.173)  (60E-3, 1.212)  (70E-3, 1.250)
▲▲▲ .OP                       ; Prints the details of the operating point
    .END                      ; End of circuit file
```

The information about the operating point, which is obtained from the output file EX7.2.OUT, is as follows:

```
****   OPERATING POINT INFORMATION    TEMPERATURE =  27.000 DEG C
   NAME              ED
V-SOURCE       1.042E+00
I-SOURCE       2.792E-02
```

Example 7-3

A Zener voltage regulator is shown in Fig. 7-14. Plot the dc transfer characteristic if the input voltage is varied from -15 V to 15 V with an increment of 0.5 V. The Zener voltages of the diodes are the same, and $V_Z = 5.2$ V; the current at the zener breakdown is $I_Z = 0.5$ μA. The model parameters are IS = 0.5UA, RS = 1, BV = 5.20, and IBV = 0.5UA. The operating temperature is 50°C. V_{in} has a normal voltage of 10 V (dc). The details of the operating point are to be printed.

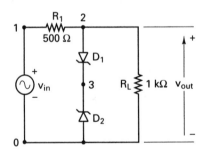

Figure 7-14 Zener regulator.

Solution A Zener diode is implemented by setting the model parameters BV = $V_Z = 5.2$ V, and IBV = $I_Z = 0.5$ μA. The listing of the circuit file follows.

Example 7-3 Zener regulator

```
▲   VIN 1  0  DC  10V      ; Dc input voltage is overridden during dc sweep.
▲▲  R1  1  2   500
    D1  2  3  DNAME        ; Diode with model name DNAME
    D2  0  3  DNAME        ; Diode with mdoel name DNAME
    RL  2  0   1K
```

```
*   Model DNAME defines the parameters of zener diodes.
.MODEL DNAME   D (IS=0.5UA RS=1 BV=5.20 IBV=0.5UA)
.TEMP    50                    ; Operating temperature 50 degrees
.OPTIONS NOPAGE NOECHO   ; Options
.DC  VIN   -15   15V  0.5V ; DC sweep from -15 to 15 V with 0.5 V increment
.PRINT  DC   V(2)    ; Plots the load voltage in the output file
.PROBE                      ; Graphical waveform analyzer
.OP
.END                        ; End of circuit file
```

The information about the operating point, which is obtained from the output file EX7.3.OUT, is as follows:

****   OPERATING POINT INFORMATION   TEMPERATURE = 50.000 DEG C

	NAME	D1	D2
	MODEL	DNAME	DNAME
I_D	ID	3.19E-03	-3.19E-03
V_D	VD	1.56E-01	-5.45E+00
R_D	REQ	8.69E+00	8.76E+00
C_D	CAP	0.00+00	0.00E+00

The dc transfer characteristic is shown in Fig. 7-15.

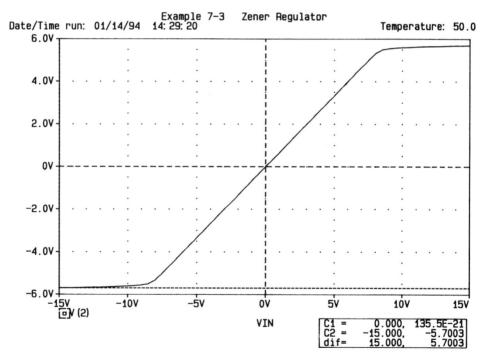

Figure 7-15   Dc transfer characteristic for Example 7-3.

Example 7-4

A diode waveform-shaping circuit is shown in Fig. 7-16, where the output is taken from node 2. Plot the transfer characteristic between V_{in} and V(2) for values of V_{in} in the range of -15 V to 30 V in steps of 0.5 V. The model parameters of the diodes are IS=100E$-$15, RS=16, BV=100, and IBV=100E$-$15. V_{in} has a normal voltage of 10 V (dc). The details of the operating point are to be printed.

Solution   The listing of the circuit file follows.

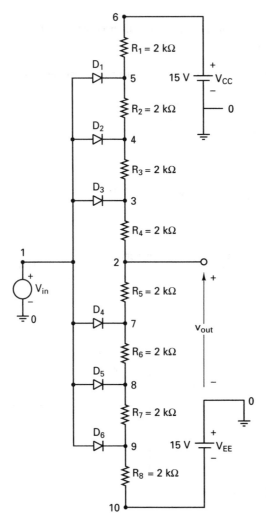

Figure 7-16   Diode waveform-shaping circuit.

Example 7-4   Diode waveform-shaping circuit

```
▲ *     The input voltage that is overridden by the dc sweep voltage is assumed
  *     to be 10 V.
  VIN   1   0   DC  −10V
  VCC   6   0   DC   15V
  VEE   10  0   DC  −15V
```

```
▲▲ R1   6   5   2K
   R2   5   4   2K
   R3   4   3   2K
   R4   3   2   2K
   R5   2   7   2K
   R6   7   8   2K
   R7   8   9   2K
   R8   9  10   2K
   *    All diodes have the same model name of DMOD.
   D1   1   5   DMOD
   D2   1   4   DMOD
   D3   1   3   DMOD
   D4   1   7   DMOD
   D5   1   8   DMOD
   D6   1   9   DMOD
   *    Diode model DMOD defines diode parameters.
   .MODEL DMOD D (IS=100E−15 RS=16 BV=100 IBV=100E−15)
▲▲▲ *    Dc sweep from −15 V to 30 V with 0.5 V increment
   .DC   VIN   −15   30 0.5
   *    Plot the results of dc sweep: (V2) versus VIN.
   .PLOT   DC   V(2)
   *    Graphic post-processor
   .OP                              ; Prints the details of the operating point
   .PROBE
 .END
```

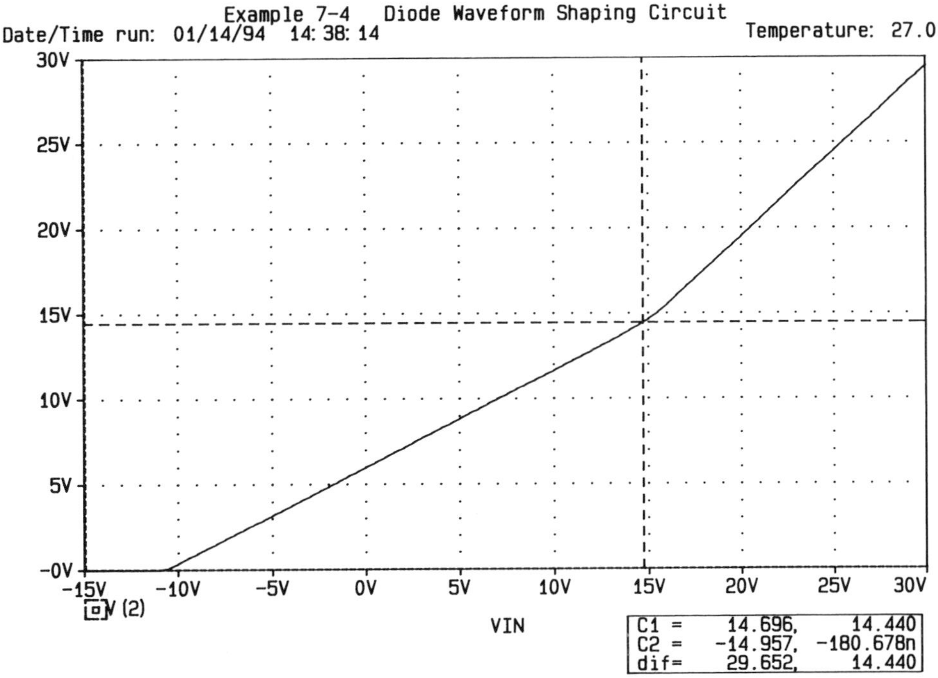

Example 7-4   Diode Waveform Shaping Circuit

Figure 7-17   Dc transfer characteristic for Example 7-4.

The information about the operating point, which is obtained from the output file EX7.4.OUT, is as follows.

****	OPERATING POINT INFORMATION			TEMPERATURE = 27.000 DEG C		
	NAME	D1	D2	D3	D4	D5
	MODEL	D1N914	D1N914	D1N914	D1N914	D1N914
I_D	ID	−4.26E−12	−3.43E−12	−2.59E−12	−9.25E−13	6.08E−14
V_D	VD	−4.16E+00	−3.33E+00	−2.49E+00	−8.25E−01	1.05E−02
R_D	REQ	1.00E+12	1.00E+12	1.00E+12	1.00E+12	1.47E+11
C_D	CAP	0.00E+00	0.00E+00	0.00E+00	0.00E+00	0.00E+00

The results of the simulations are shown in Fig. 7-17.

7.8 EXAMPLES OF TRANSIENT AND AC ANALYSIS

The transient response is determined for diode circuits in which the input signal is time-variant, usually sinusoidal. PSpice first determines the transient bias point, which is different from the regular (dc) point. The initial values of the circuit nodes are taken into account in calculating the transient bias point, along with the small-signal parameters of the nonlinear elements. The capacitors and inductors, which may have initial values, therefore, remain as parts of the circuit. The various nodes can be assigned to initial voltages by the .IC statement.

PSpice determines the small-signal model parameters from the transient bias point. The following examples illustrate the transient analysis of diode circuits.

Example 7-5

A full-wave rectifier is shown in Fig. 7-18, where the output is taken between terminals 4 and 3. Plot the transient response of the output voltage V(4, 3) for the time duration of 0 to 20 ms in steps of 0.1 ms. The peak voltage of the transformer primary is 120 V, 60 Hz. The ratio of primary to secondary windings is 10:1. The model parameters are the default values. Calculate and print the coefficients of the Fourier series.

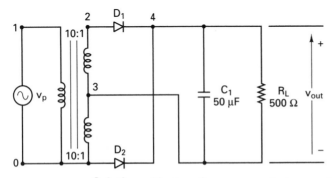

Figure 7-18 Rectifier with single-phase center-tapped transformer.

Solution The transformer secondaries may be considered as a voltage-controlled voltage source, as shown in Fig. 7-19. The primary is represented by a voltage source with a very high resistance (e.g., 10 MΩ).

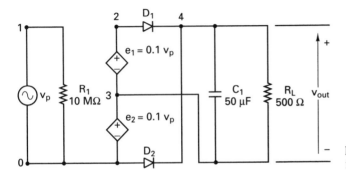

Figure 7-19 The equivalent circuit for Figure 7-18.

The listing of the circuit file follows.

Example 7-5 Rectifier with single-phase center-tapped transformer

```
 *     Primary is modeled as a voltage source of 120 V peak at 60 Hz
 *     with zero offset voltage.
 VP   1  0  SIN (0 120 60HZ)
 *     Primary winding is assumed to have a very high resistance:
 *     R1 = 10 MΩ.
 R1   1  0  10GOHM
 *     Secondary winding is assumed as a voltage-controlled voltage source
 *     with a voltage gain of 0.1.
 E1   2  3  1  0  0.1.
 E2   3  0  1  0  0.1
 C1   4  3  50UF
 RL   4  3  500
 *     Diode D1 with model name DIODE
 D1   2  4  DIODE
 D2   0  4  DIODE
 *     Diode model with default values
 .MODEL DIODE D
 *     Transient analysis from 0 to 20 ms with 0.1-ms increment
   .TRAN  0.1MS 20MS
   *     Plot the results of transient analysis for voltage across nodes
   *     4 and 3.
   .PLOT  TRAN  V(4,3)
   .FOUR  60HZ  V(4,3)
   *     Graphic post-processor
   .OP                      ; Prints details of the operating point
   .PROBE
 .END
```

The transient response for Example 7-5 is shown in Fig. 7-20. The coefficients of the Fourier analysis follow.

```
****     FOURIER ANALYSIS                        TEMPERATURE =  27.000 DEG C
FOURIER COMPONENTS OF TRANSIENT RESPONSE V (4,3)
DC COMPONENT =  1.00866SE+01
```

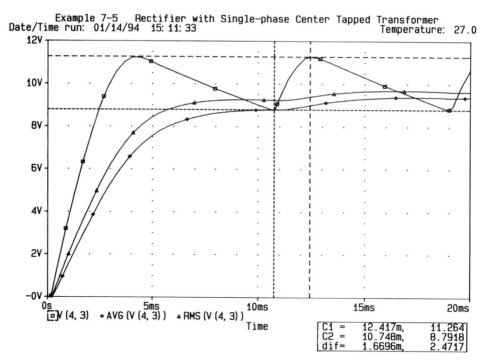

C1 =	12.417m,	11.264
C2 =	10.748m,	8.7918
dif=	1.6696m,	2.4717

□ V (4, 3)   ◆ AVG (V (4, 3))   ▲ RMS (V (4, 3))

Figure 7-20 Output voltage for Example 7-5.

HARMONIC NO	FREQUENCY (HZ)	FOURIER COMPONENT	NORMALIZED COMPONENT	PHASE (DEG)	NORMALIZED PHASE (DEG)
1	6.000E+01	2.112E-03	1.000E+00	-2.607E+00	0.000E+00
2	1.200E+02	9.843E-01	4.661E+02	1.182E+01	1.443E+01
3	1.800E+02	7.890E-04	3.736E-01	2.452E+01	2.713E+01
4	2.400E+02	3.833E-01	1.815E+02	2.131E+01	2.392E+01
5	3.000E+02	1.167E-03	5.528E-01	8.314E+01	8.575E+01
6	3.600E+02	1.669E-01	7.902E+01	3.761E+01	4.022E+01
7	4.200E+02	1.756E-03	8.314E-01	1.076E+02	1.102E+02
8	4.800E+02	7.140E-02	3.381E+01	6.687E+01	6.948E+01
9	5.400E+02	1.895E-03	8.971E-01	1.279E+02	1.305E+02

TOTAL HARMONIC DISTORTION = 5.075342E+04 PERCENT

JOB CONCLUDED

TOTAL JOB TIME   22.62

Example 7-6

A clamping circuit is shown in Fig. 7-21, where the output is taken from node 2. Plot the transient response of the output voltage V(2) for the time duration of 0 to 3 ms in steps of 20 μs. The initial capacitor voltage is -15 V. The model parameters of the diode are the default values.

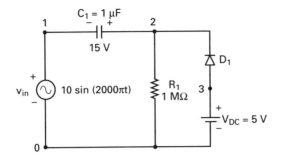

Figure 7-21 Diode clamper circuit.

Solution The listing of the circuit file follows.

Example 7-6 Diode clamper circuit

```
▲ *     Input voltage of 10 V peak at 1 kHz and zero offset voltage is
  *     connected between nodes 1 and 0.
▲▲ VIN  1  0  SIN  (0 10  1KHZ)
    *     Capacitor with an initial voltage of −15 V
    C1  1  2  1UF IC=−15V
    R1  2  0  1MEG
    VDC  3  0  DC 5
    *     Diode with model name DIODE is connected between nodes 3 and 2.
    D1  3  2  DIODE
    *     Diode model with default values of parameters
    .MODEL DIODE  D
▲▲▲ *     Transient analysis for 0 to 3 ms with 20-μs increment with UIC
        .TRAN  20US  3MS  UIC
        *     Plot transient voltages at nodes 1 and 2.
        .PLOT TRAN  V(2)  V(1)
        .OP                        ; Prints details of the operating point
        .PROBE
    .END
```

The information about the operating point, which is obtained from the output file EX7.6.OUT, is as follows.

****	OPERATING POINT INFORMATION		TEMPERATURE = 27.000 DEG C
	NAME	D1	
	MODEL	DIODE	
I_D	ID	4.48E−06	
V_D	VD	5.15E−01	
R_D	REQ	5.77E+03	
C_D	CAP	0.00E+00	

The input and output voltages of the diode clamper circuit are shown in Fig. 7-22. With the .PROBE command, there is no need for the .PLOT statement.

Example 7-7

A diode circuit is shown in Fig. 7-23(a). The ac input voltage is $v_{in} = 10 \times 10^{-3}\sin$ $(2\pi \times 10^3 t)$. Print the bias point and the small-signal parameters of the diode. Plot

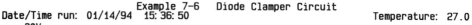

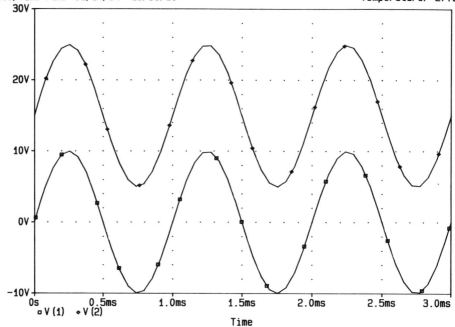

Figure 7-22   Output of diode clamper circuit for Example 7-6.

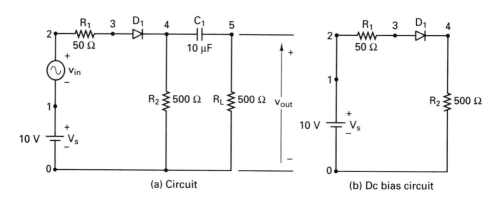

(a) Circuit   (b) Dc bias circuit

Figure 7-23   Diode circuit.

the output voltage from 0 to 200 μs with 2-μs increments. If the frequency of the ac voltage is varied from 1 Hz to 1 kHz, plot the magnitude and phase angle of the output voltage. The model parameters are IS=100E−15, RS=16, CJO=2PF, TT=12NS, BV=100, and IBV=100E−15.

Solution   The equivalent circuit for calculating the bias point and small-signal parameters is shown in Fig. 7-23(b). The listing of the circuit file follows.

Example 7-7 Diode circuit

▲ * DC voltage of 10 V
　VS 1 0 DC 10V
　* AC voltage of 10 mV peak at 10 kHz and zero offset voltage for
　* transient analysis and 10 mV peak for ac analysis
　VIN 2 1 AC 10MV SIN (0 10M 10KHZ)
▲▲ R1 2 3 50
　R2 4 0 500
　C1 4 5 10UF
　RL 5 0 500
　* Diode with model name DMOD is connected between nodes 3 and 4.
　D1 3 4 DMOD
　* Diode model defines the model parameters.
　.MODEL DMOD D (IS=100E−15 RS=16 CJO=2PF TT=12NS BV=100 IBV=100E−15)
▲▲▲ * Transient analysis for 0 to 200 μs with 2-μs increments.
　.TRAN 2US 200US
　* Ac analysis from 1 Hz to 1 kHz with 10 points per decade
　.AC DEC 10 1HZ 1kHZ
　* Plot transient voltages
　.PLOT TRAN V(5) V(2,1)
　* Magnitude and phase plots of output voltage at node 5
　.PLOT AC VDB(5) VP(5)
　* Printing of the small-signal parameters and the operating point.
　.OP   ; Prints the details of the operating point
　.PROBE
.END

Note that with the .PROBE command, there is no need for the .PLOT statement. The transient response is shown in Fig. 7-24, and the frequency response is shown in Fig. 7-25. The details of the dc bias point and the small-signal parameters are given next.

****	SMALL-SIGNAL BIAS SOLUTION					TEMPERATURE = 27.000 DEG C		
NODE	VOLTAGE	NODE	VOLTAGE	NODE	VOLTAGE	NODE		VOLTAGE
(1)	10.0000	(2)	10.0000	(3)	9.1756	(4)		8.2438
(5)	0.0000							

```
     VOLTAGE SOURCE CURRENTS
     NAME          CURRENT
     VS            −1.649E−02
     VIN           −1.649E−02
     TOTAL POWER DISSIPATION  1.65E−01  WATTS
```

****	OPERATING POINT INFORMATION	TEMPERATURE = 27.000 DEG C
NAME	D1	
MODEL	DMOD	
ID	1.65E−02	
VD	9.32E−01	
REQ	1.57E+00	
CAP	7.65E−09	

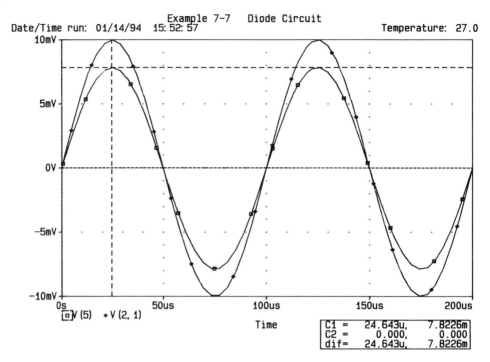

Figure 7-24 Transient response for Example 7-7.

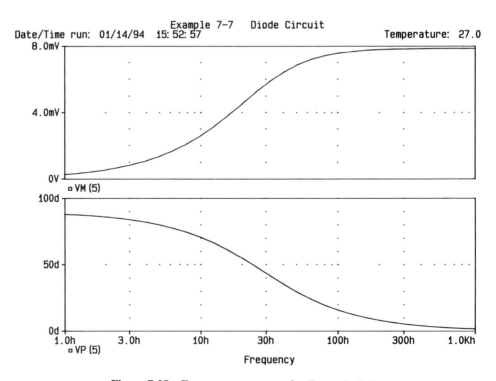

Figure 7-25 Frequency response for Example 7-7.

Example 7-8

Use PSpice to perform the worst-case analysis for the transient response of the diode circuit in Fig. 7-23. The model parameters are $R = 1$ for resistors, $C = 1$ for the capacitor, and I_S for the diode. The circuit parameters having uniform deviations are

$R_1 = 50 \pm 10\%$

$R_2 = 500 \; \Omega \pm 15\%$

$R_L = 500 \; \Omega \pm 20\%$

$C_1 = 10 \; \mu\text{F} \pm 15\%$

$I_S = 100\text{E} - 15 \pm \text{DEV} = 10\%$

The maximum value of the output voltage is to be printed.

Solution   The listing of the circuit file follows.

Example 7-8   Worst-case analysis of diode circuit

```
▲ VS  1   0   DC  10V                ; Dc voltage of 10 V
  VIN 2   1   AC  10MV  SIN (0 10M 10KHZ) ; sinusoidal voltage of 10 mV
▲▲ R1 2   3   RMOD1   50
   R2  4   0   RMOD2   500
   RL  5   0   RMODL   500
   C1  4   5   CMOD   10UF
   .MODEL RMOD1  RES  (R=1  DEV=10%)
   .MODEL RMOD2  RES  (R=1  DEV=20%)
   .MODEL RMODL  RES  (R=1  DEV=25%)
   .MODEL CMOD   CAP  (C=1  DEV=15%)
   D1   3 4  D1N914                  ; Diode with model name D1N914
   *   Diode model defines the model parameters
   .MODEL D1N914 D(IS=100E-15 DEV=10% RS=1 CJO=2PF TT=12NS BV=100
   +IBV=100E-15)
▲▲▲ .TRAN  2US  200US ; Transient analysis for 0 to 200 μs with 2 μs
                      ; increment
    .WCASE   TRAN  V(5)   MAX
    .OP                ; Prints details of the operating point
    .PROBE             ; Graphics post-processor
  .END                 ; End of circuit file
```

The results of the worst-case analysis, which are obtained from the output file EX7.8.OUT, are shown below.

****	SORTED DEVIATIONS OF V(5)	TEMPERATURE = 27.000 DEG C

WORST-CASE SUMMARY

RUN	MAXIMUM VALUE
ALL DEVICES	8.6367E-03 at T = 25.1210E-06
	(104.82% of Nominal)
NOMINAL	8.2395E-03 at T = 25.1210E-06

SUMMARY

The statements for diodes are

```
D⟨name⟩  NA  NK  DNAME  [(area) value]
.MODEL DNAME  D (P1=A1  P2=A2  P3=A3 ... PN=AN)
```

REFERENCES

1. *PSpice Manual.* Irvine, Calif.: MicroSim Corporation, 1992.
2. P. Antognetti, *Power Integrated Circuits.* New York: McGraw-Hill, 1986.
3. A. Laha and D. Smart, "A Zener diode model with application to SPICE2," *IEEE Journal of Solid-State Circuits,* Vol. SC-16, No. 1, pp. 21–22.
4. M. S. Ghausi, *Electronic Devices and Circuits: Discrete and Integrated.* New York: Holt, Rinehart, and Winston, 1985, pp. 3–7.
5. William H. Hayt, Jr., and Gerold W. Neudeck, *Electronic Circuit Analysis and Design.* Boston, Mass.: Houghton Mifflin, 1984.

PROBLEMS

7-1. For the diode circuit in Fig. P7-1, print the bias point and the small-signal parameters of the diode. Use the default values of model parameters.

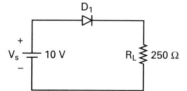

Figure P7-1

7-2. If the input voltage to the circuit in Fig. 7-14 is $v_{in} = 15 \sin (2000\pi\ t)$, plot the transient response of the output voltage for a time duration of 0 to 2 ms with a time increment of 10 μs. Print the details of the transient analysis bias point. The Zener voltages of the diodes are the same, $V_Z = 5.2$ V, and the current at the zener breakdown is $I_Z = 0.5$ μA. The model parameters are IS = 0.5UA, RS = 6, CJO = 2PF, TT = 12NS, BV = 5.20, and IBV = 0.5UA. The operating temperature is 50°C.

7-3. If the input voltage of the circuit in Fig. 7-16 is $V_{in} = 10$ V DC, print the details of the dc operating point. Print the voltage gain (V_{out}/V_{in}), the input resistance, and the output resistance.

7-4. A full-wave bridge rectifier is shown in Fig. P7-4. Plot the transient response of the output voltage for the time duration of 0 to 20 ms in steps of 0.1 ms. The model parameters are the default values. Print the details of the transient analysis bias point and the coefficients of the Fourier series.

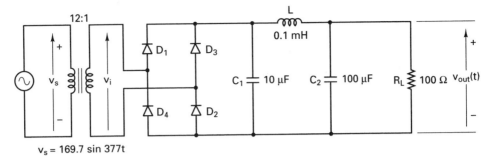

Figure P7-4

7-5. For the diode circuit in Fig. P7-5, plot the dc transfer characteristic between v_{in} and v_{out} for values of v_{in} in the range of -18 V to 18 V in steps of 0.5 V. The model parameters of the diodes are the default values.

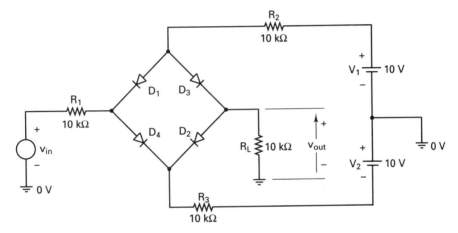

Figure P7-5

7-6. For the diode circuit in Fig. P7-6, plot the input current against the input voltage for values of V_{in} in the range of -10 V to 10 V in steps of 0.25 V. The model parameters are IS=0.5UA, RS=6, BV=5.20, and IBV=0.5UA. The operating temperature is 50°C.

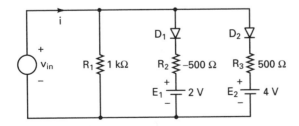

Figure P7-6

7-7. Repeat Example 7-6 if the direction of the diode D_1 is reversed.

7-8. Repeat Example 7-5 if the diodes are represented by voltage-controlled switches.

The model parameters of the switches are RON = 0.25, ROFF = 1E + 6, VON = 0.25, and VOFF = 0.

7-9. A demodulator circuit is shown in Fig. P7-9. Plot the transient output voltage for the time duration of 0 to 100 μs with an increment of 0.5 μs. The model parameters are IS = 0.5UA, RS = 6, CJO = 2PF, TT = 12NS, BV = 5.20, and IBV = 0.5UA. The input voltage is given by

$$v_{in} = 10[1 + 0.5 \sin(2\pi \times 10 \times 10^3 t)]\sin(2\pi \times 20 \times 10^6 t)$$

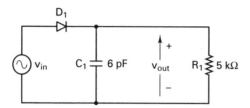

Figure P7-9

7-10. Use PSpice to perform a Monte Carlo analysis for six runs and for the transient response of Problem 7-4. The model parameters are $R = 1$ for resistors, L for inductor and $C = 1$ for the capacitors. The circuit parameters having uniform deviations are

$R_L = 100 \ \Omega \pm 5\%$
$C_1 = 10 \ \mu\text{F} \pm 15\%$
$C_2 = 100 \ \mu\text{F} \pm 10\%$
$L = 0.1 \ \text{mH} \pm 20\%$

(a) The greatest difference from the nominal run is to be printed.
(b) The maximum value of the output voltage is to be printed.
(c) The minimum value of the output voltage is to be printed.
(d) The first occurrence of the output voltage crossing below 5 V is to be printed.

7-11. Use PSpice to perform the worst-case analysis for Problem 7-10.

7-12. Use PSpice to perform a Monte Carlo analysis for five runs and for the dc sweep of Problem 7.5. The model parameter for resistors is $R = 1$. The circuit parameters having uniform deviations are

$R_1 = 10 \ \text{k}\Omega \pm 20\%$
$R_2 = 10 \ \text{k}\Omega \pm 5\%$
$R_3 = 10 \ \text{k}\Omega \pm 10\%$
$R_L = 10 \ \Omega \pm 15\%$

(a) The greatest difference from the nominal run is to be printed.
(b) The maximum value of the output voltage is to be printed.
(c) The minimum value of the output voltage is to be printed.
(d) The first occurrence of the output voltage crossing below -2 V is to be printed.

7-13. Use PSpice to perform the worst-case analysis for Problem 7-12.

7-14. Use PSpice to perform a Monte Carlo analysis for five runs and for the dc sweep of Problem 7-6. The model parameters for resistors are $R = 1$ and I_S for diodes. The circuit parameters having uniform deviations are

$R_1 = 1 \text{ k}\Omega \pm 20\%$

$R_2 = 500 \ \Omega \pm 15\%$

$R_3 = 500 \ \Omega \pm 20\%$

$I_S = 0.5\text{E} - 6 + 10\%$

(a) The greatest difference from the nominal run is to be printed.

(b) The maximum value of the input current is to be printed.

(c) The minimum value of the input current is to be printed.

7-15. Use Pspice to perform the worst-case analysis for Problem 7-14.

Bipolar Junction Transistors

8-1 INTRODUCTION

A bipolar junction transistor (BJT) may be specified by a device statement in conjunction with a model statement. Similar to diode models, the BJT model incorporates an extensive range of characteristics: for example, dc and small-signal behavior, temperature dependency, and noise generation. The model parameters take into account temperature effects, various capacitances, and the physical properties of semiconductors.

8-2 BJT MODEL

PSpice generates a complex model for BJTs. The model equations that are used by PSpice are described in Gummel and Poon [1] and Getreu [2]. If a complex model is not necessary, the model parameters can be ignored by the users, and PSpice assigns default values to the parameters.

The PSpice model, which is based on the integral charge-control model of Gummel and Poon [1, 6], is shown in Fig. 8-1. The small-signal and static models that are generated by PSpice are shown in Figs. 8-2 and 8-3, respectively.

The model statement for NPN transistors has the general vorm

```
.MODEL QNAME NPN (P1=A1 P2=A2 P3=A3 ...PN=AN)
```

and the general form for PNP transistors is

```
.MODEL QNAME PNP (P1=A1 P2=A2 P3=A3 ...PN=AN)
```

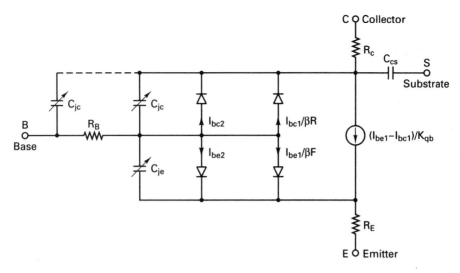

Figure 8-1 PSpice BJT model.

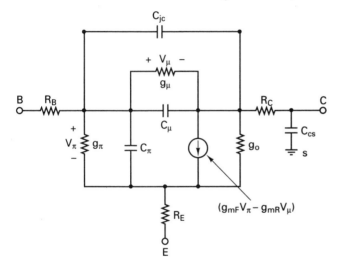

Figure 8-2 Small-signal BJT model.

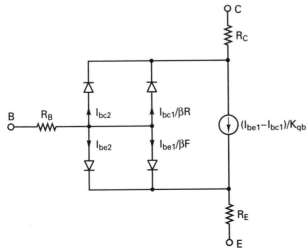

Figure 8-3 Static BJT model.

where QNAME is the name of the BJT model. NPN and PNP are the type symbols for NPN and PNP transistors, respectively. QNAME, which is the model name, can begin with any character, and its word size is normally limited to eight characters. P1, P2, . . . and A1, A2, . . . are the parameters and their values, respectively. Table 8-1 shows the model parameters of BJTs. If certain parameters are not specified, PSpice assumes the simple Ebers-Moll model [3], which is shown in Fig. 8-4(a).

TABLE 8-1 MODEL PARAMETERS OF BJTS

Name	Area	Model parameters	Units	Default	Typical
IS	*	*p-n* saturation current	Amps	1E–16	1E–16
BF		Ideal maximum forward beta		100	100
NF		Forward current emission coefficient		1	1
VAF(VA)		Forward Early voltage	Volts	∞	100
IKF(IK)		Corner for forward beta high-current roll-off	Amps	∞	10M
ISE(C2)		Base-emitter leakage saturation current	Amps	0	1000
NE		Base-emitter leakage emission coefficient		1.5	2
BR		Ideal maximum reverse beta		1	0.1
NR		Reverse current emission coefficient		1	
VAR(VB)		Reverse Early voltage	Volts	∞	100
IKR	*	Corner for reverse beta high-current roll-off	Amps	∞	100M
ISC(C4)		Base-collector leakage saturation current	Amps	0	1
NC		Base-collector leakage emission coefficient		2	2
RB	*	Zero-bias (maximum) base resistance	Ohms	0	100
RBM		Minimum base resistance	Ohms	RB	100
IRB		Current at which RB falls halfway to RBM	Amps	∞	
RE	*	Emitter ohmic resistance	Ohms	0	1
RC	*	Collector ohmic resistance	Ohms	0	10
CJE	*	Base-emitter zero-bias *p-n* capacitance	Farads	0	2P
VJE(PE)		Base-emitter built-in potential	Volts	0.75	0.75
MJE(ME)		Base-emitter *p-n* grading factor		0.33	0.33
CJC	*	Base-collector zero-bias *p-n* capacitance	Farads	0	1P
VJC(PC)		Base-collector built-in potential	Volts	0.75	0.75
MJC(MC)		Base-collector *p-n* grading factor		0.33	0.33
XCJC		Fraction of C_{bc} connected internal to R_B		1	
CJS(CCS)		Collector-substrate zero-bias *p-n* capacitance	Farads	0	2PF

TABLE 8-1 CONTINUED

Name	Area	Model parameters	Units	Default	Typical
VJS(PS)		Collector-substrate built-in potential	Volts	0.75	
MJS(MS)		Collector-substrate p-n grading factor		0	
FC		Forward-bias depletion capacitor coefficient		0.5	
TF		Ideal forward transit time	Seconds	0	0.1NS
XTF		Transit-time bias dependence coefficient		0	
VTF		Transit-time dependency on V_{bc}	Volts	∞	
ITF		Transit-time dependency on I_c	Amps	0	
PTF		Excess phase at $1/(2\pi*\text{TF})\text{Hz}$	Degrees	0	30°
TR		Ideal reverse transit time	Seconds	0	10NS
EG		Bandgap voltage (barrier height)	Electron-volts	1.11	1.11
XTB		Forward and reverse beta temperature coefficient		0	
XTI(PT)		IS temperature-effect exponent		3	
KF		Flicker noise coefficient		0	6.6E–16
AF		Flicker noise exponent		1	1

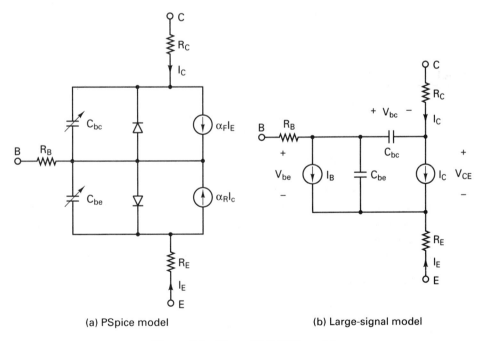

(a) PSpice model   (b) Large-signal model

Figure 8-4 Ebers-Moll BJT model.

As with diodes, an *area factor* is used to determine the number of equivalent parallel BJTs of a specified model. The model parameters that are affected by the area factor are marked by an asterisk (*) in Table 8-1. RC, RE, and RB represent the contact and bulk resistances per unit area of the collector, emitter, and base, respectively. The bipolar transistor is modeled as an intrinsic device. The area value, which is the relative device area, is specified in the .MODEL statement (Section 8.3), and changes the actual resistance values. The area has a default value of 1.

Some parameters have alternate names, such as VAF and VA. One can use either name, VAF or VA. These are indicated by parentheses in Table 8-1.

The parameters ISE (C2) and ISC (C4) can be either greater than or less than 1. If they are less than 1, they represent the absolute currents. If they are greater than 1, they act as the multipliers of IS instead of absolute currents. That is, the value of ISE becomes ISE*IS for ISE > 1 and that of ISC becomes ISC*IS for ISC > 1.

The dc model is defined (1) by parameters BF, C2, IK, and NE, which determine the forward-current gain, (2) by BR, C4, IKR, and VC, which determine the reverse-current gain characteristics, (3) by VA and VB, which determine the output conductance for forward and reverse regions, and (4) by the reverse saturation current IS.

Base-charge storage is modeled (1) by forward and reverse transit times TF and TF, and nonlinear depletion-layer capacitances, which are determined by CJE, PE, and ME for a b-e junction, and (2) by CJC, PC, and MC for a b-c junction. CCS is a constant collector-substrate capacitance.

The temperature dependence of the saturation current is determined by the energy gap EG and the saturation-current temperature exponent PT.

8-3 BJT STATEMENTS

The symbol for a bipolar junction transistor (BJT) is Q. The name of a bipolar transistor must start with Q, and it takes the general form

```
Q⟨name⟩  NC  NB  NE  NS  QNAME  [(area) value]
```

where NC, NB, NE, and NS are the collector, base, emitter, and substrate nodes, respectively. QNAME could be any name of up to eight characters. The substrate node is optional; if not specified, it defaults to ground. Positive current is the current that flows into a terminal. That is, the current flows from the collector node through the device to the emitter node for an NPN BJT.

Some Statements for BJTs

```
QIN   5  7  8   2N2222
Q5    2  4  5   2N2907   1.5
QX    1  4  9   NMOD
.MODEL 2N2222 NPN(IS=3.108E-15 XTI=3 EG=1.11 VAF=131.5 BF=217.5
```

```
+  NE=1.541 ISE=190.7E-15 IKF=1.296 XTB=1.5 BR=6.18 NC=2 ISC=0 IKR=0
+  RC=1 CJC=14.57E-12 VJC=.75 MJC=.3333 FC=.5 CJE=26.08E-12 VJE=.75
+  MJE=.3333 TR=51.35E-9 TF=451E-12 ITF=.1 VTF=10 XTF=2)
.MODEL 2N2907 PNP(IS=9.913E-15 XTI=3 EG=1.11 VAF=90.7 BF=197.8
+  NE=2.264 ISE=6.191E-12 IKF=.7322 XTB=1.5 BR=3.369 NC=2 ISC=0 IKR=0
+  RC=1 CJC=14.57E-12 VJC=.75 MJC=.3333 FC=.5 CJE=20.16E-12 VJE=.75
+  MJE=.3333 TR=29.17E-9 TF=405.7E-12 ITF=.4 VTF=10 XTF=2)
.MODEL NMOD NPN
```

Note. A + (plus) sign at the first column indicates the continuation of the statement preceding it.

8-4 BJT PARAMETERS

The data sheet for NPN transistor of type 2N2222A is shown in Fig. 8-5. SPICE parameters are not quoted directly in the data sheet. Some versions of SPICE (e.g., PSpice) support device library files that give the model parameters. The library file EVAL.LIB contains the list of devices and their model statements in the student's version of PSpice. The software PARTS of PSpice can generate SPICE models from the data-sheet parameters of diodes. SPICE model parameters are also supplied by some manufacturers. However, some parameters that significantly influence the performance of a transistor can be determined from the data sheet [7, 8].

The diode characteristic described by Eq. (7-1) can be applied to an NPN transistor by selecting appropriate subscripts, as in the following equation:

$$I_C = I_S[e^{V_{BE}/\eta V_T} - 1] \qquad (8\text{-}1)$$

Using Eq. (7-8), the difference in the base-emitter voltages can be expressed by

$$V_{BE2} - V_{BE1} = 2.3 \, \eta V_T \log \left(\frac{I_{C2}}{I_{C1}}\right) \qquad (8\text{-}2)$$

From the data sheet for V_{BD} versus I_C, we get $V_{BE1} = 0.6$ V at $I_{C1} = 0.2$ mA, and $V_{BE2} = 0.7$ V at $I_{C2} = 20$ mA. Assuming $V_T = 25.8$ mV $= 0.0258$, we can apply Eq. (8-2) to find the *emission coefficient* η as follows:

$$0.7 - 0.6 = 2.3 \, \eta \times 0.0258 \log \left(\frac{20}{0.2}\right)$$

which gives $\eta = 0.843$. Since we did not include any contact and bulk resistance RE of the emitter, we got the value $\eta < 1$; its practical value is $\eta \geq 1$. Let us assume $\eta = 1$. For $\eta = 1$, and $V_{BE2} = 0.7$ V at $I_{C2} = 20$ mA, we can apply Eq. (8-1) to find the saturation current I_S

$$20 \text{ mA} = I_S[e^{0.7/(1 \times 0.0258)} - 1]$$

which gives $I_S = 3.295E-14$ A. The dc current gain at 150 mA is $h_{FE} = 100$ to 300. Taking the geometric mean gives $BF = \sqrt{(100 \times 300)} \approx 173$.

2N2222 2N2222A

TO-18

TL/G/10100-9

PN2222 PN2222A

TO-92

E
B
C

TL/G/10100-1

MMBT2222 MMBT2222A

C
B
E
TO-236
(SOT-23)

TL/G/10100-5

MPQ2222*

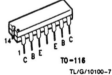

14
1
C B E
C B
E B C

TO-116
TL/G/10100-7

NPN General Purpose Amplifier

Electrical Characteristics T_A = 25°C unless otherwise noted

Symbol	Parameter		Min	Max	Units
OFF CHARACTERISTICS					
$V_{(BR)CEO}$	Collector-Emitter Breakdown Voltage (Note 1) ($I_C = 10$ mA, $I_B = 0$)	2222	30		V
		2222A	40		
$V_{(BR)CBO}$	Collector-Base Breakdown Voltage ($I_C = 10$ μA, $I_E = 0$)	2222	60		V
		2222A	75		
$V_{(BR)EBO}$	Emitter Base Breakdown Voltage ($I_E = 10$ μA, $I_C = 0$)	2222	5.0		V
		2222A	6.0		
I_{CEX}	Collector Cutoff Current ($V_{CE} = 60$V, $V_{EB(OFF)} = 3.0$V)	2222A		10	nA
I_{CBO}	Collector Cutoff Current ($V_{CB} = 50$V, $I_E = 0$)	2222		0.01	μA
	($V_{CB} = 60$V, $I_E = 0$)	2222A		0.01	
	($V_{CB} = 50$V, $I_E = 0$, $T_A = 150$°C)	222		10	
	($V_{CB} = 60$V, $I_E = 0$, $T_A = 150$°C)	2222A		10	
I_{EBO}	Emitter Cutoff Current ($V_{EB} = 3.0$V, $I_C = 0$)	2222A		10	nA
I_{BL}	Base Cutoff Current ($V_{CE} = 60$V, $V_{EB(OFF)} = 3.0$)	2222A		20	nA
ON CHARACTERISTICS					
h_{FE}	DC Current Gain ($I_C = 0.1$ mA, $V_{CE} = 10$V)		35		
	($I_C = 1.0$ mA, $V_{CE} = 10$V)		50		
	($I_C = 10$ mA, $V_{CE} = 10$V)		75		
	($I_C = 10$ mA, $V_{CE} = 10$V, $T_A = -55$°C)		35		
	($I_C = 150$ mA, $V_{CE} = 10$V) (Note 1)		100	300	
	($I_C = 150$ mA, $V_{CE} = 1.0$V) (Note 1)		50		
	($I_C = 500$ mA, $V_{CE} = 10$V) (Note 1)	2222	30		
		2222A	40		

Note 1: Pulse Test: Pulse Width ≤ 300 μs, Duty Cycle ≤ 2.0%.

*16-SOIC version also available. Contact factory.

Figure 8-5 Data sheet for transistor of type 2N2222A (Courtesy of National Semiconductor, Inc.).

NPN General Purpose Amplifier (Continued)

Electrical Characteristics T_A = 25°C unless otherwise noted (Continued)

Symbol	Parameter		Min	Max	Units
ON CHARACTERISTICS (Continued)					
$V_{CE(sat)}$	Collector-Emitter Saturation Voltage (Note 1)				
	(I_C = 150 mA, I_B = 15 mA)	2222		0.4	
		2222A		0.3	V
	(I_C = 500 mA, I_B = 50 mA)	2222		1.6	
		2222A		1.0	
$V_{BE(sat)}$	Base-Emitter Saturation Voltage (Note 1)				
	(I_C = 150 mA, I_B = 15 mA)	2222	0.6	1.3	
		2222A	0.6	1.2	V
	(I_C = 500 mA, I_B = 50 mA)	2222		2.6	
		2222A		2.0	
SMALL-SIGNAL CHARACTERISTICS					
f_T	Current Gain—Bandwidth Product (Note 3)				
	(I_C = 20 mA, V_{CE} = 20V, f = 100 MHz)	2222	250		MHz
		2222A	300		
C_{obo}	Output Capacitance (Note 3)				
	(V_{CB} = 10V, I_E = 0, f = 100 kHz)			8.0	pF
C_{ibo}	Input Capacitance (Note 3)				
	(V_{EB} = 0.5V, I_C = 0, f = 100 kHz)	2222		30	pF
		2222A		25	
$rb'C_C$	Collector Base Time Constant				
	(I_E = 20 mA, V_{CB} = 20V, f = 31.8 MHz)	2222A		150	ps
NF	Noise Figure				
	(I_C = 100 μA, V_{CE} = 10V, R_S = 1.0 kΩ, f = 1.0 kHz)	2222A		4.0	dB
$Re(h_{ie})$	Real Part of Common-Emitter High Frequency Input Impedance				
	(I_C = 20 mA, V_{CE} = 20V, f = 300 MHz)			60	Ω

SWITCHING CHARACTERISTICS

Symbol	Parameter			Min	Max	Units
t_D	Delay Time	(V_{CC} = 30V, $V_{BE(OFF)}$ = 0.5V, I_C = 150 mA, I_{B1} = 15 mA)	except MPQ2222		10	ns
t_R	Rise Time				25	ns
t_S	Storage Time	(V_{CC} = 30V, I_C = 150 mA, I_{B1} = I_{B2} = 15 mA)	except MPQ2222		225	ns
t_F	Fall Time				60	ns

Note 1: Pulse Test: Pulse Width < 300 μs, Duty Cycle ≤ 2.0%.

Note 2: For characteristics curves, see Process 19.

Note 3: f_T is defined as the frequency at which $|h_{fe}|$ extrapotates to unity.

Note 4: 2N also available in JAN/TX/V series.

Figure 8-5 *Continued*

SMALL SIGNAL CHARACTERISTICS (f = 1.0 kHz)

Symbol	Parameter	Conditions	Typ	Units
h_{ie}	Input Resistance	I_C = 10 mA, V_{CE} = 10V	700	Ω
h_{oe}	Output Conductance	I_C = 10 mA, V_{CE} = 10V	120	μmhos
h_{fe}	Small Signal Current Gain	I_C = 10 mA, V_{CE} = 10V	240	
h_{re}	Voltage Feedback Ratio	I_C = 10 mA, V_{CE} = 10V	460	$\times 10^{-6}$

TYPICAL COMMON EMITTER CHARACTERISTICS (f = 1.0 kHz)

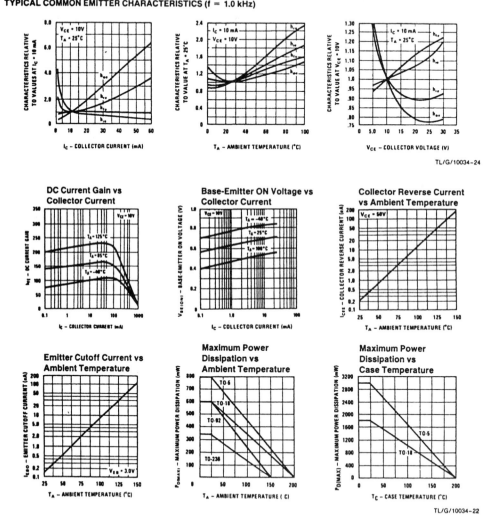

Figure 8-5 *Continued*

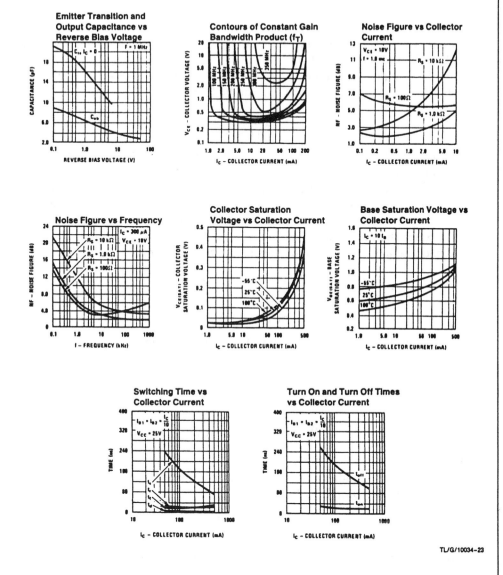

Figure 8-5 *Continued*

TL/G/10034-23

The input capacitance at the base-emitter junction is $C_{ibo} = 25$ pF at $V_{EB} = 0.5$ V, $I_C = 0$ (reverse-biased). Since $C_{je} = C_{ibo}$, then C_{jeo} can be found from

$$C_{je} = \frac{C_{jeo}}{(1 + V_{EB}/V_{je})^{M_{je}}} \tag{8-3}$$

where $M_{je} = \text{MJE} \approx 1/3$, and $V_{je} = \text{VJE} \approx 0.75$ V. Equation (8-3) gives $C_{jeo} = \text{CJE} = 29.6$ pF at $V_{BE} = 0$ V.

The output capacitance is $C_{obo} = 8$ pF at $V_{CB} = 10$ V, $I_E = 0$ (reverse-biased). Since $C_\mu = C_{obo}$, then $C_{\mu o}$ can be found from

$$C_\mu = \frac{C_{\mu o}}{(1 + V_{CB}/V_{jc})^{M_{jc}}} \tag{8-4}$$

where $M_{jc} = \text{MJC} \approx 1/3$, and $V_{jc} \approx \text{VJC} = 0.75$ V. From Eq. (8-4), $C_{\mu o} = \text{CJC} = 19.4$ pF at $V_{CB} = 0$ V.

The transition frequency $f_{T(\min)} = 300$ MHz at $V_{CE} = 20$ V, $I_C = 20$ mA. The transition period is $\tau_T = 1/2\pi f_T = 1/(2\pi \times 300 \text{ MHz}) = 530.5$ ps. Assuming $V_{BE} = 0.7$ V, $V_{CB} \approx V_{CE} - V_{BE} = 20 - 0.7 = 19.3$ V, and Eq. (8-4) gives $C_\mu = 6.49$ pF.

Since the transition frequency $f_{T(\min)} = 300$ MHz is specified at $I_C = 20$ mA, we need to find the transconductance g_m (at $I_C = 20$ mA), which is given by

$$g_m = \frac{I_C}{V_T} \tag{8-5}$$

$$= 20 \text{ mA}/25.8 \text{ mV} = 775.2 \text{ mA/V}$$

The transition period τ_T is related to forward transit time τ_F by

$$\tau_T = \tau_F + \frac{C_{je}}{g_m} + \frac{C_\mu}{g_m} \tag{8-6}$$

or

$$530.5 \text{ ps} = \tau_F + \frac{25 \text{ pF}}{0.7752} + \frac{6.49 \text{ pF}}{0.7752}$$

which gives $\tau_F = \text{TF} = 489.88$ ps.

The output conductance h_{oe} of a transistor is related to the collector current I_C and the *Early voltage* V_A by

$$\frac{1}{h_{oe}} = \frac{V_A}{I_C} \tag{8-7}$$

From the data sheet, $h_{oe} = 120$ μmhos at $I_C = 10$ mA and $V_{CE} = 10$ V. From Eq. (8-7), the Early voltage V_A becomes

$$V_A = \text{VA} = I_C/h_{oe} = 10 \text{ mA}/120 \text{ } \mu\text{mhos} = 83.3 \text{ V}$$

The reverse transit time can be approximated to $\tau_R = \text{TR} = 10 \tau_F = 4.9$ ns.

The model statement for transistor 2N2222A is

```
.MODEL   Q2N2222A NPN (IS=3.295E−14 BF=173 VA=83.3V CJE=29.6PF CJC=19.4PF
+        TF=489.88PS TR=4.9NS)
```

This model is used to plot the characteristics of the BJT as illustrated in Example 8-1. It may be necessary to modify the parameter values to conform to the actual characteristics.

Note. If a model parameter is not specified in the model statement, SPICE assumes its default value. However, it should be noted that some default values represent ideal conditions (e.g., TR = 0 and VAF = ∞). More accurate results can be obtained by using the typical values (e.g., TR = 0.1NS and VAF = 100V) rather than the default ones.

8-5 EXAMPLES OF BJT CIRCUITS

The PSpice simulation of BJT circuits requires specifying the BJT model parameters. If the model parameters are not specified, PSpice assumes the default values that are given in Table 8-1. The following examples illustrate the PSpice simulation of BJT circuits.

Example 8-1

For the NPN BJT transistor of Fig. 8-6, plot the output characteristics (I_C versus V_{CE}) if V_{CE} is varied from 0 to 10 V in steps of 0.02 V and I_B is varied from 0 to 1 mA in steps of 200 μA. Use the model parameters that were determined in Section 8-4. Print the details of the small-signal parameters at the operating point for I_B = 1 mA and V_{CE} = 12 V.

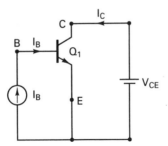

Figure 8-6   A circuit with an NPN BJT.

Solution   The listing of the circuit file follows.

Example 8-1   NPN BJT characteristics

```
▲ IB    0   1   DC   1MA              ; Base current
  VCE   2   0   DC   12V              ; Collector-emitter voltage
▲▲ Q1     2   1   0      Q2N2222A        ; BJT statement
    .MODEL   Q2N2222A   NPN  (IS=2.105E-16 BF=173 VA=83.3V CJE=29.6PF CJC=19.4PF
    +        TF=489.88PS TR=4.9NS)      ; Model parameters
▲▲▲ .DC  VCE   0   10V   0.02V   IB   0   1MA   200UA ; Dc sweep for VCE and IB
    .PROBE                             ; Graphics post-processor
  .END
```

The output characteristics, which are plots of I_C versus V_{CE}, are shown in Fig. 8-7. The students are encouraged to compare the transistor characteristics with those obtained by using the model parameters of transistor 2N2222A, which is listed

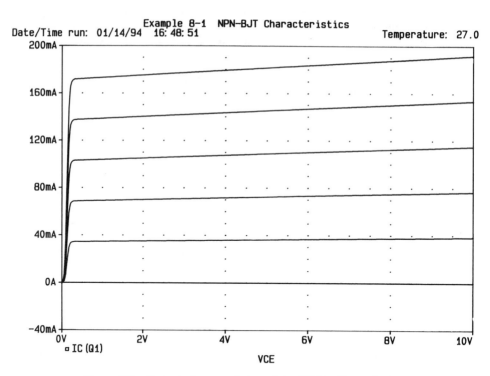

Figure 8-7 Output characteristics of the BJT in Example 8.1.

in the PSpice library file EVAL.LIB. This can be done by replacing the .MODEL statement in the above circuit file by the .LIB EVAL.LIB statement.

The small-signal parameters of the transistor at the operating point, which are obtained from the output file EX8.1.OUT, are as follows.

```
****    OPERATING POINT INFORMATION          TEMPERATURE =  27.000 DEG C
        NAME        Q1
        MODEL       Q2N2222A
        IB          1.00E-03
        IC          1.96E-01
        VBE         8.88E-01
        VBC         -1.11E+01
        VCE         1.20E+01
        BETADC      1.96E+02
        GM          7.58E+00
        RPI         2.59E+01
        RX          0.00E+00
        RO          4.82E+02
        CBE         3.77E-09
        CBC         7.80E-12
        CBX         0.00E+00
        CJS         0.00E+00
        BETAAC      1.96E+02
        FT          3.19E+08
```

Example 8-2

A bipolar transistor circuit is shown in Fig. 8-8(a), where the output is taken from node 4. Calculate and print the sensitivity of the collector current with respect to all parameters. Print the details of the bias point. The equivalent circuit for transistor Q_1 is shown in Fig. 8-8(b).

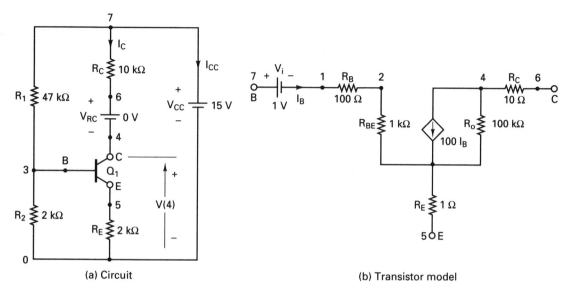

(a) Circuit

(b) Transistor model

Figure 8-8 Bipolar transistor circuit.

Solution The listing of the circuit file follows.

Example 8-2   Biasing sensitivity of bipolar transistor amplifier

```
▲ *  Supply voltage is 15 V DC.
   VCC  7 0  DC 15V
   *   A dummy voltage source of 0 V to measure the collector current
   VRC  6  4  DC  0V
▲▲ R1  7  3   47K
   R2  3  0   2K
   RC  7  6   10K
   RE     5  0  2K
   *   Subcircuit call for transistor model QMOD, and the substrate is
   *   connected to ground by default.
   XQ1 4  3  5  QMOD
   *   Subcircuit definition for QMOD
   .SUBCKT   QMOD   6  7  5
   RB   1  2   100
   RE   3  5   1
   RC   4  6   10
   RBE  2  3   1K
   RO   4  3   100K
```

```
    *    A dummy voltage source of 0 V to measure the controlling current
    VI  7  1  DC 0V
    F1  4  3  VI  20
    *    End of subcircuit definition
    .ENDS  QMOD
▲ ▲ ▲ .OPTIONS NOPAGE NOECHO
    *    Sensitivity of collector current (which is the current through voltage
    *    source VRC)
    .SENS  I(VRC)
.END
```

The .SENS command does not require a .PRINT command for printing the output. The output for the sensitivity analysis and the bias point follow.

****   SMALL-SIGNAL BIAS SOLUTION   **TEMPERATURE = 27.000 DEG C**

NODE	VOLTAGE	NODE	VOLTAGE	NODE	VOLTAGE	NODE	VOLTAGE
(3)	.5960	(4)	12.1520	(5)	.5864	(6)	12.1520
(7)	15.0000	(XQ1.1)	.5960	(XQ1.2)	.5952	(XQ1.3)	.5867
(XQ1.4)	12.1500						

VOLTAGE SOURCE CURRENTS

NAME	CURRENT
VCC	−5.912E−04
VRC	2.848E−04
XQ1.VI	8.456E−06

TOTAL POWER DISSIPATION 8.87E−03 WATTS

****   DC SENSITIVITY ANALYSIS   **TEMPERATURE = 27.000 DEG C**

DC SENSITIVITIES OF OUTPUT I(VRC)

ELEMENT NAME	ELEMENT VALUE	ELEMENT SENSITIVITY (AMPS/UNIT)	NORMALIZED SENSITIVITY (AMPS/PERCENT)
R1	4.700E+04	−5.481E−09	−2.576E−06
R2	2.000E+03	1.252E−07	2.505E−06
RC	1.000E+04	−3.134E−10	−3.134E−08
RE	2.000E+03	−1.288E−07	−2.576E−06
XQ1.RB	1.000E+02	−3.705E−09	−3.705E−09
XQ1.RE	1.000E+00	−1.288E−07	−1.288E−09
XQ1.RC	1.000E+01	−3.134E−10	−3.134E−11
XQ1.RBE	1.000E+03	−3.705E−09	−3.705E−08
XQ1.RO	1.000E+05	−1.273E−10	−1.273E−07
VCC	1.500E+01	1.898E−05	2.848E−06
VRC	0.000E+00	−1.101E−06	0.000E+00
XQ1.VI	0.000E+00	−4.381E−04	0.000E+00

JOB CONCLUDED

TOTAL JOB TIME   2.52

Example 8-3

A bipolar Darlington pair amplifier is shown in Fig. 8-9. Calculate and print the voltage gain, the input resistance, and the output resistance. The input voltage is 5 V. The model parameters of the bipolar transistors are BF=100, BR=1, RB=5, RC=1, RE=0, VJE=0.8, and VA=100.

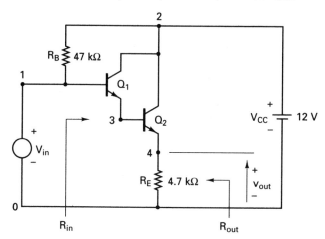

Figure 8-9 Darlington pair amplifier.

Solution The listing of the circuit file follows.

Example 8-3   Darlington pair

```
▲ .OPTIONS  NOPAGE  NOECHO
   VCC  2  0  DC  12V
   VIN  1  0  DC  5V
▲▲ *   BJTs with model QM
   Q1   2  1  3  QM
   Q2   2  3  4  QM
   RB   2  1  47K
   RE   4  0  4.7K
   *    Model QM for NPN BJTs
   .MODEL QM  NPN (BF=100 BR=1 RB=5 RC=1 RE=0 VJE=0.8 VA=100)
▲▲▲ *   Transfer-function analysis to calculate dc gain, input
   *    resistance, and output resistance
   .TF  V(4)   VIN
.END
```

The results of the transfer-function analysis follow.

****	**SMALL-SIGNAL BIAS SOLUTION**				**TEMPERATURE =**	**27.000 DEG C**	
NODE	**VOLTAGE**	**NODE**	**VOLTAGE**	**NODE**	**VOLTAGE**	**NODE**	**VOLTAGE**
(1)	5.0000	(2)	12.0000	(3)	4.3560	(4)	3.5909

```
VOLTAGE SOURCE CURRENTS
NAME          CURRENT

VCC          -9.129E-04

VIN           1.489E-04

TOTAL POWER DISSIPATION   1.02E-02  WATTS
```

        V(4)/VIN = 9.851E-01
        INPUT RESISTANCE AT VIN = 4.696E+04
        OUTPUT RESISTANCE AT V(4) = 6.679E+01
        JOB CONCLUDED
        TOTAL JOB TIME   2.97

Example 8-4

A bipolar transistor amplifier circuit is shown in Fig. 8-10. The output is taken from node 6. Calculate and plot the magnitude and phase of the voltage gain for frequencies from 1 Hz to 10 kHz with a decade increment and with 10 points per decade. The input voltage for ac analysis is 10 mV. Calculate and plot the transient response of voltages at nodes 4 and 6 for an input voltage of $v_{in} = 0.01 \sin(2\pi \times 1000t)$ and for a duration of 0 to 2 ms in steps of 50 μs. The details of ac and transient analysis operating points should be printed. The model parameters of the PNP BJT are IS=2E−16, BF=50, BR=1, RB=5, RC=1, RE=0, TF=0.2NS, TR=5NS, CJE=0.4PF, VJE=0.8, ME=0.4, CJC=0.5PF, VJC=0.8, CCS=1PF, and VA=100.

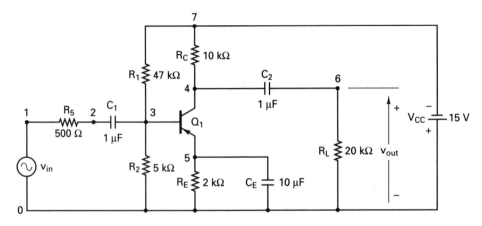

Figure 8-10 Bipolar transistor amplifier circuit.

Solution The listing of the circuit file follows.

Example 8-4 Bipolar transistor amplifier

▲ * Input voltage is 10 mV peak for ac analysis and for transient response:
  * It is 10 mV peak at 1 kHz with zero-offset value.
    VIN 1 0 AC 10MV SIN(0 10MV 1KHZ)
    VCC 0 7 DC 15V
▲▲ RS 1 2 500
    R1 7 3 47K
    R2 3 0 5K
    RC 7 4 10K
    RE 5 0 2K
    RL 6 0 20K
    C1 2 3 1UF
    C2 4 6 1UF
    CE 5 0 10UF

```
*  Transistor Q1 with model QM
Q1  4  3  5  0  QM
*  Model QM for PNP transistors
.MODEL   QM  PNP (IS=2E-16 BF=50 BR=1 RB=5 RC=1 RE=0 TF=0.2NS TR=5NS
+         CJE=0.4PF VJE=0.8 ME=0.4 CJC=0.5PF VJC=0.8 CCS=1PF VA=100)
```
▲▲▲ `*  Plot the results of transient analysis for voltages at nodes 4, 6, and 1`
```
   .PLOT  TRAN  V(4) V(6) V(1)
   *   Plot the results of the ac analysis for the magnitude and phase angle
   *   of output voltage at node 6.
   .PLOT  AC  VM(6)  VP(6)
   .OPTIONS  NOPAGE NOECHO
   *   Transient analysis for 0 to 2 ms with 50-µs increment
   *   Print details of transient analysis operating point.
   .TRAN/OP  50US  2MS
   *   AC analysis from 1 Hz to 10 KHz with a decade increment and 10 points
   *   per decade
   .AC  DEC  10  1HZ  10KHZ
   *   Print the details of the ac analysis operating point.
   .OP
   .PROBE
.END
```

Note. .PLOT statements generate graphical plots in the output file. If the .PROBE command is included, there is no need for the .PLOT command.

The determination of the operating point is the first step in analyzing a circuit with nonlinear devices (e.g., bipolar transistors). The equivalent circuit for determining the ac analysis (or dc analysis) bias point of the amplifier in Fig. 8-10 is shown in Fig. 8-11, where the capacitors are open-circuited. The details of the bias point follow.

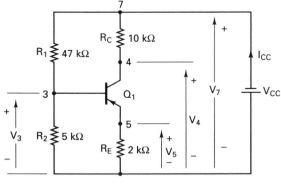

Figure 8-11 Equivalent circuit for dc bias calculation.

****	SMALL-SIGNAL BIAS SOLUTION				TEMPERATURE =	27.000 DEG C	
NODE	VOLTAGE	NODE	VOLTAGE	NODE	VOLTAGE	NODE	VOLTAGE
(1)	0.0000	(2)	0.0000	(3)	−1.4280	(4)	−11.5240
(5)	−.7016	(6)	0.0000	(7)	−15.0000		

VOLTAGE SOURCE CURRENTS
NAME   CURRENT

VIN   0.000E+00

VCC   −6.364E−04

TOTAL POWER DISSIPATION   9.55E−03   WATTS

      Once the dc bias point is determined, PSpice generates a small-signal model of the BJT. This model is similar to that in Fig. 8-12. PSpice replaces the transistor with this circuit model. It should be noted that this model is valid only at the operating point. The details of the operating point and model values follow.

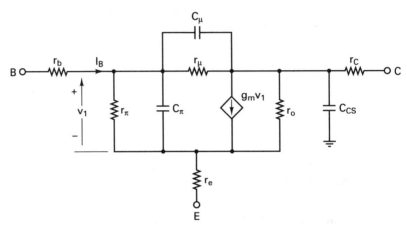

Figure 8-12   Small-signal equivalent circuit of bipolar transistors.

**** OPERATING POINT INFORMATION		TEMPERATURE = 27.000 DEG C

**** BIPOLAR JUNCTION TRANSISTORS

NAME	Q1
MODEL	QM
IB	−3.16E−06
IC	−3.48E−04
VBE	−7.26E−01
VBC	1.01E+01
VCE	−1.08E+01
BETADC	1.10E+02
GM	1.34E−02
RPI	8.19E+03
RX	5.00E+00
RO	3.17E+05
CBE	3.39E−12
CBC	2.11E−13
CBX	0.00E+00
CJS	1.00E−12
BETAAC	1.10E+02
FT	5.94E+08

Prior to the transient analysis, PSpice determines the small-signal parameters of the nonlinear devices and the potentials of the various nodes. The method for the calculation of the transient analysis bias point differs from that of the dc analysis bias point because, in transient analysis, all the nodes have to be assigned initial values, and the nonlinear sources may have transient values at the beginning of transient analysis. The capacitors, which may have initial values, therefore remain as parts of the circuit. The equivalent circuit for determining the transient analysis bias point for the circuit in Fig. 8-10 is shown in Fig. 8-13. Since the capacitors in Fig. 8-10 do not have any initial values, the bias points for dc and transient analysis are the same. There, the small-signal parameters are also the same. The details of the transient analysis bias point and the small-signal parameters are given next to compare with those of dc analysis.

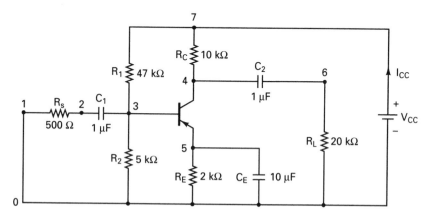

Figure 8-13 Equivalent circuit for the transient analysis bias point.

****	INITIAL TRANSIENT SOLUTION					TEMPERATURE =	27.000 DEG C	
NODE	VOLTAGE	NODE	VOLTAGE	NODE	VOLTAGE	NODE	VOLTAGE
(1)	0.0000	(2)	0.0000	(3)	-1.4280	(4)	-11.5240
(5)	-.7016	(6)	0.0000	(7)	-15.0000		

VOLTAGE SOURCE CURRENTS
NAME   CURRENT

VIN   0.000E+00
VCC   -6.364E-04
TOTAL POWER DISSIPATION   9.55E-03   WATTS

**** OPERATING POINT INFORMATION   TEMPERATURE =   27.000 DEG C
**** BIPOLAR JUNCTION TRANSISTORS

NAME   Q1
MODEL   QM
IB   -3.16E-06
IC   -3.48E-04

VBE	−7.26E−01
VBC	1.01E+01
VCE	−1.08E+01
BETADC	1.10E+02
GM	1.34E−02
RPI	8.19E+03
RX	5.00E+00
RO	3.17E+05
CBE	3.39E−12
CBC	2.11E−13
CBX	0.00E+00
CJS	1.00E−12
BETAAC	1.10E+02
FT	5.94E+08

The frequency and transient responses are shown in Figs. 8-14 and 8-15, respectively.

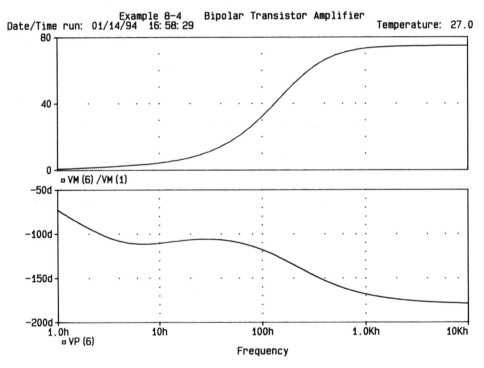

Figure 8-14 Frequency response for Example 8-4.

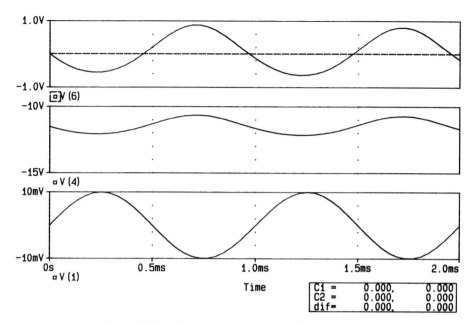

Figure 8-15   Transient response for Example 8-4.

Example 8-5

If the transistor in Fig. 8-10 is replaced by the equivalent circuit of Fig. 8-16, repeat Example 8-4. There is no need to print the details of the operating point.

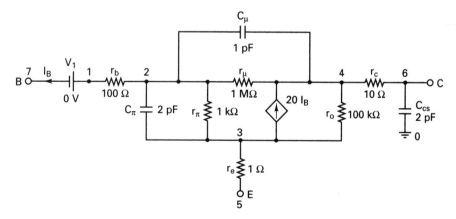

Figure 8-16   Subcircuit for PNP bipolar transistor.

Solution   The listing of the circuit file follows.

Example 8-5   Bipolar transistor amplifier

▲ * Input voltage is 10 mV peak for ac analysis and for transient response:

```
*  It is 10 mV peak at 1 kHz with zero-offset value.
VIN  1  0  AC  1  SIN(0 0.01 1KHZ)
VCC  0  7  DC  15V
▲▲ RS  1  2  500
   R1  7  3  47K
   R2  3  0  2K
   RC  7  4  10K
   RE  5  0  2K
   RL  6  0  20K
   C1  2  3  1UF
   C2  4  6  1UF
   CE  5  0  10UF
   *  Calling subcircuit for transistor model TRANS
   XQ1  4  3  5  TRANS
   *  Subcircuit definition for TRANS
   .SUBCKT  TRANS  6  7  5
   RB  1  2  100
   RE  3  5  1
   RC  4  6  10
   RPI 2  3  1K
   CPI 2  3  2PF
   RU  2  4  1MEG
   CU  2  4  1PF
   RO  4  3  100K
   CCS 6  0  2PF
   *  A dummy voltage source of 0 V through which the controlling current flows
   VI  1  7  DC 0V
   * The collector current is controlled by the current through source VI.
   F1  3  4  VI  20
   *  End of subcircuit definition
   .ENDS  TRANS
▲▲▲ .OPTIONS  NOPAGE  NOECHO
   *    Transient analysis for 0 to 2 ms with 50-μs increment
   .TRAN  50US  2MS
   *  Ac analysis from 1 Hz to 10 KHz with a decade increment and
   *  10 points per decade
   .AC  DEC  10  1HZ  10KHZ
   *  Plot the results of transient analysis for voltages at nodes 4, 6, and 1.
   .PLOT  TRANS  V(4) V(6) V(1)
   *  Plot the results of ac analysis for the magnitude and phase angle
   *  of voltage at node 6.
   .PLOT  AC  VM(6)  VP(6)
   .PROBE
.END
```

The frequency and transient responses are shown in Figs. 8-17 and 8-18, respectively. The .PLOT statements generate graphical plots in the output file. If the .PROBE command is included, there is no need for the .PLOT command.

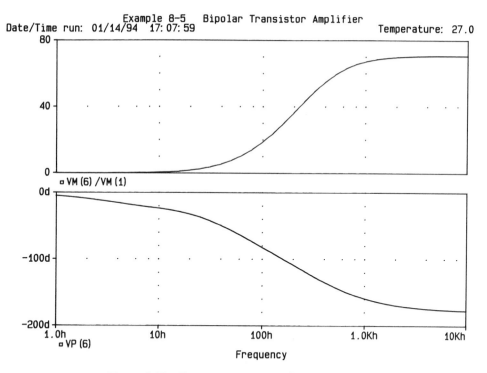

Figure 8-17 Frequency response for Example 8-5.

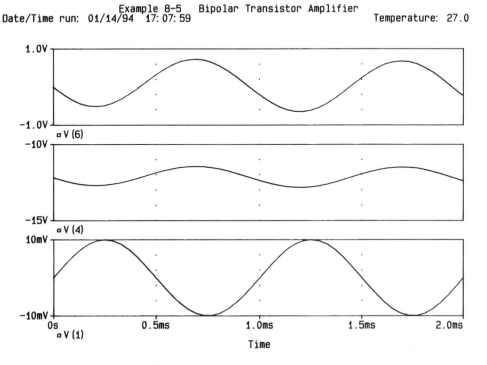

Figure 8-18 Transient response for Example 8-5.

Example 8-6

A two-stage bipolar transistor amplifier is shown in Fig. 8-19. The output is taken from node 9. Plot (a) the magnitude and phase angle of the voltage gain and (b) the magnitude of input impedance for frequencies from 10 Hz to 10 MHZ with a decade increment and 10 points per decade. The peak input voltage is 1 mV. The model parameters of the BJTs are IS=2E−16, BF=50, BR=1, RB=5, RC=1, RE=0, CJE=0.4PF, VJE=0.8, ME=0.4, CJC=0.5PF, VJC=0.8, CCS=1PF, and VA=100.

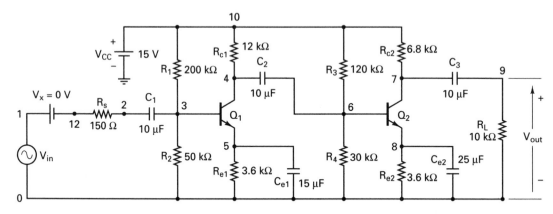

Figure 8-19 Two-stage BJT amplifier.

Solution The listing of the circuit file follows.

Example 8-6   Two-stage BJT amplifier

```
▲ VCC  10   0   DC   15V
  *    Input voltage is 1 mV peak for frequency response.
  VIN   1   0   AC   1MV
  *   A dummy voltage source of 0 V to measure the input current
  VX    1   12  DC   0V
▲▲ RS   12  2    150
   C1   2   3    10UF
   R1   10  3    200K
   R2   3   0    50K
   *  Transistors Q1 and Q2 have model name QM.
   Q1   4   3   5   0   QM
   Q2   7   6   8   0   QM
   RC1  10  4    12K
   RE1  5   0    3.6K
   CE1  5   0    15UF
   C2   4   6    10UF
   R3   10  6    120K
   R4   6   0    30K
   RC2  10  7    6.8K
   RE2  8   0    3.6K
   CE2  8   0    25UF
```

```
C3    7    9    10UF
RL    9    0    10K
*  Model statement for NPN transistors whose model name is QM
.MODEL   QM   NPN  (IS=2E-16 BF=50 BR=1 RB=5 RC=1 RE=0 CJE=0.4PF
+               VJE=0.8 ME=0.4 CJC=0.5PF VJC=0.8 CCS=1PF VA=100)
▲▲▲ *  Ac analysis from 10 Hz to 10 MHz with a decade increment and 10
      *  points per decade
      .AC   DEC   10   10HZ   10MEGHZ
      .PLOT  AC   VM(9)   VP(9)
      .PROBE
.END
```

The results of the frequency response are shown in Fig. 8-20. If the .PROBE command is included, there is no need for the .PLOT command.

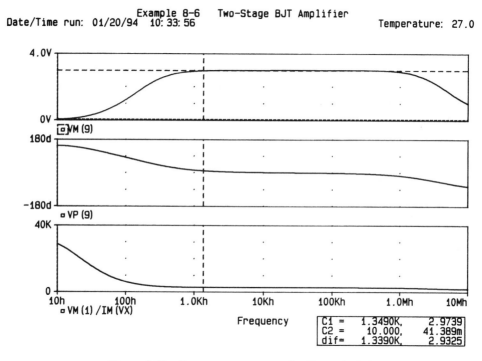

Figure 8-20   Frequency response for Example 8-6.

Example 8-7

A two-stage amplifier with shunt-series feedback is shown in Fig. 8-21. Plot (a) the magnitude and phase angle of voltage gain and (b) the magnitude of the input impedance if the frequency is varied from 100 Hz to 100 MHz in decade steps with 10 points per decade. The peak input voltage is 10 mV. The model parameters of the BJTs are IS=2E-16, BF=50, BR=1, RB=5, RC=1, RE=0, CJE=0.4PF, VJE=0.8, ME=0.4, CJC=0.5PF, VJC=0.8, CCS=1PF, and VA=100.

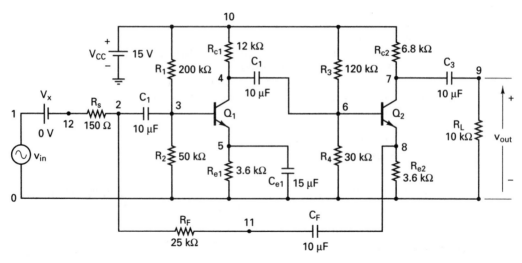

Figure 8-21 Two-stage BJT amplifier with shunt-series feedback.

Solution

Example 8-7   Two-stage BJT amplifier with shunt-series feedback

▲ VCC 10   0 DC 15V
  *   Input voltage of 10 mV peak for frequency response
  VIN 1   0 AC 10MV
  *   A dummy voltage source of 0 V
  VX   1   12 DC 0V
▲▲ RS   12   2   150
  C1   2   3   10UF
  R1   10   3 200K
  R2   3 0   50K
  *   Substrate of BJTs with model QM is connected to node 0.
  Q1   4   3   5   0   QM
  Q2   7 6 8 0 QM
  RC1 10   4   12K
  RE1 5   0   3.6K
  CE1 5   0   15UF
  C2   4   6   10UF
  R3   10   6   120K
  R4   6   0   30K
  RC2 10   7   6.8K
  RE2 8   0   3.6K
  CF   11   8   10UF
  RF   2   11   25K
  C3   7 9   10UF
  RL   9 0   10K
  * Model statement for NPN transistors with model name QM
  .MODEL   QM NPN (IS=2E-16 BF=50 BR=1 RB=5 RC=1 RE=0 CJE=0.4PF
  +   VJE=0.8 ME=0.4 CJC=0.5PF VJC=0.8 CCS=1PF VA=100)
▲▲▲ *   Ac analysis for 10 Hz to 100 MHz with a decade increment and 10
  *   points per decade

```
. AC   DEC   10   10   10MEGHZ
. PLOT   AC   VM (9)   VP (9)
. END
```

The results of the frequency response are shown in Fig. 8-22. If the .PROBE command is included, there is no need for the .PLOT command.

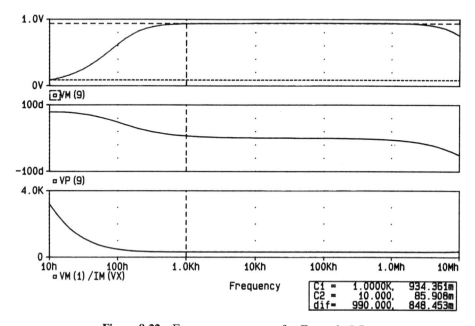

Example 8-7   Two-stage BJT amplifier with shunt-series feedback
Date/Time run: 01/20/94   10: 45: 58   Temperature: 27.0

Figure 8-22   Frequency response for Example 8-7.

Example 8-8

An astable multivibrator is shown in Fig. 8-23. The output is taken from nodes 1 and 2. Plot the transient responses of voltages at nodes 1 and 2 from 0 to 15 μs in steps of 0.1 μs. The initial voltages of nodes 1 and 3 are 0. The CPU time should

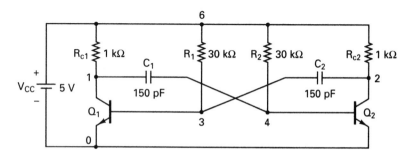

Figure 8-23   Astable multivibrator.

be limited to 1.22E2 s. The model parameters of the BJTs are IS=2E−16, BF=50, BR=1, RB=5, RC=1, RE=0, TF=0.2NS, and TR=5NS.

Solution Due to the regenerative nature of the circuit, the solution may not converge, and the simulation will continue for a very long time. The CPU time is limited so that the circuit does not run for a long time. The run time should be less than the CPU time itself if the circuit converges. The listing of the circuit file follows.

Example 8-8   Astable multivibrator

```
▲ VCC   6   0   DC   5V
▲▲ RC1  6   1   1K
   RC2  6   2   1K
   R1   6   3   30K
   R2   6   4   30K
   C1   1   4   150PF
   C2   2   3   150PF
   *  Q1 and Q2 with model QM and substrate connected to ground by
   +default
   Q1   1   3   0   QM
   Q2   2   4   0   QM
   *  Model statement for NPN transistors
   .MODEL QM NPN (IS=2E−16 BF=50 BR=1 RB=5 RC=1 RE=0 TF=0.2NS TR=5NS)
   *    CPU time is limited.
   .OPTIONS  NOPAGE  NOECHO  CPTIME=1.2E2
   *    Node voltages are set to defined values to break the tie-in
   +condition.
   .NODESET V(1)=0   V(3)=0
▲▲▲ *    Transient analysis from 0 to 10 μs with 0.1-μs increment
     .TRAN/OP   0.1US   10US
     *    Plot the results of transient analysis: voltages at nodes 2 and +4.
     .PLOT   TRAN   V(1)   V(2)
     .OPTIONS   ABSTOL=1.0N   RELTOL=10M   VNTOL=1M   ITL5=40000
     .PROBE
.END
```

The transient responses are shown in Fig. 8-24. If the .PROBE command is included, there is no need for the .PLOT command.

Example 8-9

A TTL inverter circuit is shown in Fig. 8-25(a). The output is taken from node 4. Plot the dc transfer characteristic V(4) versus V_{in} if the input voltage is varied from 0 to 2 V with a step of 0.01 V. If the input is a pulse voltage with a period of 60 μs, as shown in Fig. 8-25(b), plot the transient response of voltage at node 4 from 0 to 80 ns in steps of 1 ns. The model parameters of the BJTs are BF=50, RB=70, RC=40, CCS=2PF, TF=0.1NS, TR=10NS, VJC=0.85, and VAF=50.

Solution The listing of the circuit file follows.

Example 8-9   TTL inverter

```
▲ *    Pulsed input voltage
   VIN  1   0   PULSE (0   5   1NS   1NS   1NS   38NS   60NS)
   VCC  6   0   DC   5V
```

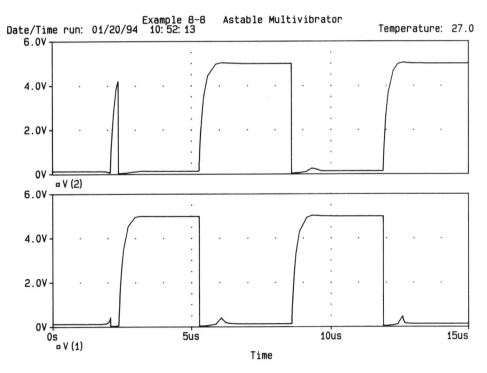

Figure 8-24   Transient responses for Example 8-8.

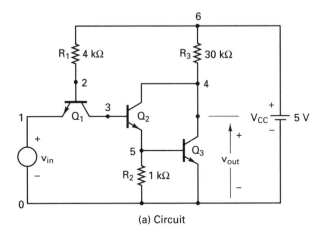

(a) Circuit

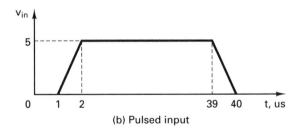

(b) Pulsed input

Figure 8-25   A TTL inverter.

```
▲▲ *   BJTs with model QN and substrate connected to ground by default
    Q1   3   2   1   QN
    Q2   4   3   5   QN
    Q3   4   5   0   QN
    *   Model for NPN BJTs with model QN
    .MODEL QN NPN (BF=50 RB=70 RC=40 CCS=2PF TF=0.1NS TR=10NS VJC=0.85
    +VAF=50)
    R1   6   2   4K
    R2   5   0   1K
    R3   6   4   1K
▲▲▲ *   DC sweep for 0 to 2 with 0.01 V increment
    .DC  VIN  0  2  0.01
    *   Transient analysis for 0 to 80 ns with 1-ns increment
    .TRAN 0.5NS 80NS
    *   Plot the results of dc sweep: voltage at node 4 versus VIN.
    .PLOT  DC  V(4)
    *   Plot the results of transient analysis: voltage at nodes 4 and 1.
    .PLOT  TRAN  V(4)  V(1)
    .PROBE
  .END
```

The results of the dc sweep and transient analyses are shown in Figs. 8-26 and 8-27, respectively. If the .PROBE command is included, there is no need for the .PLOT command.

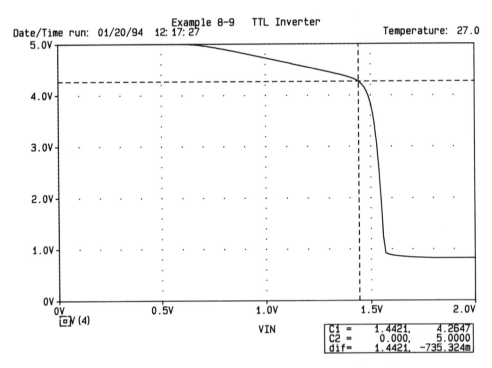

Figure 8-26 Dc transfer characteristic for Example 8-9.

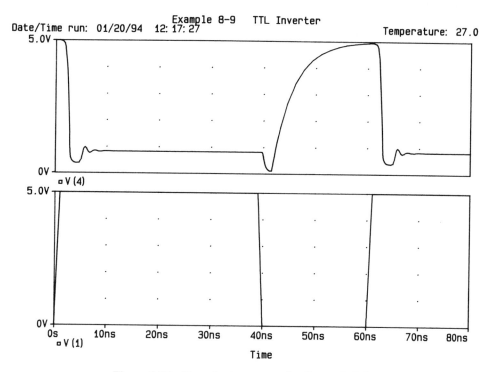

Figure 8-27   Transient response for Example 8-9.

Example 8-10

A TTL inverter circuit is shown in Fig. 8-28(a). Plot the dc transfer characteristic between nodes 1 and 9 for values of V_{in} in the range of 0 to 2 V in steps of 0.01 V. If the input is a pulsed waveform of period 80 μs, as shown in Fig. 8-28(b), plot the transient response from 0 to 80 ns with steps of 1 ns. The model parameters of the BJTs are BF=50, RB=70, RC=40, TF=0.1NS, TR=10NS, VJC=0.85, and VAF=50. The model parameters of the diodes are RS=40, TT=0.1NS.
Solution   The listing of the circuit file follows.

Example 8-10   TTL inverter

```
▲ *  Pulse input voltage
   VIN  1  0  PULSE (0  3.5V  1NS 1NS  1NS  38NS  80NS)
   VCC  13  0  5V
▲▲ RS  1  2  50
   RB1 13  3  4K
   RC2 13  5  1.4K
   RE2  6  0  1K
   RC3 13  7  100
   RB5 13  10  4K
   *  BJTs with model QNP and substrate connected to ground by default
   Q1  4  3  2  QNP
   Q2  5  4  6  QNP
   Q3  7  5  8  QNP
   Q4  9  6  0  QNP
```

Sec. 8-5   Examples of BJT Circuits

217

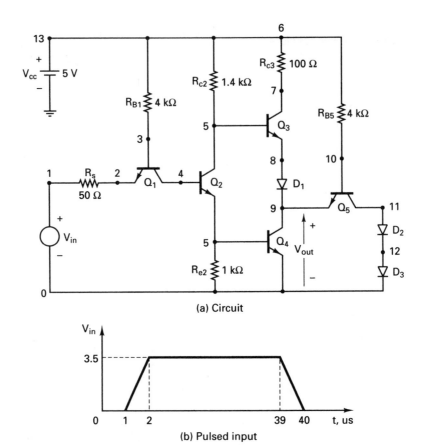

(a) Circuit

(b) Pulsed input

Figure 8-28 TTL inverter.

```
Q5   11 10 9   QNP
*  Diodes with model DIODE
D1   8  9    DIODE
D2   11 12   DIODE
D3   12  0   DIODE
*  Model of NPN transistors with model QNP
.MODEL QNP NPN (BF=50 RB=70 RC=40 TF=0.1NS TR=10NS VJC=0.85 VAF=50)
*  Diodes with model DIODE
.MODEL DIODE D (RS=40 TT=0.1NS)
*  Dc sweep from 0 to 2 V with 0.01 V increment
 .DC  VIN  0  2  0.01
  *  Transient analysis from 0 to 80 ns with 1-ns increment
  .TRAN  1NS  80NS
  *  Plot the results of dc sweep: voltage at node 9 against VIN.
  .PLOT  DC  V(9)
  *  Plot the results of transient analysis: voltage at node 9.
  .PLOT  TRAN  V(9)
  .PROBE
 .END
```

The results of the dc sweep and transient analyses are shown in Figs. 8-29 and 8-30, respectively. If the .PROBE command is included, there is no need for the .PLOT command.

Example 8-11

The circuit diagram of an OR/NOR gate is shown in Fig. 8-31(a). The inputs to nodes 1 and 4 are pulses of period 60 μs, as shown in Fig. 8-31(b). Plot the transient responses of voltages at nodes 12, 13, and 1 from 0 to 100 ns in steps of 1 ns. The model parameters of the BJTs are BF=50, RB=70, RC=40, TF=0.1NS, TR=10NS, VJC=0.85, and VAF=50. The parameters of the diodes are RS=40, and TT=0.1NS.

Solution The listing of the circuit file follows.

Example 8-11   OR/NOR logic gate

```
▲ *  Pulsed input voltages
  VA   1  0  PULSE (0  −5  1NS  1NS  1NS  38NS  60NS)
  VB   4  0  PULSE (0  −5  1NS  1NS  1NS  38NS  60NS)
  VEE  0  14  DC  5.2V
▲ ▲ *    BJTs with model QN and substrate connected to ground by default
    Q1   5  4  3   QN
    Q2   7  8  3   QN
    Q3   2  1  3   QN
    Q4   0  9  8   QN
    Q5   0  2  13  QN
    Q6   0  7  12  QN
```

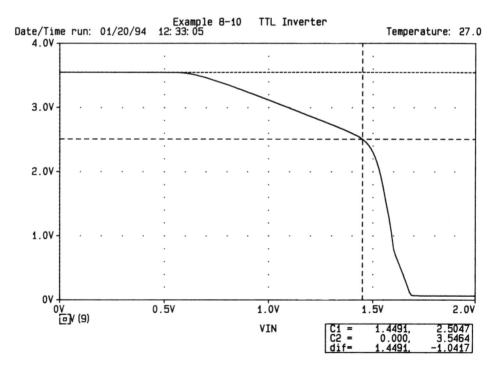

Example 8-10   TTL Inverter
Date/Time run: 01/20/94 12: 33: 05   Temperature: 27.0

	C1 =	1.4491,	2.5047
	C2 =	0.000,	3.5464
	dif=	1.4491,	−1.0417

Figure 8-29 Dc transfer characteristic for Example 8-10.

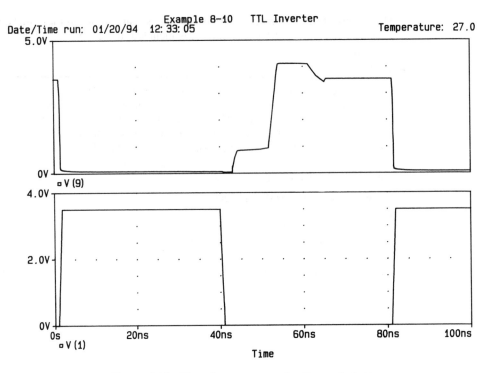

Example 8-10   TTL Inverter

Figure 8-30   Transient response for Example 8-10.

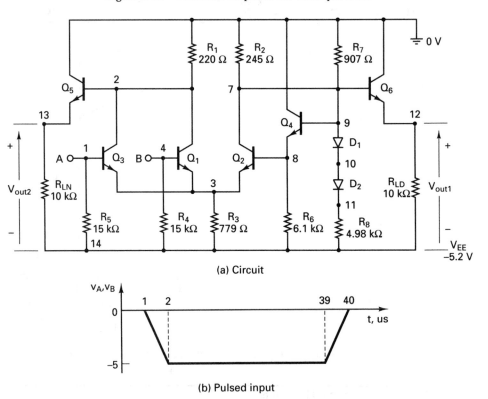

(a) Circuit

(b) Pulsed input

Figure 8-31   OR/NOR logic gate.

```
.MODEL QN NPN (BF=50 RB=70 RC=40 TF=0.1NS TR=10NS VJC=0.85 VAF=50)
* Diodes with model DIODE
D1    9   10  DIODE
D2    10  11  DIODE
.MODEL DIODE  D (RS=40   TT=0.1NS)
R1    0   2   220
R2    0   7   245
R3    3   14  779
R4    4   14  15K
R5    1   14  15K
R6    8   14  6.1K
R7    0   9   907
R8    11  14  4.98K
RLO   12  14  10K
RLN   13  14  10K
```
▲▲▲ * Transient analysis from 0 to 80 ns with 1-ns increment
```
   .TRAN   0.5NS   120NS
    * Plot the results of transient analysis: voltages at nodes 12 and +13.
    .PLOT TRAN V(12) V(13) V(1)
    .PROBE
.END
```

The results of the transient analysis are shown in Fig. 8-32. If the .PROBE command is included, there is no need for the .PLOT command.

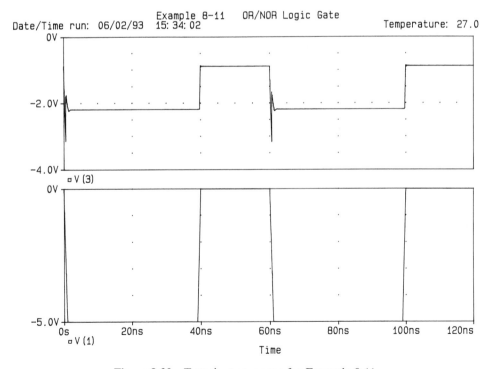

Figure 8-32 Transient response for Example 8-11.

SUMMARY

The statements for BJTS are

```
Q(name) NC  NB  NE  NS  QNAME  [(area) value]
.MODEL  QNAME  NPN  (P1=V1 P2=V2 P3=V3 ....... PN=VN)
.MODEL  QNAME  PNP  (P1=V1 P2=V2 P3=V3 ....... PN=VN)
```

REFERENCES

1. H. K. Gummel and H. C. Poon, "An integral charge control model for bipolar transistors," *Bell System Technical Journal*, Vol. 49, January 1970, pp. 827–852.
2. Ian Getreu, *Modeling the bipolar transistor*—Part # 062-2841-00. Beaverton, Ore.: Tektronix, Inc., 1979.
3. J. J. Ebers and J. J. Moll, "Large-signal behavior of junction transistors," *Proc. IRE*, Vol. 42, December 1954, pp. 1161–1172.
4. L. W. Nagel, *SPICE2—A computer program to simulate semiconductor circuits*, Memorandum no. ERL-M520, May 1975, Electronics Research Laboratory, University of California, Berkeley.
5. A. S. Grove, *Physics and Technology of Semiconductor Devices*. New York: Wiley, 1967.
6. R. B. Schilling, "A bipolar transistor model for device and circuit design," *RCA Review*, Vol. 32, September 1971, pp. 339–371.
7. S. Natarajan, "An effective approach to obtain model parameters for BJTs and FETs from data books," *IEEE Transactions on Education*, Vol. 35, No. 2., 1992, pp. 164–169.
8. M. H. Rashid, *SPICE For Power Electronics and Electric Power*. Englewood Cliffs, N.J.: Prentice Hall, 1993, Chapter 11.

PROBLEMS

8-1. For Example 8-2, calculate the coefficients of a Fourier series for the output voltage.

8-2. For Example 8-6, calculate the equivalent input and output noise.

8-3. For example 8-7, plot the output impedance and the current gain.

8-4. For Fig. 8-28, calculate the input and output noise for frequencies from 1 Hz to 10 kHz.

8-5. For Fig. 8-28, calculate and plot the frequency response of the output voltage from 10 Hz to 10 MHz in decade steps with 10 points per decade. Assume the peak input voltage is 5 V. The model parameters of the BJTs are BF=50, RB=70, RC=40, TF=0.1NS, TR=10NS, VJC=0.85, and VAF=50. The model parameters of the diodes are RS=40, and TT=0.1NS.

8-6. For the circuit in Fig. P8-6, calculate and plot (a) the magnitude and phase angle of voltage gain, (b) the magnitude of input impedance, and (c) the magnitude of output impedance. The frequency is varied from 1 Hz to 10 MHz in decade steps with 10 points per decade. The peak input voltage is 10 mV. The model parameters of the

BJT are IS=2E−16, BF=50, BR=1, RB=5, RC=1, RE=0, CJE=0.4PF, VJE=0.8, ME=0.4, CJC=0.5PF, VJC=0.8, CCS=1PF, and VA=100.

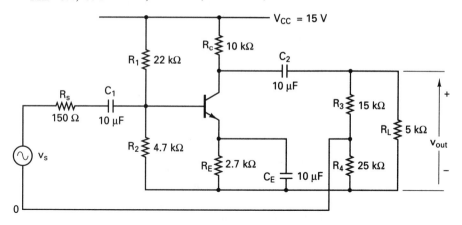

Figure P8-6

8-7. Repeat Problem 8-6 for the circuit in Fig. P8-7.

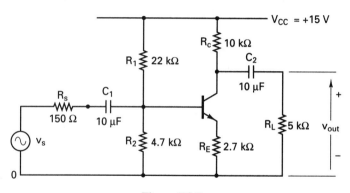

Figure P8-7

8-8. Repeat Problem 8-6 for the circuit in Fig. P8-8.

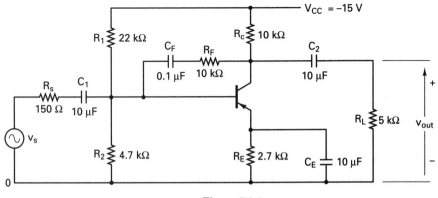

Figure P8-8

8-9. Repeat Problem 8-6 for the circuit in Fig. P8-9.

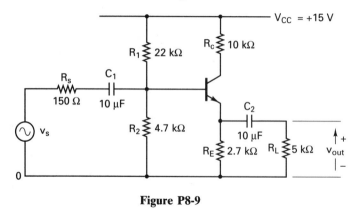

Figure P8-9

8-10. Repeat Problem 8-6 for the circuit in Fig. P8-10. Calculate the input and output noise.

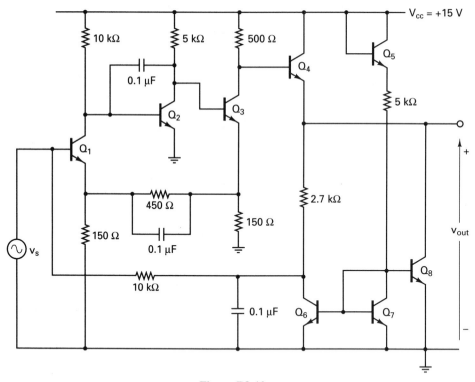

Figure P8-10

8-11. For the circuit in Fig. P8-11, calculate and print the dc transfer function (the voltage gain, the input resistance, and the output resistance) between the output current and the input voltage V_{EE}. The model parameters of the BJTs are BF=100, BR=1, RB=5, RC=1, RE=0, VJE=0.8, and VA=100.

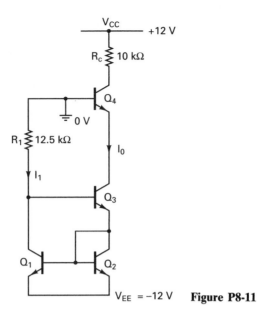

Figure P8-11

8-12. For the circuit in Fig. P8-12, calculate and print the voltage gain, the input resistance, and the output resistance. The input voltage is 5 V dc. The model parameters of the BJTs are BF=100, BR=1, RB=5, RC=1, RE=0, VJE=0.8, and VA=100.

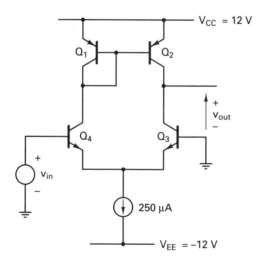

Figure P8-12

8-13. Use PSpice to perform a Monte Carlo analysis for six runs and for the dc sweep of Problem 8-11. The model parameter is $R=1$ for resistors. The circuit and transistor parameters having uniform deviations are

$R_1 = 12.5$ kΩ ± 5%
$R_C = 10$ kΩ ± 15%
$B_F = 100 ± 50$
$V_A = 100 ± 20$

(a) The greatest difference from the nominal run is to be printed.

(b) The maximum value of the output voltage is to be printed.

(c) The minimum value of the output voltage is to be printed.

(d) The first occurrence of the output voltage crossing below 5 V is to be printed.

8-14. Use PSpice to perform the worst-case analysis for Problem 8-13.

8-15. Use PSpice to perform a Monte Carlo analysis for five runs and for the dc sweep of Problem 8-12. The transistor parameters having uniform deviations are

$$B_F = 100 \pm 50$$
$$V_A = 100 \pm 20$$

(a) The greatest difference from the nominal run is to be printed.

(b) The maximum value of the output voltage is to be printed.

(c) The minimum value of the output voltage is to be printed.

8-16. Use PSpice to perform the worst-case analysis for Problem 8-15.

Field-Effect Transistors

9-1 INTRODUCTION

A field-effect transistor (FET) may be specified by a device statement. PSpice generates complex models for FETs. These models are quite complex and incorporate an extensive range of device characteristics (e.g., dc and small-signal behavior, temperature dependency, and noise generation). If such complex models are not necessary, users can ignore many model parameters, and PSpice assigns default values to the parameters. The FETs are of three types:

> Junction field-effect transistors (JFETs)
> Metal-oxide silicon field-effect transistors (MOSFETs)
> Gallium arsenide MESFETs

9-2 JUNCTION FIELD-EFFECT TRANSISTORS

The PSpice JFET model is based on the FET model of Schichman and Hodges [1]. The model of an n-channel JFET is shown in Fig. 9-1. The small-signal model and the static (or dc) model, which are generated by PSpice, are shown in Figs. 9-2 and 9-3, respectively. The model parameters for a JFET device and the default values assigned by PSpice are given in Table 9-1. The model equations of JFETs that are used by PSpice are described in Schichman and Hodges [1], Vladimirescu and Liu [3], and the *PSpice Manual* [7].

The model statement of an n-channel JFET has the general form

```
.MODEL  JNAME  NJF (P1=A1 P2=A2 P3=A3 ... PN=AN)
```

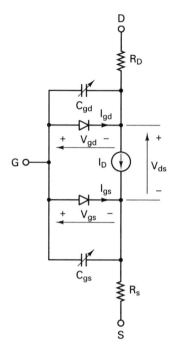

Figure 9-1 PSpice n-channel JFET.

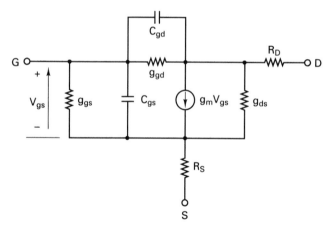

Figure 9-2 Small-signal n-channel JFET model.

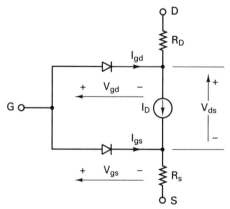

Figure 9-3 Static n-channel JFET model.

TABLE 9-1 MODEL PARAMETERS OF JFETS

Name	Area	Model parameters	Units	Default	Typical
VTO		Threshold voltage	Volts	-2	-2
BETA	*	Transconductance coefficient	Amps/Volts^2	1E−4	1E−3
LAMBDA		Channel-length modulation	Volts^{-1}	0	1E−4
RD	*	Drain ohmic resistance	Ohms	0	100
RS	*	Source ohmic resistance	Ohms	0	100
IS	*	Gate *p-n* saturation current	Amps	1E−14	1E−14
PB		Gate *p-n* potential	Volts	1	0.6
CGD	*	Gate-drain zero-bias *p-n* capacitance	Farads	0	5PF
CGS	*	Gate-source zero-bias *p-n* capacitance	Farads	0	1PF
FC		Forward-bias depletion capacitance coefficient		0.5	
VTOTC		VTO temperature coefficient	Volts/°C	0	
BETATCE		BETA exponential temperature coefficient	percent/°C	0	
KF		Flicker noise coefficient		0	
AF		Flicker noise exponent		1	

and for a *p*-channel JFET, the statement has the form

```
.MODEL   JNAME   PJF  (P1=A1 P2=A2 P3=A3 ... PN=AN)
```

where JNAME is the model name; it can begin with any character and its word size is normally limited to eight characters. NJF and PJF are the type symbols of *n*-channel and *p*-channel JFETs, respectively. P1, P2, . . . and A1, A2, . . . are the parameters and their values, respectively.

As with diodes and BJTs, an *area factor* is used to determine the number of equivalent parallel JFETs. The model parameters that are affected by the area factor are marked by an asterisk (*) in Table 9-1. The [(area) value] scales BETA, RD, RS, CGD, CGS, and IS; it defaults to 1.

RD and RS represent the contact and bulk resistances per unit area of the drain and source, respectively. The JFET is modeled as an intrinsic device. The area value, which is the relative device area, is specified in the .MODEL statement and changes the actual resistance values. The default value of the area is 1.

The dc characteristics that are represented by the nonlinear current source I_D are defined (1) by parameters VTO and BETA, which determine the variation of the drain current with the gate voltage, (2) by LAMBDA, which determines the output conductance, and (3) by IS, which determines the reserve saturation current of the two gate junctions. VTO is negative for depletion-mode JFETs, both for *n*-channel and *p*-channel types, and it is positive for enhancement-mode JFETs. VTO does not identify whether the JFET is *n*-channel or *p*-channel.

The symbol for a JFET is *J*. The name of a JFET must start with *J* and it takes the general form of

```
J⟨name⟩ ND  NG  JNAME  [⟨area⟩ value]
```

where ND, NG and NS are the drain, gate, and source nodes, respectively.

Some JFET Statements

```
JIM  5   6  8   JNAME
.MODEL JNAME  NJF
J15  3   9  12   SWITCH  1.5
.MODEL SWITCH NJF  (IS=100E-14 RD=10 RS=10 BETA=1E-3 VTO=-5)
JQ   1   5   9   JMOD
.MODEL JMOD PJF  (IS=100E-14 RD=10 RS=10 BETA=1E-3 CGD=5PF CGS=1PF
+VTO=5)
```

9-3 JFET PARAMETERS

The library file EVAL.LIB of the student version of PSpice supports models for n-channel JFETs: J2N3819 and J2N4393. As an example, we shall generate approximate values of some parameters [8, 9] from the data sheet of the n-channel JFET of type 2N5459, shown in Fig. 9-4.

I_{DSS} = 4 to 16 mA at V_{GS} = 0 V, and V_{DS} = 15 V. Taking the geometric mean,

$$I_{DSS} = \sqrt{(4 \times 16)} = 8 \text{ mA}$$

The threshold voltage, $V_{Th} = V_{GS(\text{off})} = -2$ to -8 V. Taking the geometric mean,

$$V_{Th} = -\sqrt{(2 \times 8)} = -4 \text{ V}$$

That is, VTO = -4 V (for depletion-mode).

The transconductance coefficient is given by

$$\text{BETA} = \frac{I_{DSS}}{V_{Th}^2} \tag{9-1}$$

$$= \frac{8 \text{ mA}}{(-4)^2} = 0.5 \text{ mA/V}^2$$

The gate reverse current $I_{GSS} = -\text{IS} = -1$ nA at $V_{GS} = -15$ V, and $V_{DS} = 0$.

The common-source reverse transfer capacitance, $C_{rss} = C_{gd} = 1.5$ to 3 pF at V_{DS} = 15 V, and V_{GS} = 0 V. Taking the geometric mean,

$$C_{rss} = C_{gd} = \sqrt{(1.5 \times 3)} \text{ pF} = 2.12 \text{ pF}$$

At $V_{DG} = V_{DS} - V_{GS} = 15 - 0 = 15$ V, C_{gdo} can be found from

$$C_{gd} = \frac{C_{gdo}}{(1 + V_{DG}/V_{\text{off}})^{1/3}} \tag{9-2}$$

where $V_{\text{off}} = 0.75$ V. Equation (9-2) gives, $C_{gdo} = 5.85$ pF.

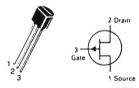

2N5460
thru
2N5465

CASE 29-04, STYLE 7
TO-92 (TO-226AA)

2 Drain

3 Gate

1 Source

**JFET
AMPLIFIER**

P-CHANNEL — DEPLETION

MAXIMUM RATINGS

Rating	Symbol	2N5460 2N5461 2N5462	2N5463 2N5464 2N5465	Unit
Drain-Gate Voltage	V_{DG}	40	60	Vdc
Reverse Gate-Source Voltage	V_{GSR}	40	60	Vdc
Forward Gate Current	$I_{G(f)}$	10		mAdc
Total Device Dissipation @ T_A = 25°C Derate above 25°C	P_D	310 2.82		mW mW/°C
Junction Temperature Range	T_J	− 65 to + 135		°C
Storage Channel Temperature Range	T_{stg}	− 65 to + 150		°C

ELECTRICAL CHARACTERISTICS (T_A = 25°C unless otherwise noted.)

Characteristic		Symbol	Min	Typ	Max	Unit
OFF CHARACTERISTICS						
Gate-Source Breakdown Voltage (I_G = 10 μAdc, V_{DS} = 0)	2N5460, 2N5461, 2N5462 2N5463, 2N5464, 2N5465	$V_{(BR)GSS}$	40 60	— —	— —	Vdc
Gate Reverse Current (V_{GS} = 20 Vdc, V_{DS} = 0) (V_{GS} = 30 Vdc, V_{DS} = 0) (V_{GS} = 20 Vdc, V_{DS} = 0, T_A = 100°C) (V_{GS} = 30 Vdc, V_{DS} = 0, T_A = 100°C)	2N5460, 2N5461, 2N5462 2N5463, 2N5464, 2N5465 2N5460, 2N5461, 2N5462 2N5463, 2N5464, 2N5465	I_{GSS}	— — — —	— — — —	5.0 5.0 1.0 1.0	nAdc μAdc
Gate Source Cutoff Voltage (V_{DS} = 15 Vdc, I_D = 1.0 μAdc)	2N5460, 2N5463 2N5461, 2N5464 2N5462, 2N5465	$V_{GS(off)}$	0.75 1.0 1.8	— — —	6.0 7.5 9.0	Vdc
Gate Source Voltage (V_{DS} = 15 Vdc, I_D = 0.1 mAdc) (V_{DS} = 15 Vdc, I_D = 0.2 mAdc) (V_{DS} = 15 Vdc, I_D = 0.4 mAdc)	2N5460, 2N5463 2N5461, 2N5464 2N5462, 2N5465	V_{GS}	0.5 0.8 1.5	— — —	4.0 4.5 6.0	Vdc
ON CHARACTERISTICS						
Zero-Gate-Voltage Drain Current (V_{DS} = 15 Vdc, V_{GS} = 0, f = 1.0 kHz)	2N5460, 2N5463 2N5461, 2N5464 2N5462, 2N5465	I_{DSS}	1.0 2.0 4.0	— — —	5.0 9.0 16	mAdc
SMALL-SIGNAL CHARACTERISTICS						
Forward Transfer Admittance (V_{DS} = 15 Vdc, V_{GS} = 0, f = 1.0 kHz)	2N5460, 2N5463 2N5461, 2N5464 2N5462, 2N5465	$|y_{fs}|$	1000 1500 2000	— — —	4000 5000 6000	μmhos
Output Admittance (V_{DS} = 15 Vdc, V_{GS} = 0, f = 1.0 kHz)		$|y_{os}|$	—	—	75	μmhos
Input Capacitance (V_{DS} = 15 Vdc, V_{GS} = 0, f = 1.0 MHz)		C_{iss}	—	5.0	7.0	pF
Reverse Transfer Capacitance (V_{DS} = 15 Vdc, V_{GS} = 0, f = 1.0 MHz)		C_{rss}	—	1.0	2.0	pF
FUNCTIONAL CHARACTERISTICS						
Noise Figure (V_{DS} = 15 Vdc, V_{GS} = 0, R_G = 1.0 Megohm, f = 100 Hz, BW = 1.0 Hz)		NF	—	1.0	2.5	dB
Equivalent Short-Circuit Input Noise Voltage (V_{DS} = 15 Vdc, V_{GS} = 0, f = 100 Hz, BW = 1.0 Hz)		e_n	—	60	115	nV/$\sqrt{Hz}$

Figure 9-4 Data sheet for the *n*-channel JFET of type 2N5459 (Courtesy of Motorola, Inc.).

The common-source input capacitance, C_{iss} = 4.5 to 7 pF at V_{DS} = 15 V, and V_{GS} = 0 V. Taking the geometric mean,

$$C_{iss} = \sqrt{(4.5 \times 7)} \text{ pF} = 5.61 \text{ pF}$$

Since C_{iss} is measured at V_{GS} = 0 V, $C_{gs} = C_{gso}$. That is,

$$C_{iss} = C_{gso} + C_{gd}$$

which gives $C_{gso} = C_{iss} - C_{gd}$ = 5.61 − 2.12 = 3.49 pF.

The output admittance $|Y_{os}|$ = 10 to 50 μmhos at V_{DS} = 15 V, and V_{GS} = 0. Taking the geometric mean,

$$|Y_{os}| = \sqrt{(10 \times 50)} \text{ } \mu\text{mhos} = 22.36 \text{ } \mu\text{mhos}$$

Since $|Y_{os}|$ is given at V_{GS} = 0, the channel-modulation length λ (LAMBDA) can be found approximately from

$$\text{LAMBDA} = \frac{|Y_{os}|}{I_{DSS}} \approx \frac{22.36 \text{ } \mu\text{mhos}}{8 \text{ mA}} = 2.395\text{E}-3$$

The PSpice model statement for the JFET of type J2N5459 is

```
.MODEL J2N5459 NJF (IS=1N   VTO=-4   BETA=0.5M   CGDO=5.85PF
+                   CGSO=3.49PF   LAMBDA=2.395E-3)
```

9-4 EXAMPLES OF JFET AMPLIFIERS

The approximate values of JFET model parameters can be determined from the data sheet. If a model parameter is not specified, PSpice assumes its default value, as indicated in Table 9-1. The following examples illustrate the PSpice simulation of JFET circuits.

Example 9-1

For the *n*-channel JFET in Fig. 9-5, plot the output characteristics if V_{DD} is varied from 0 to 12 V in steps of 0.2 V and V_{GS} is varied from 0 to −4 V in steps of 1 V. The model parameters are IS=100E−14, RD=10, RS=10, BETA=1E−3, and VTO=−5.

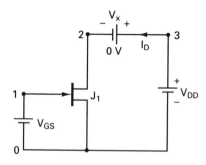

Figure 9-5 A circuit with an *n*-channel JFET.

Solution The listing of the circuit file follows.

Example 9-1 Output characteristics of an *n*-channel JFET

```
▲ *    Gate to source voltage of 0 V
  VGS  1  0  DC  0V
  *    A dummy voltage source of 0 V
  VX   3  2  DC  0V
  *    Dc supply voltage of 12 V
  VDD  3  0  DC  12V
▲▲ *   J1 with model JMOD
  J1   2  1  0  JMOD
  .MODEL  JMOD  NJF  (IS=100E-14 RD=10 RS=10 BETA=1E-3 VTO=-5)
▲▲▲ *  VDD is swept from 0 to 12 V and VGS from 0 to -4 V.
  .DC  VDD  0  12  0.2  VGS  0  -4  1
  .PLOT  DC  I(VX)
  .PROBE
.END
```

The output characteristics, which are plots of I_D versus V_{DD}, are shown in Fig. 9-6.

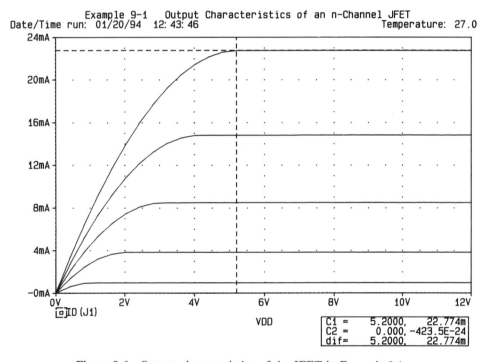

Figure 9-6 Output characteristics of the JFET in Example 9-1.

Example 9-2

For the JFET in Example 9-1, plot the input characteristic if V_{GS} is varied from 0 to −5 V in steps of 0.1 V, and $V_{DD} = 10$ V.

Solution The listing of the circuit file follows.

Example 9-2   Input characteristics of an *n*-channel JFET

```
▲ VGS  1  0  DC  0V
  VX   3  2  DC  0V
  *   Dc supply voltage of 10 V
  VDD  3  0  DC  10V
▲▲ *  J1 with model JMOD
  J1   2  1  0  JMOD
  .MODEL  JMOD  NJF  (IS=100E-14 RD=10 RS=10 BETA=1E-3 VTO=-5)
▲▲▲ *   VGS is swept from 0 to -5 V.
  .DC  VGS  0  -5V  0.1V
  .PLOT  DC  I(VX)
  .PROBE
.END
```

The input characteristic, which is a plot of I_D versus V_{GS}, is shown in Fig. 9-7.

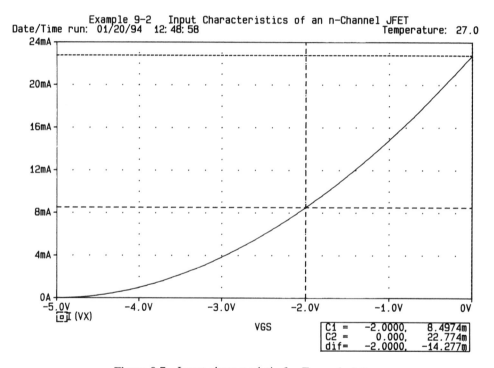

Figure 9-7   Input characteristic for Example 9-2.

Example 9-3

A JFET transistor amplifier circuit is shown in Fig. 9-8. The output is taken from node 6. If the input voltage is $v_{in} = 0.5 \sin(2000\pi t)$, use ac analysis to calculate and print the magnitudes and phase angles of the output voltage, the input current, and the load current. Plot the transient responses of the voltages at nodes 1, 4, and 6 from 0 to 1 ms in steps of 10 μs. The model parameters of the JFET are IS=100E−

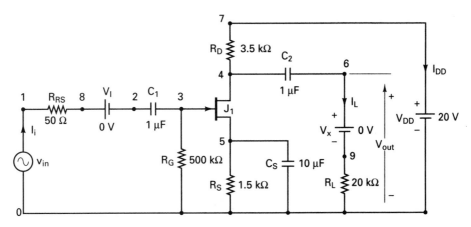

Figure 9-8 An *n*-channel JFET amplifier circuit.

14, RD=10, RS=10, BETA=1E−3, CGD=5PF, CGS=1PF, and VTO=−5. The
details of the dc analysis and transient analysis operating points should be printed.
Solution The listing of the circuit file follows.

Example 9-3   An *n*-channel JFET amplifier

```
▲ .OPTIONS  NOPAGE  NOECHO
  *  Input voltage has 0.5 V peak at 1 kHz with zero offset value for
  *  transient response and 0.5 V peak for frequency response.
  VIN  1  0  AC  0.5V  SIN (0  0.5V  1KHZ)
  VDD  7  0  DC  20V
  *  Dummy voltage source of 0 V
  VI  8  2  DC  0V
  VX  6  9  DC  0V
▲▲ RRS  1  8  50
   RG  3  0  0.5MEG
   RD  7  4  3.5K
   RS  5  0  1.5K
   RL  9  0  20K
   C1  2  3  1UF
   C2  4  6  1UF
   CS  5  0  10UF
   *  n-channel JFET with model JMOD
   J1  4  3  5  JMOD
   .MODEL JMOD NJF (IS=100E−14 RD=10 RS=10 BETA=1E−3 CGD=5PF CGS=1PF VTO=−5)
▲▲▲ *  Ac analysis at 1 kHz with a linear increment and only 1 point
    .AC  LIN  1  1KHZ   1KHZ
    *  Transient analysis with details of transient analysis operating point
    .TRAN/OP  10US  1MS
    *   Print the details of the ac analysis operating point.
    .OP
    * Print the results of the ac analysis for the magnitudes of voltages at
    * node 6 and 1 and for the magnitude of current through resistance RRS
    * and the current through VX.
```

```
          .PRINT AC   VM(6)  VP(6)  IM(RRS)  IP(RRS)
          .PRINT AC   IM(VI)  IP(VI)  IM(VX)  IP(VX)
          *  Plot transient response.
          .PLOT TRAN V(6)   V(1)
          .PROBE
    .END
```

The equivalent circuit for determining the dc bias point is shown in Fig. 9-9. The details of the dc bias are given next.

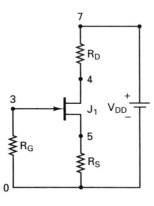

Figure 9-9 Equivalent circuit for dc bias calculation.

****	**SMALL-SIGNAL BIAS SOLUTION**					**TEMPERATURE =**	**27.000 DEG C**	
NODE	**VOLTAGE**	**NODE**	**VOLTAGE**	**NODE**	**VOLTAGE**	**NODE**	**VOLTAGE**	
(1)	0.0000	(2)	0.0000	(3)	8.694E-06	(4)	11.9300	
(5)	3.4585	(6)	0.0000	(7)	20.0000	(8)	0.0000	
(9)	0.0000							

VOLTAGE SOURCE CURRENTS

NAME	CURRENT
VIN	0.000E+00
VDD	-2.306E-03
VI	0.000E+00
VX	0.000E+00

TOTAL POWER DISSIPATION 4.61E-02 WATTS

Once the dc bias points are determined, the small-signal parameters of the JFET in Fig. 9-8 are calculated. The details of the operating points are given next.

****	**OPERATING POINT INFORMATION**	**TEMPERATURE =**	**27.000 DEG C**
**** JFETS			

NAME	J1
MODEL	JMOD
ID	2.31E-03
VGS	-3.46E+00
VDS	8.47E+00
GM	3.04E-03
GDS	0.00E+00
CGS	4.72E-13
CGD	1.39E-12

The outputs at a frequency of 1 kHz are as follows.

FREQ	VM(6)	VP(6)	IM(RRS)	IP(RRS)
1.000E+03	4.382E+00	−1.769E+02	9.990E−07	2.555E+00
FREQ	IM(VI)	IP(VI)	IM(VX)	IP(VX)
1.000E+03	9.990E−07	2.555E+00	2.191E−04	−1.769E+02

The equivalent circuit for determining the transient analysis bias point is shown in Fig. 9-10. The transient analysis bias point and the operating point are the same as those of the dc analysis because the capacitors do not have any initial voltages. The transient responses are shown in Fig. 9-11.

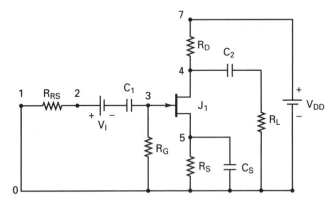

Figure 9-10 Equivalent circuit for transient analysis bias calculation.

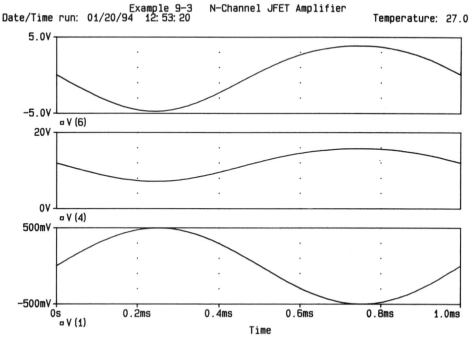

Figure 9-11 Transient responses for Example 9-3.

Example 9-4

If the JFET in Fig. 9-8 is replaced by the subcircuit model of Figure 9-12, plot the frequency response of the output voltage. The frequency is varied from 10 Hz to 100 MHz with a decade increment and 10 points per decade.

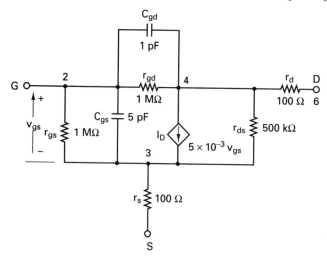

Figure 9-12 Subcircuit model for JFET of Example 9-4.

Solution The listing of the circuit file follows.

Example 9-4 An *n*-channel JFET amplifier

```
▲ .OPTIONS  NOPAGE  NOECHO
  *  Input voltage has 0.5 V peak for frequency response.
  VIN  1  0  AC  0.5V
  VDD  7  0  DC  20V
  *  Dummy voltage source of 0V
  VI  8  2  DC  0V
  VX  6  9  DC  0V
▲▲ RRS  1  8   50
   RG  3  0   0.5MEG
   RD  7  4   3.5K
   RS  5  0   1.5K
   RL  9  0   20K
   C1  2  3   1UF
   C2  4  6   1UF
   CS  5  0   10UF
  *  Calling subcircuit for TRANS
  XQ1 4  3  5   TRANS
  *  Subcircuit definition for TRANS
  .SUBCKT  TRANS  6  2  5
   RD  4  6  100
   RS  3  5  100
   RGS 2  3  1MEG
   CGS 2  3  5PF
   RGD 2  4  1MEG
   CGS 2  4  1PF
```

```
          RDS  4  3  500K
          *   Voltage-controlled current source with a gain of 5E-3
          G1  4  3  2  3  5E-3
          .ENDS  TRANS
▲▲▲ *  Ac analysis for 100 Hz to 100 MHz with a decade increment and
          *  10 points per decade
          .AC  DEC  10  10HZ  100MEGHZ
          *  Plot the results of the ac analysis for the magnitudes and phases of
          *  output voltage and the magnitudes of input and load currents.
          .PLOT  AC  VM(6)  VP(6)
          .PLOT  AC  IM(VI)  IM(VX)
          .PROBE
   .END
```

The frequency response for Example 9-4 is shown in Fig. 9-13. The .PLOT statement generates graphical plots in the output file. If the .PROBE command is included, there is no need for the .PLOT command.

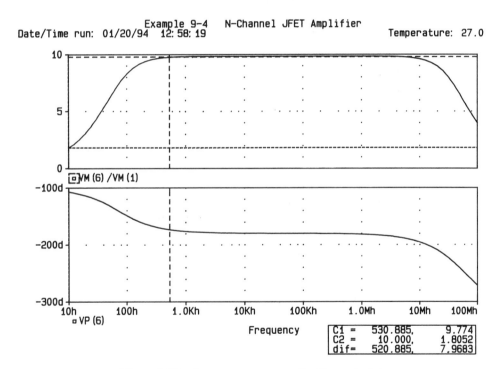

Figure 9-13 Frequency response for Example 9-4.

Example 9-5

A *p*-channel JFET bootstrapped amplifier is shown in Fig. 9-14. The output is taken from node 5. Calculate and print the voltage gain, the input resistance, and the output resistance. The model parameters of the JFET are IS$=100E-14$, RD$=10$, RS$=10$, BETA$=1E-3$, and VTO$=5$.

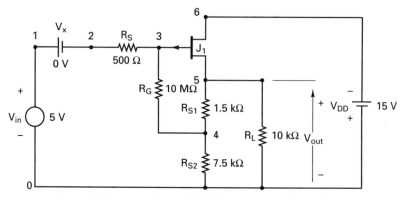

Figure 9-14 A *p*-channel JFET bootstrapped amplifier.

Solution The listing of the circuit follows.

Example 9-5   Bootstrapped JFET amplifier

```
▲ VDD   0    6    15V
   *    Input voltage of 5 V dc
   VIN   1    0    DC   5V
   *    A dummy voltage source of 0 V
   VX    1    2    DC   0V
▲▲ RS    2    3    500
   RG    3    4    10MEG
   RS1   5    4    1.5K
   RS2   4    0    7.5K
   RL    5    0    10K
   *    p-channel JFET of model JMOD
   JX    6    3    5    JMOD
   *    Model statement for p-channel JFET
   .MODEL JMOD PJF (IS=100E-14 RD=10 RS=10 BETA=1E-3 VTO=5)
▲▲▲ *    Transfer function analysis between the output and input voltages
   .TF   V(5)   VIN
.END
```

The results of the transfer function analysis are

```
****        SMALL-SIGNAL CHARACTERISTICS
      V(5)/VIN = -9.257E-01
      INPUT RESISTANCE AT VIN =  4.375E+07
      OUTPUT RESISTANCE AT V(5) =  3.521E+02
         JOB CONCLUDED
         TOTAL JOB TIME          9.50
```

9-5 METAL OXIDE SILICON FIELD-EFFECT TRANSISTORS

The PSpice model of an *n*-channel MOSFET [1–3] is shown in Fig. 9-15. The small-signal model and the static (or dc) model generated by PSpice are shown in

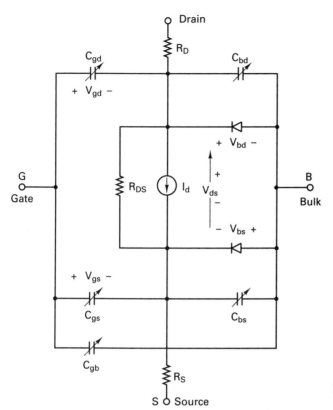

Figure 9-15 PSpice *n*-channel MOSFET model.

Figs. 9-16 and 9-17, respectively. The model parameters for a MOSFET device and the default values assigned by PSpice are given in Table 9-2. The model equations of MOSFETs that are used by PSpice are described in Schichman and Hodges [1], Vladimirescu and Liu [3], and the *PSpice Manual* [7].

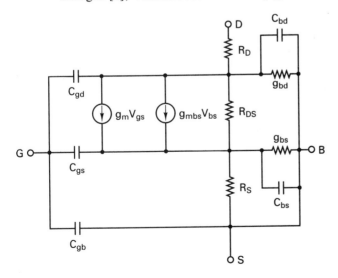

Figure 9-16 Small-signal *n*-channel MOSFET model.

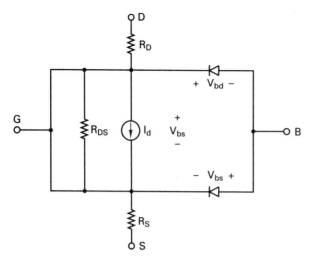

Figure 9-17 Static *n*-channel MOSFET model.

TABLE 9-2 MODEL PARAMETERS OF MOSFETS

Name	Model parameters	Units	Default	Typical
LEVEL	Model type (1, 2, or 3)		1	
L	Channel length	meters	DEFL	
W	Channel width	meters	DEFW	
LD	Lateral diffusion length	meters	0	
WD	Lateral diffusion width	meters	0	
VTO	Zero-bias threshold voltage	Volts	0	0.1
KP	Transconductance	Amps/Volts^2	2E−5	2.5E−5
GAMMA	Bulk threshold parameter	Volts$^{1/2}$	0	0.35
PHI	Surface potential	Volts	0.6	0.65
LAMBDA	Channel-length modulation (LEVEL = 1 or 2)	Volts^{-1}	0	0.02
RD	Drain ohmic resistance	Ohms	0	10
RS	Source ohmic resistance	Ohms	0	10
RG	Gate ohmic resistance	Ohms	0	1
RB	Bulk ohmic resistance	Ohms	0	1
RDS	Drain-source shunt resistance	Ohms	∞	
RSH	Drain-source diffusion sheet resistance	Ohms/square	0	20
IS	Bulk *p-n* saturation current	Amps	1E−14	1E−15
JS	Bulk *p-n* saturation current/area	Amps/meters^2	0	1E−8
PB	Bulk *p-n* potential	Volts	0.8	0.75
CBD	Bulk-drain zero-bias *p-n* capacitance	Farads	0	5PF
CBS	Bulk-source zero-bias *p-n* capacitance	Farads	0	2PF
CJ	Bulk *p-n* zero-bias bottom capacitance/length	Farads/meters^2	0	
CJSW	Bulk *p-n* zero-bias perimeter capacitance/length	Farads/meters	0	
MJ	Bulk *p-n* bottom grading coefficient		0.5	

(continued)

TABLE 9-2 *Continued*

Name	Model parameters	Units	Default	Typical
MJSW	Bulk *p-n* sidewall grading coefficient		0.33	
FC	Bulk *p-n* forward-bias capacitance coefficient		0.5	
CGSO	Gate-source overlap capacitance/channel width	Farads/meters	0	
CGDO	Gate-drain overlap capacitance/channel width	Farads/meters	0	
CGBO	Gate-bulk overlap capacitance/channel length	Farads/meters	0	
NSUB	Substrate doping density	$1/\text{centimeter}^3$	0	
NSS	Surface-state density	$1/\text{centimeter}^2$	0	
NFS	Fast surface-state density	$1/\text{centimeter}^2$	0	
TOX	Oxide thickness	meters	∞	
TPG	Gate material type: $+1$ = opposite of substrate, -1 = same as substrate, 0 = aluminum		$+1$	
XJ	Metallurgical junction depth	meters	0	
UO	Surface mobility	$\text{centimeters}^2/\text{Volts} \cdot \text{seconds}$	600	
UCRIT	Mobility degradation critical field (LEVEL = 2)	Volts/centimeter	1E4	
UEXP	Mobility degradation exponent (LEVEL = 2)		0	
UTRA	(Not used) mobility degradation transverse field coefficient			
VMAX	Maximum drift velocity	meters/second	0	
NEFF	Channel charge coefficient (LEVEL = 2)		1	
XQC	Fraction of channel charge attributed to drain		1	
DELTA	Width effect on threshold		0	
THETA	Mobility modulation (LEVEL = 3)	Volts^{-1}	0	
ETA	Static feedback (LEVEL = 3)		0	
KAPPA	Saturation field factor (LEVEL = 3)		0.2	
KF	Flicker noise coefficient		0	$1E-26$
AF	Flicker noise exponent		1	1.2

The model statement for *n*-channel MOSFETs has the general form

```
.MODEL   MNAME   NMOS   (P1=A1   P2=A2   P3=A3 ... PN=AN)
```

and the statement for *p*-channel MOSFETs has the form

```
.MODEL   MNAME   PMOS   (P1=A1   P2=A2   P3=A3 ... PN=AN)
```

where MNAME is the model name; it can begin with any character, and its word size is normally limited to eight characters. NMOS and PMOS are the type symbols of *n*-channel and *p*-channel MOSFETs, respectively. P1, P2, . . . and A1, A2, . . . are the parameters and their values, respectively.

In Table 9-2 L and W are the channel length and width, respectively. AD and AS are the drain and source diffusion areas. L is decreased by twice LD to get the effective channel length. Similarly, W is decreased by twice WD to get the effective channel width. L and W can be specified for the device, the model, or in the .OPTION statement. PSpice sets priority in selecting their values. The value specified for the device supersedes that for the model, which supersedes that in the .OPTION statement.

AD and AS are the drain and source diffusion areas. PD and PS are the drain and source diffusion perimeters. The drain *p-n*-saturation current can be specified either by IS in an absolute value, or by JS, which is multiplied by AD and AS. The zero-bias depletion capacitances can be specified (1) by CBD and CBS in absolute values, (2) by CJ, which is multiplied by AD and AS, or (3) by CJSW, which is multiplied by PD and PS.

Contact and bulk resistances are included in series with the drain, source, gate, and bulk (substrate). The MOSFET is modeled as an intrinsic device. RDS is a shunt resistance in parallel with the drain-source channel. These ohmic resistances can be specified in absolute values of RD, RS, RG, and RB. Alternatively, one could specify these resistances by RSH, which is multiplied by NRD, NRS, NRG, and NRB, respectively. NRD, NRS, NRG, and NRB are the relative resistivities of the drain, source, gate, and substrate in squares.

PD, PS, NRG, and NRB default to 0. NRD and NRS default to 1. Defaults for L, W, AD, and AS may be set in the .OPTIONS statement. The default value of AD or AS is 0. The default value of L or W is 100 μm.

The dc characteristics are defined by parameters VTO, KP, LAMBDA, PHI, and GAMMA, which are computed by PSpice by using the fabrication-process parameters NSUB, TOX, NSS, NFS, TPG, and so on. The values of VTO, KP, LAMDA, PHI, and GAMMA, which are specified in the model statement, supersede the values calculated by PSpice based on fabrication-process parameters. *VTO is positive for enhancement type n-channel MOSFETs and for depletion type p-channel MOSFETs. VTO is negative for enhancement type p-channel MOSFETs and for depletion type n-channel MOSFETs.*

PSpice incorporates three MOSFET device models. The LEVEL parameter selects among different models for the intrinsic MOSFET. If LEVEL = 1, the Schichman-Hodges model [1] is used. If LEVEL = 2, an advanced version of the Schichman-Hodges model, which is a geometry-based analytical model and incorporates extensive second-order effects [3], is used. If LEVEL = 3, a modified version of the Schichman-Hodges model, which is a semiempirical short-channel model [3], is used.

The LEVEL 1 model, which employs fewer fitting parameters, gives approximate results. However, it is useful for a quick and rough estimate of the circuit performances and it is normally adequate for the analysis of basic electronic circuits. The LEVEL 2 model, which can take into consideration various parameters, requires a great amount of CPU time for the calculations and could cause convergence problems. The LEVEL 3 model introduces a smaller error as compared to the LEVEL 2 model, and the CPU time is also approximately 25% less. The LEVEL 3 model is designed for MOSFETs with short channels.

The symbol for a metal-oxide silicon field-effect transistor (MOSFET) is M. The name of a MOSFET must start with M and takes the general form

```
M⟨name⟩  ND   NG   NS   NB   MNAME
+        [L=⟨value⟩]  [W=⟨value⟩]
+        [AD=⟨value⟩]  [AS=⟨value⟩]
+        [PD=⟨value⟩]  [PS=⟨value⟩]
+        [NRD=⟨value⟩]  [NRS=⟨value⟩]
+        [NRG=⟨value⟩]  [NRB=⟨value⟩]
```

where ND, NG, NS, and NB are the drain, gate, source, and bulk (or substrate) nodes, respectively. MNAME is the model name, and it can begin with any character; its word size is normally limited to eight characters. Positive current is the current that flows into a terminal. That is, the current flows from the drain node through the device to the source node for an n-channel MOSFET.

Some MOSFET Statements

```
M1    4   2   7   0 MMOD   L=10U W=20U
.MODEL   MMOD   NMOS
M13  15   3   0   0   IRF150
.MODEL   IRF150   NMOS   (LEVEL=3 TOX=.10U L=3.0U LD=.5U W=2.0 WD=0
+XJ=1.2U
+        NSUB=4E14 IS=2.1E-14 RB=0 RD=.01 RS=.03 RDS=1E6 VTO=3.25
+        U0=550 THETA=.1 ETA=0 VMAX=1E6 CBS=1P CBD=4000P PB=.7 MJ=.5
+        RG=4.9 CGSO=1690P CGDO=365P CGBO=1P)
M2A   0   2  20  20   IRF9130
.MODEL   IRF9130   PMOS   (LEVEL=3 TOX=.1U L=3.0U LD=.5U W=1.3 WD=0
+XJ=1.2U
+        NSUB=4E14 IS=2.1E-14 RB=0 RD=.03 RS=.2 RDS=5E5 VTO=-3.7
+        U0=600 THETA=.1 ETA=0 VMAX=1E6 CBS=1P CBD=2000P PB=.7 MJ=.5
+        RG=5 CGSO=520P CGDO=180P CGBO=1P)
MA    0   2  15  15   PMOD   L=20U W=20U AD=100U AS=200U PD=50U
+                            PS=50U NRD=10 NRS=20 NRG=10
.MODEL PMOD   PMOS
```

9-6 MOSFET PARAMETERS

The data sheet for the n-channel MOSFET of type IRF150 is shown in Fig. 9-18. The library file EVAL.LIB of the student version of PSpice supports models for the n-type MOSFET of type IRF150 and the p-type MOSFET of type IRF9140. As an example, we shall generate approximate values of some parameters [8, 9] from the data sheet of IRF150.

INTERNATIONAL RECTIFIER

HEXFET® TRANSISTORS IRF150
IRF151
IRF152
IRF153

N-Channel

100 Volt, 0.055 Ohm HEXFET

The HEXFET® technology is the key to International Rectifier's advanced line of power MOSFET transistors. The efficient geometry and unique processing of the HEXFET design achieve very low on-state resistance combined with high transconductance and great device ruggedness.

The HEXFET transistors also feature all of the well established advantages of MOSFETs such as voltage control, freedom from second breakdown, very fast switching, ease of paralleling, and temperature stability of the electrical parameters.

They are well suited for applications such as switching power supplies, motor controls, inverters, choppers, audio amplifiers, and high energy pulse circuits.

Features:

- Fast Switching
- Low Drive Current
- Ease of Paralleling
- No Second Breakdown
- Excellent Temperature Stability

Product Summary

Part Number	V_{DS}	$R_{DS(on)}$	I_D
IRF150	100V	0.055Ω	40A
IRF151	60V	0.055Ω	40A
IRF152	100V	0.08Ω	33A
IRF153	60V	0.08Ω	33A

CASE STYLE AND DIMENSIONS

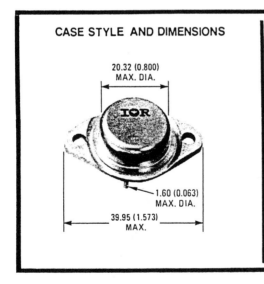

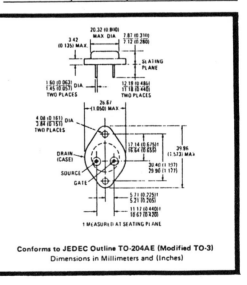

Conforms to JEDEC Outline TO-204AE (Modified TO-3)
Dimensions in Millimeters and (Inches)

Figure 9-18 Data sheet for MOSFET type IRF150 (Courtesy of International Rectifier).

IRF150, IRF151, IRF152, IRF153 Devices

Absolute Maximum Ratings

	Parameter	IRF150	IRF151	IRF152	IRF153	Units
V_{DS}	Drain · Source Voltage ①	100	60	100	60	V
V_{DGR}	Drain · Gate Voltage (R_{GS} = 20 kΩ) ①	100	60	100	60	V
I_D @ T_C = 25°C	Continuous Drain Current	40	40	33	33	A
I_D @ T_C = 100°C	Continuous Drain Current	25	25	20	20	A
I_{DM}	Pulsed Drain Current ③	160	160	132	132	A
V_{GS}	Gate · Source Voltage	± 20				V
P_D @ T_C = 25°C	Max. Power Dissipation	150 (See Fig. 14)				W
	Linear Derating Factor	1.2 (See Fig. 14)				W/K
I_{LM}	Inductive Current, Clamped	(See Fig. 15 and 16) L = 100µH				A
		160	160	132	132	
T_J T_{stg}	Operating Junction and Storage Temperature Range	−55 to 150				°C
	Lead Temperature	300 (0.063 in. (1.6mm) from case for 10s)				°C

Electrical Characteristics @ T_C = 25°C (Unless Otherwise Specified)

	Parameter	Type	Min.	Typ.	Max.	Units	Test Conditions
BV_{DSS}	Drain · Source Breakdown Voltage	IRF150 IRF152	100	–	–	V	V_{GS} = 0V
		IRF151 IRF153	60	–	–	V	I_D = 250µA
$V_{GS(th)}$	Gate Threshold Voltage	ALL	2.0	–	4.0	V	V_{DS} = V_{GS}, I_D = 250µA
I_{GSS}	Gate-Source Leakage Forward	ALL	–	–	100	nA	V_{GS} = 20V
I_{GSS}	Gate-Source Leakage Reverse	ALL	–	–	-100	nA	V_{GS} = ·20V
I_{DSS}	Zero Gate Voltage Drain Current	ALL	–	–	250	µA	V_{DS} = Max. Rating, V_{GS} = 0V
			–	–	1000	µA	V_{DS} = Max. Rating x 0.8, V_{GS} = 0V, T_C = 125°C
$I_{D(on)}$	On-State Drain Current ②	IRF150 IRF151	40	–	–	A	V_{DS} > $I_{D(on)}$ x $R_{DS(on)}$ max.· V_{GS} = 10V
		IRF152 IRF153	33	–	–	A	
$R_{DS(on)}$	Static Drain-Source On-State Resistance ②	IRF150 IRF151	–	0.045	0.055	Ω	V_{GS} = 10V, I_D = 20A
		IRF152 IRF153	–	0.06	0.08	Ω	
g_{fs}	Forward Transconductance ②	ALL	9.0	11	–	S (℧)	V_{DS} > $I_{D(on)}$ x $R_{DS(on)}$ max.· I_D = 20A
C_{iss}	Input Capacitance	ALL	–	2000	3000	pF	V_{GS} = 0V, V_{DS} = 25V, f = 1.0 MHz See Fig. 10
C_{oss}	Output Capacitance	ALL	–	1000	1500	pF	
C_{rss}	Reverse Transfer Capacitance	ALL	–	350	500	pF	
$t_{d(on)}$	Turn-On Delay Time	ALL	–	–	35	ns	V_{DD} = 24V, I_D = 20A, Z_o = 4.7Ω
t_r	Rise Time	ALL	–	–	100	ns	See Figure 17.
$t_{d(off)}$	Turn-Off Delay Time	ALL	–	–	125	ns	(MOSFET switching times are essentially
t_f	Fall Time	ALL	–	–	100	ns	independent of operating temperature.)
Q_g	Total Gate Charge (Gate-Source Plus Gate-Drain)	ALL	–	63	120	nC	V_{GS} = 10V, I_D = 50A, V_{DS} = 0.8 Max. Rating. See Fig. 18 for test circuit. (Gate charge is essentially
Q_{gs}	Gate-Source Charge	ALL '	–	27	–	nC	independent of operating temperature.)
Q_{gd}	Gate-Drain ("Miller") Charge	ALL	–	36	–	nC	
L_D	Internal Drain Inductance	ALL	–	5.0	–	nH	Measured between the contact screw on header that is closer to source and gate pins and center of die.
L_S	Internal Source Inductance	ALL	–	12.5	–	nH	Measured from the source pin, 6 mm (0.25 in.) from header and source bonding pad.

Modified MOSFET symbol showing the internal device inductances.

Thermal Resistance

			Min.	Typ.	Max.	Units	
R_{thJC}	Junction-to-Case	ALL	–	–	0.83	K/W	
R_{thCS}	Case-to-Sink	ALL	–	0.1	–	K/W	Mounting surface flat, smooth, and greased.
R_{thJA}	Junction-to-Ambient	ALL	–	–	30	K/W	Free Air Operation

Figure 9-18 *Continued*

Source-Drain Diode Ratings and Characteristics

							Modified MOSFET symbol showing the integral reverse P-N junction rectifier.
I_S	Continuous Source Current (Body Diode)	IRF150 IRF151	–	–	40	A	
		IRF152 IRF153	–	–	33	A	
I_{SM}	Pulse Source Current (Body Diode) ③	IRF150 IRF151	–	–	160	A	
		IRF152 IRF153	–	–	132	A	
V_{SD}	Diode Forward Voltage ②	IRF150 IRF151	–	–	2.5	V	T_C = 25°C, I_S = 40A, V_{GS} = 0V
		IRF152 IRF153	–	–	2.3	V	T_C = 25°C, I_S = 33A, V_{GS} = 0V
t_{rr}	Reverse Recovery Time	ALL	–	600	–	ns	T_J = 150°C, I_F = 40A, dI_F/dt = 100A/μs
Q_{RR}	Reverse Recovered Charge	ALL	–	3.3	–	μC	T_J = 150°C, I_F = 40A, dI_F/dt = 100A/μs
t_{on}	Forward Turn-on Time	ALL	Intrinsic turn-on time is negligible. Turn-on speed is substantially controlled by $L_S + L_D$.				

① T_J = 25°C to 150°C. ② Pulse Test: Pulse width ≤ 300μs, Duty Cycle ≤ 2%. ③ Repetitive Rating: Pulse width limited by max. junction temperature. See Transient Thermal Impedance Curve (Fig. 5).

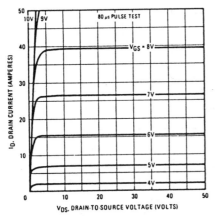

Fig. 1 – Typical Output Characteristics

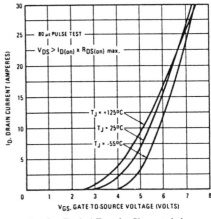

Fig. 2 – Typical Transfer Characteristics

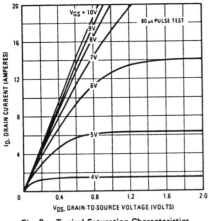

Fig. 3 – Typical Saturation Characteristics

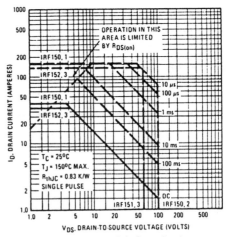

Fig. 4 – Maximum Safe Operating Area

Figure 9-18 *Continued*

IRF150, IRF151, IRF152, IRF153 Devices

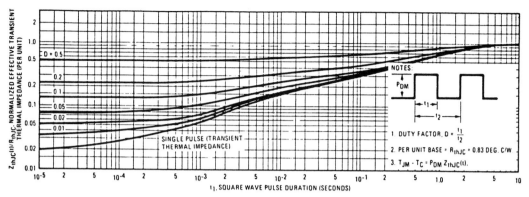

Fig. 5 — Maximum Effective Transient Thermal Impedance, Junction-to-Case Vs. Pulse Duration

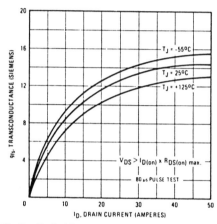

Fig. 6 — Typical Transconductance Vs. Drain Current

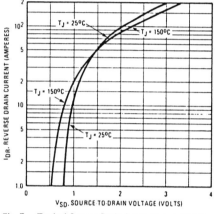

Fig. 7 — Typical Source-Drain Diode Forward Voltage

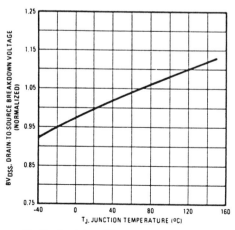

Fig. 8 — Breakdown Voltage Vs. Temperature

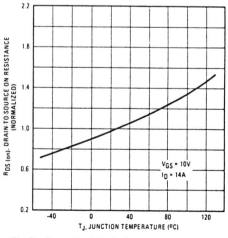

Fig. 9 — Normalized On-Resistance Vs. Temperature

Figure 9-18 *Continued*

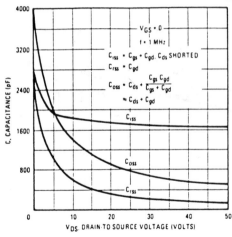

Fig. 10 — Typical Capacitance Vs. Drain-to-Source Voltage

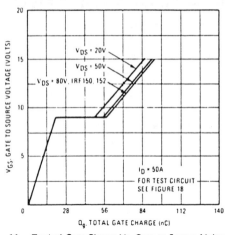

Fig. 11 — Typical Gate Charge Vs. Gate-to-Source Voltage

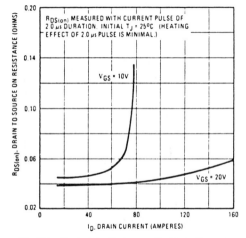

Fig. 12 — Typical On-Resistance Vs. Drain Current

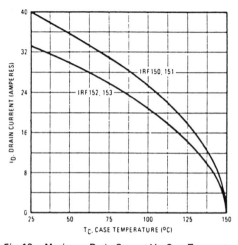

Fig. 13 — Maximum Drain Current Vs. Case Temperature

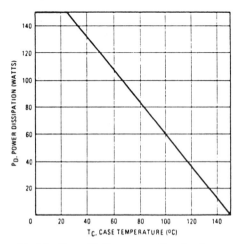

Fig. 14 — Power Vs. Temperature Derating Curve

Figure 9-18 *Continued*

From the data sheet we get $I_{DSS} = 250\ \mu A$ at $V_{GS} = 0$ V, $V_{DS} = 100$ V. $V_{Th} = 2$ to 4 V. Geometric mean, $V_{Th} = VTO = \sqrt{2 \times 4} = 2.83$ V. The constant K_p can be found from

$$I_D = K_p(V_{GS} - V_{Th})^2 \qquad (9\text{-}3)$$

For $I_D = I_{DSS} = 250\ \mu A$, and $V_{Th} = 2.83$ V, Eq. (9-3) gives $K_p = 250\ \mu A/2.83^2 = 31.2\ \mu A/V^2$. K_p is related to channel length L and channel width W by

$$K_p = \frac{\mu_a C_o}{2}\left(\frac{W}{L}\right) \qquad (9\text{-}4)$$

where C_o is the capacitance per unit area of the oxide layer, a typical value for a power MOSFET being 3.5×10^{-11} F/cm^2 at a thickness of 0.1 nm (assumed), and μ_a is the surface mobility of electrons, 600 cm^2/(V $\cdot$ s).

The ratio W/L can be found from Eq. (9-4),

$$\frac{W}{L} = \frac{2K_p}{\mu_a C_o} = \frac{2 \times 31.2 \times 10^{-6}}{600 \times 3.5 \times 10^{-11}} = 3000$$

Let $L = 1$ nm and $W = 3\ \mu m$. $C_{rss} = 350 - 500$ pF at $V_{GS} = 0$, $V_{DS} = 25$ V. Geometric mean, $C_{rss} = C_{gd} = \sqrt{350 \times 500} = 418.3$ pF at $V_{DG} = 25$ V.

For a MOSFET, the values of C_{gs} and C_{gd} remain relatively constant with changing V_{GS} or V_{DS}. They are determined mainly by the thickness and type of the insulating oxide. Although, the curves of the capacitances versus drain-source voltage show some variations, we will assume constant capacitances. Thus, $C_{gdo} = 418.3$ pF. $C_{iss} = 2000$ to 3000 pF. Geometric mean $C_{iss} = \sqrt{2000 \times 3000} = 2450$ pF. Since C_{iss} is measured at $V_{GS} = 0$ V, $C_{gs} = C_{gso}$. That is,

$$C_{iss} = C_{gso} + C_{gd}$$

which gives $C_{gso} = C_{iss} - C_{sd} = 2450 - 418.3 = 2032$ pF $= 2.032$ nF. Thus the PSpice model statement for MOSFET IRF150 is

```
.MODEL IRF150 NMOS (VTO=2.83 KP=31.2U L=1N W=3U CGDO=0.418N CGSO=2.032N)
```

The model can be used to plot the characteristics of the MOSFET. It may be necessary to modify the parameter values to conform with the actual characteristics. It should be noted that the parameters would differ from those given in the PSpice library, because their values are dependent on the constants used in derivations. Students are encouraged to run a circuit file with the PSpice library model and compare the results obtained with the above model statement.

9-7 EXAMPLES OF MOSFET AMPLIFIERS

The large number of parameters involved is an indication of the complexity of modeling a MOSFET. An accurate modeling requires a SPICE library file of a MOSFET. The parameters that are determined from the data sheet in Section 9-6 are the approximate values only. If a model parameter is not available, its typical value as indicated in Table 9-2 should be used. The following examples illustrate the PSpice simulation of MOSFET circuits.

Example 9-6

An *n*-channel enhancement-type MOSFET amplifier with series-shunt feedback is shown in Fig. 9-19. Plot the magnitude of output voltage. The frequency is varied from 10 Hz to 100 MHz in decade steps with 10 points per decade. The peak input voltage is 100 mV. The model parameters of the MOSFET are VTO=1, KP=6.5E−3, CBD=5PF, CBS=2PF, RD=5, RS=2, RB=0, RG=0, RDS=1MEG, CGSO=1PF, CGDO=1PF, and CGBO=1PF. Print the details of the bias and operating points.

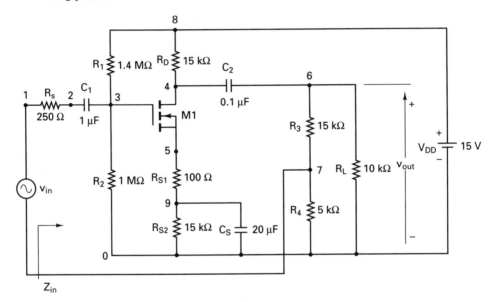

Figure 9-19 A MOSFET feedback amplifier.

Solution The listing of the circuit file follows.

Example 9-6 A MOSFET feedback amplifier
```
▲ *    Input voltage of 100 mV peak for frequency response
   VIN   1  7  AC   100mV
   VDD   8  0  15V
▲ ▲ RS    1  2   250
   C1    2  3   1UF
   R1    8  3   1.4MEG
   R2    3  0   1MEG
   RD    8  4   15K
   RS1   5  9   100
   RS2   9  0   15K
   CS    9  0   20UF
   C2    4  6   0.1UF
   R3    6  7  15K
```

```
       R4    7  0  5K
       RL    6  0  10K
       *    MOSFET M1 with model MQ is connected to 4 (drain), 3 (gate), 5
       *    (source) and 5 (substrate).
       M1 4  3  5  5  MQ
       *    Model for MQ
       .MODEL MQ NMOS (VTO=1 KP=6.5E-3 CBD-5PF CBS=2PF RD=5 RS=2 RB=0
       +    RG=0 RDS=1MEG CGS0=1PF CGD0=1PF CGB0=1PF)
▲ ▲ ▲  *    Ac analysis for 10 Hz to 100 MHz with a decade increment and 10
       *    points per decade
       .AC   DEC  10  10HZ   100MEGHZ
       *    Plot the results of ac analysis: voltage at node 6.
       .PLOT  AC  VM(6)
       *    Print the details of the dc operating point.
       .OP
       .PROBE
.END
```

The details of the bias and operating points are given next.

**** SMALL-SIGNAL BIAS SOLUTION   TEMPERATURE = 27.000 DEG C

NODE	VOLTAGE	NODE	VOLTAGE	NODE	VOLTAGE	NODE	VOLTAGE
(1)	0.0000	(2)	0.0000	(3)	6.2500	(4)	10.1000
(5)	4.9323	(6)	0.0000	(7)	0.0000	(8)	15.0000
(9)	4.8997						

VOLTAGE SOURCE CURRENTS

NAME	CURRENT
VIN	0.000E+00
VDD	−3.329E−04

TOTAL POWER DISSIPATION   4.99E−03   WATTS

**** OPERATING POINT INFORMATION   TEMPERATURE = 27.000 DEG C
**** MOSFETS

NAME	M1
MODEL	MQ
ID	3.32E−04
VGS	1.32E+00
VDS	5.17E+00
VBS	0.00E+00
VTH	1.00E+00
VDSAT	3.17E−01
GM	2.06E−03
GDS	1.00E−06
GMB	0.00E+00
CBD	1.83E−12
CBS	2.00E−12
CGSOV	1.00E−16
CGDOV	1.00E−16
CGBOV	1.00E−16

CGS	0.00E+00
CGD	0.00E+00
CGB	0.00E+00

JOB CONCLUDED

TOTAL JOB TIME   17.41

The frequency response for Example 9-6 is shown in Fig. 9-20. If the .PROBE command is included, there is no need for the .PLOT command.

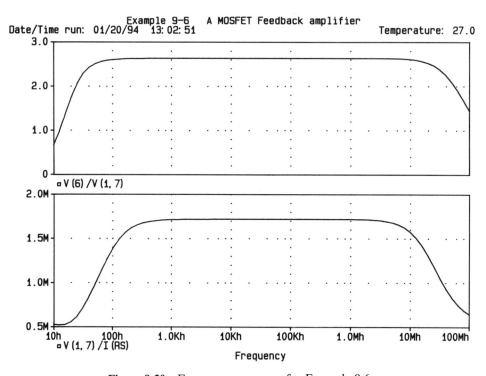

Figure 9-20   Frequency response for Example 9-6.

Example 9-7

For Fig. 9-19, plot the magnitude response of output impedance. The frequency is varied from 10 Hz to 100 MHz in decade steps with 10 points per decade.

Solution   The output impedance of the MOSFET feedback amplifier in Fig. 9-19 can be determined by short-circuiting the input source and connecting a test current source between terminals 0 and 6, as shown in Fig. 9-21. Let the peak value of the test current be 1 mA. The voltage at node 6 is a measure of the output impedance:

$Z_{out} = V(6)/1$ mA $= V(6)$ kΩ.

The listing of the circuit file follows.

Example 9-7   Output impedance of a MOSFET feedback amplifier

▲ * Input source VIN is shorted.

    VIN 1 9 AC 0V

   * Test current of 1 mA peak for frequency response

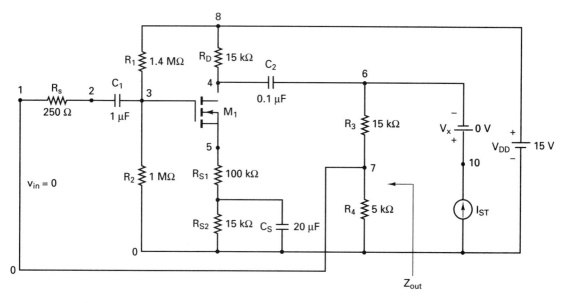

Figure 9-21 Equivalent circuit for output impedance calculation (Example 9-7).

```
IST   0   10   AC   1MA
*   A dummy source of 0 V dc
VX    10   6   DC 0V
VDD   8    0   15V
RS    1    2   250
C1    2    3   1UF
R1    8    3   1.4MEG
R2    3    0   1MEG
RD    8    4   15K
RS1   5    9   100
RS2   9    0   15K
CS    9    0   20UF
C2    4    6   0.1UF
R3    6    7   15K
R4    7    0   5K
RL    6    0   10K
*     M1 with model MQ, whose substrate is connected to node 5
M1 4  3   5   5   MQ
*     Model for n-channel MOSFET with model name MQ
.MODEL MQ NMOS (VTO=1 KP=6.5E-3 CBD=5PF CBS=2PF RD=5 RS=2 RB=0
+     RG=0 RDS=1MEG CGSO=1PF CGDO=1PF CGBO=1PF)
*     Ac analysis for 10 Hz to 100 MHz with a decade increment and 10
*     points per decade
.AC   DEC   10   10HZ   10MEGHZ
*     Plot the results of the ac analysis: voltage at node 6.
.PLOT   AC   VM(6)
.PROBE
.END
```

The frequency response of the output impedance for Example 9-7 is shown in Fig. 9-22. If the .PROBE command is included, there is no need for the .PLOT command.

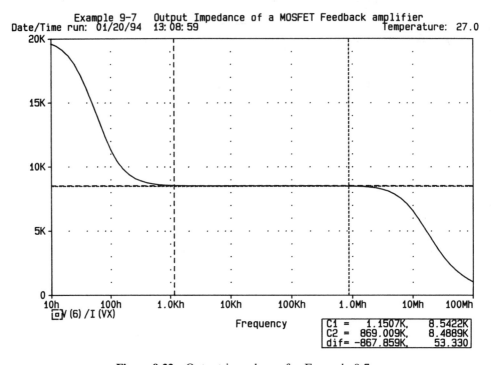

Figure 9-22 Output impedance for Example 9-7.

Example 9-8

A CMOS inverter circuit is shown in Fig. 9-23(a). The output is taken from node 3. The input voltage is shown in Fig. 9-23(b). Plot the transient response of the output voltage from 0 to 80 μs in steps of 2 μs. If the input voltage is 5 V, calculate

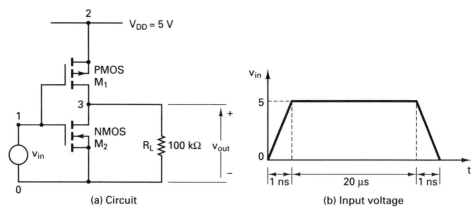

Figure 9-23 A CMOS inverter (Example 9-8).

the voltage gain, the input resistance, and the output resistance. Print the small-signal parameters of the MOS and NMOS. The model parameters of the PMOS are L=1U, W=20U, VTO=−2, KP=4.5E−4, CBD=5PF, CBS=2PF, RD=5, RS=2, RB=0, RG=0, RDS=1MEG, CGSO=1PF, CGDO=1PF, and CGBO=1PF. The model parameters of the NMOS are L=1U, W=5U, VTO=2, KP=4.5E−5, CBD=5PF, CBS=2PF, RD=5, RS=2, RB=0, RG=0, RDS=1MEG, CGSO=1PF, CGDO=1PF, and CGBO=1PF.

Solution   The listing of the circuit file follows.

Example 9-8   A CMOS inverter

```
▲ VDD   2   0   5V
  *  The input voltage is 5 V for dc analysis and pulse waveform for
  *  transient analysis.
  VIN  1  0  DC  5V  PULSE (0  5V  0  1NS  1NS  20US 40US)
▲▲ RL    3   0  100K
    *   PMOS with model PMOD
    M1   3  1  2  2  PMOD  L=1U  W=20U
    .MODEL PMOD PMOS (VTO=−2 KP=4.5E−4 CBD=5PF CBS=2PF RD=5 RS=2 RB=0
    +  RG=0 RDS=1MEG CGSO=1PF CGDO=1PF CGBO=1PF)
    M2   3  1  0  0  NMOD  L=1U  W=5U
    *   NMOS with model NMOD
    .MODEL NMOD NMOS (VTO=2 KP=4.5E−5 CBD=5PF CBS=2PF RD=5 RS=2 RB=0
    +  RG=0 RDS=1MEG CGSO=1PF CGDO=1PF CGBO=1PF)
    *   Transient analysis from 0 to 80μs in steps of 1 μs
▲▲▲ .TRAN   1US   80US
      *  Transfer-function analysis
      .TF  V(3)  VIN
      *  Print details of operating points.
      .OP
      .PLOT TRAN  V(3)  V(1)
      .PROBE
  .END
```

```
    ****     OPERATING POINT INFORMATION        TEMPERATURE =  27.000 DEG C
    **** MOSFETS
    NAME        M1          M2
    MODEL       MQ          NMOD
    ID          −5.00E−06   1.53E−11
    VGS         0.00E+00    5.00E+00
    VDS         −5.00E+00   2.27E−08
    VBS         0.00E+00    0.00E+00
    VTH         −2.00E+00   2.00E+00
    VDSAT       0.00E−00    3.00E+00
    GM          0.00E−00    5.09E−12
    GDS         1.00E−06    6.76E−04
    GMB         0.00E+00    0.00E+00
    CBD         1.86E−12    5.00E−12
    CBS         2.00E−12    2.00E−12
    CGSOV       2.00E−17    5.00E−18
    CGDOV       2.00E−17    5.00E−18
```

CGBOV	1.00E-18	1.00E-18
CGS	0.00E+00	0.00E+00
CGD	0.00E+00	0.00E+00
CGB	0.00E+00	0.00E+00

```
****    SMALL-SIGNAL CHARACTERISTICS
        V(3)/VIN = -7.422E-09
        INPUT RESISTANCE AT VIN = 1.000E+20
        OUTPUT RESISTANCE AT V(3) = 1.462E+03
           JOB CONCLUDED
           TOTAL JOB TIME          48.33
```

The frequency response for Example 9-8 is shown in Fig. 9-24. If the .PROBE command is included, there is no need for the .PLOT command. The .OPTIONS statement is essential to avoid convergence problems, and it is often necessary to adjust the .OPTIONS parameters (e.g., ABSTOL, RELTOL, and VNTOL) in order to let the simulation converge.

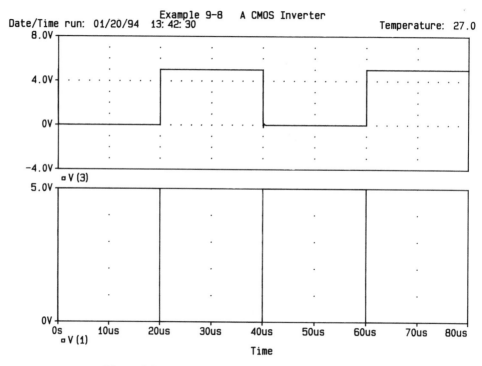

Figure 9-24 Frequency response for Example 9-8.

9-8 GALLIUM ARSENIDE MESFETs

The PSpice model of an *n*-channel GaAsFET (gallium arsenide FET) is shown in Fig. 9-25 [5, 6]. The small-signal model, which is generated by PSpice, is shown in Fig. 9-26. The model parameters for a GaAsFET device and the default values

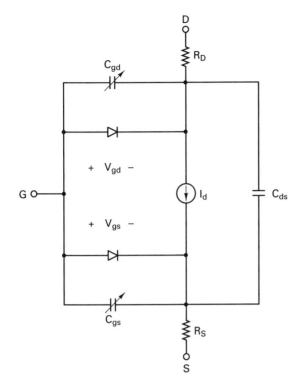

Figure 9-25 PSpice *n*-channel GaAsFET model.

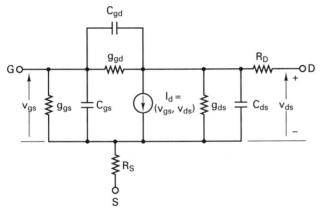

Figure 9-26 Small-signal *n*-channel GaAsFET model.

assigned by PSpice are listed in Table 9-3. The model equations of GaAsFETs that are used by PSpice are described in Curtice [5], Sussman-Fort et al. [6], and the *PSpice Manual* [7].

The model statement of *n*-channel GaAsFETs has the general form

```
.MODEL  BNAME  GASFET  (P1=A1  P2=A2  P3=A3 ... PN=AN)
```

where GASFET is the type symbol of *n*-channel GaAsFETs. BNAME is the model name. It can begin with any character and its word size is normally limited

TABLE 9-3 MODEL PARAMETERS OF GaAs MESFETs

Name	Area	Model parameters	Units	Default	Typical
VTO		Threshold voltage	Volts	-2.5	-2.0
ALPHA		Tan h constant	Volts^{-1}	2.0	1.5
BETA		Transconductance coefficient	Amps/Volts^2	0.1	25U
LAMBDA		Channel-length modulation	Volts	0	1E-10
RG	*	Gate ohmic resistance	Ohms	0	1
RD	*	Drain ohmic resistance	Ohms	0	1
RS	*	Source ohmic resistance	Ohms	0	1
IS		Gate p-n saturation current	Amps	1E-14	
M		Gate p-n grading coefficient		0.5	
N		Gate p-n emission coefficient		1	
VBI		Threshold voltage	Volts	1	0.5
CGD		Gate-drain zero-bias p-n capacitance	Farads	0	1FF
CGS		Gate-source zero-bias p-n capacitance	Farads	0	6FF
CDS		Drain-source capacitance	Farads	0	0.3FF
TAU		Transit time	seconds	0	10PS
FC		Forward-bias depletion capacitance coefficient		0.5	
VTOTC		VTO temperature coefficient	Volts/°C	0	
BETATCE		BETA exponent temperature coefficient	%/°C	0	
KF		Flicker noise coefficient		0	
AF		Flicker noise exponent		1	

to eight characters. P1, P2, . . . and A1, A2, . . . are the parameters and their values, respectively.

RD, RS, and RG represent the contact and bulk resistances per unit area of the drain, source, and gate, respectively. The GaAsFET is modeled as an intrinsic device. The area value, which is the relative device area, is specified in the .MODEL statement and changes the actual resistance values. The default value of the area is 1.

The symbol for a gallium arsenide MESFET (GaAs MESFET or GaAsFET) is *B*. The name of a GaAs MESFET must start with *B*, and it takes the general form

```
B⟨name⟩ ND  NG  NS  BNAME  [(area) value]
```

where ND, NG, ND are the drain, gate, and source nodes, respectively. BNAME, which is the model name, can begin with any character, and its word size is normally limited to eight characters. Positive current flows into a terminal.

Some GaAs MESFET Statements

```
BIX  2  5  7  NMOD
.MODEL NMOD GASFET
BIM 15  1  0  GMOD
.MODEL GMOD GASFET (VTO=-2.5 BETA=60U VBI=0.5 ALPHA=1.5 TAU=10PS)
B5   7  9  3    NMOM 1.5
.MODEL MNOM GASFET (VTO=-2.5 BETA=32U VBI=0.5 ALPHA=1.5)
```

Example 9-9

A GaAsFET inverter with active load is shown in Fig. 9-27(a). The input voltage is a pulse waveform, as shown in Fig. 9-27(b). Plot the transient response of the output voltage for a time duration of 240 ps in steps of 2 ps. Plot the dc transfer characteristic if the input voltage is varied from −2.5 V to 1 V in steps of 0.1 V. The model parameters of the GaAsFET are VTO=−2, BETA=60U, VBI=0.5, ALPHA=1.5, and TAU=10PS, and those of B2 are VTO=−2, BETA=3U, VBI=0.5, and ALPHA=1.5. Calculate the dc voltage gain, the input resistance, and the output resistance. Print the small-signal parameters for the dc analysis.

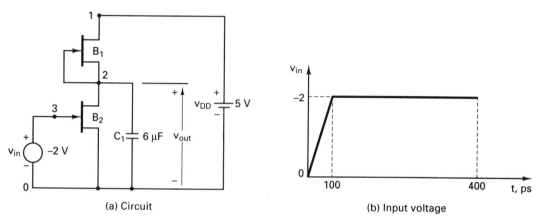

(a) Circuit   (b) Input voltage

Figure 9-27 GaAsFET inverter with active load.

Solution The listing of the circuit file follows.

Example 9-9   A GaAsFET inverter with active load

▲ VDD   1   0   5V
  *   Pulsed input voltage
  VIN   3   0   DC   −2V   PWL (0   0   100PS   −2V   1NS   −2V)
▲▲ *   GaAsFET, which is connected to 1 (drain), 2 (gate) and
  *   2 (source), has a model of GF1.
  B1   1   2   2   GF1
  B2   2   4   0   GF2
  C1   2   0   6F   IC=0V
  RS   3   4   50
  *   Model for GF1
  .MODEL GF1 GASFET (VTO=−2.5 BETA=65U VBI=0.5 ALPHA=1.5 TAU=10PS)
  .MODEL GF2 GASFET (VTO=−2.5 BETA=32.5U VBI=0.5 ALPHA=1.5)
▲▲▲ *   Transient analysis for 0 to 240 ps with 2-ps increment
  .TRAN   2PS   240PS   UIC
  .DC VIN   −2.5   1   0.1
  *   Plot the results of transient analysis.
  .PLOT TRAN   V(3)   V(2)
  .PLOT   DC   V(2)
  *   Dc transfer characteristics
  .TF   V(2)   VIN

```
*      Small-signal parameters for dc analysis
.OP
.PROBE
.END
```

****   SMALL-SIGNAL BIAS SOLUTION   TEMPERATURE = 27.000 DEG C

NODE	VOLTAGE	NODE	VOLTAGE	NODE	VOLTAGE	NODE	VOLTAGE
(1)	5.0000	(2)	4.8787	(3)	−1.0000	(4)	−1.0000

VOLTAGE SOURCE CURRENTS

NAME	CURRENT
VDD	−7.313E−05
VIN	6.899E−12

TOTAL POWER DISSIPATION 3.66E−04 WATTS

****   OPERATING POINT INFORMATION   TEMPERATURE = 27.000 DEG C

**** GASFETS

NAME	B1	B2
MODEL	GF1	GF2
ID	7.31E−05	7.31E−05
VGS	0.00E+00	−1.00E+00
VDS	1.21E−01	4.88E+00
GM	5.85E−05	9.75E−05

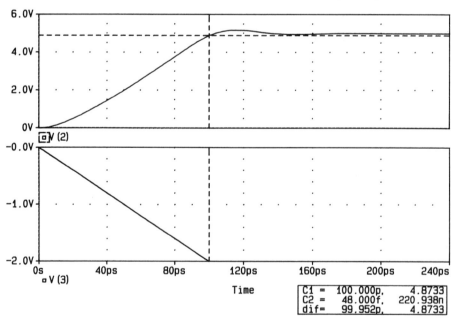

Example 9-9   A GaAsFET inverter with active load
Date/Time run: 01/20/94 13:56:18   Temperature: 27.0

Figure 9-28 Transient response for Example 9-9.

GDS	5.90E-04	1.93E-10
CGS	0.00E+00	0.00E+00
CGD	0.00E+00	0.00E+00
CDS	0.00E+00	0.00E+00

****   SMALL-SIGNAL CHARACTERISTICS
      V(2)/VIN = -1.654E-01
      INPUT RESISTANCE AT VIN = 4.618E+11
      OUTPUT RESISTANCE AT V(2) = 1.696E+03

The transient response and the dc transfer characteristics for Example 9-9 are shown in Fig. 9-28. If the .PROBE command is included, there is no need for the .PLOT command.

SUMMARY

The model statements for FETs can be summarized as follows:

```
B⟨name⟩ ND  NG   NS   BNAME  [(area) value]
.MODEL  BNAME    GASFET (P1=V1 P2=V2 P3=V3 ...PN=VN)
J⟨name⟩ ND  NG   NS   JNAME  [(area) value]
.MODEL  JNAME    NJF  (P1=V1 P2=V2 P3=V3 ...PN=VN)
.MODEL  JNAME    PJF  (P1=V1 P2=V2 P3=V3 ...PN=VN)
N⟨name⟩ ND  NG   NS   NB    MNAME
+        [L=⟨value⟩]   [W=⟨value⟩]
+        [AD=⟨value⟩]  [AS=⟨value⟩]
+        [PD=⟨value⟩]  [PS=⟨value⟩]
+        [NRD=⟨value⟩] [NRS=⟨value⟩]
+        [NRG=⟨value⟩] [NRB=⟨value⟩]
.MODEL  MNAME    NMOS (P1=V1 P2=V2 P3=V3 ...PN=VN)
.MODEL  MNAME    PMOS (P1=V1 P2=V2 P3=V3 ...PN=VN)
```

REFERENCES

1. H. Schichman and D. A. Hodges, "Modeling and simulation of insulated gate field effect transistor switching circuits," *IEEE Journal of Solid-State Circuits,* Vol. SC-3, September 1968, pp. 285–289.

2. J. F. Meyer, "MOS models and circuit simulation," *RCA Review,* Vol. 32, March 1971, pp. 42–63.

3. A. Vladimirescu and Sally Liu, *The simulation of MOS integrated circuits using SPICE2,* Memorandum no. M80/7, February 1980, University of California, Berkeley.

4. L. M. Dang, "A simple current model for short channel IGFET and its application to circuit simulation," *IEEE Journal of Solid-State Circuits,* Vol. SC-14, No. 2, 1979, pp. 358–367.

5. W. R. Curtice, "A MESFET model for use in the design of GaAs integrated circuits," *IEEE Transactions on Microwave Theory and Techniques,* Vol. MTT-23, May 1980, pp. 448–456.

6. S. E. Sussman-Fort, S. Narasimhan, and K. Mayaram, "A complete GaAs MESFET computer model for SPICE," *IEEE Transactions on Microwave Theory and Techniques,* Vol. MTT-32, April 1984, pp. 471–473.

7. *PSpice Manual.* Irvine, Calif.: MicroSim Corporation, 1992.

8. S. Natarajan, "An effective approach to obtain model parameters for BJTs and FETs from data books," *IEEE Transactions on Education,* Vol. 35, No. 2, 1992, pp. 164–169.

9. M. H. Rashid, *SPICE for Power Electronics and Electric Power.* Englewood Cliffs, N.J.: Prentice Hall, 1993.

PROBLEMS

9-1. For the amplifier circuit in Fig. 6-8, calculate and plot the frequency responses of the output voltage and the input current. The frequency is varied from 10 Hz to 10 MHz in decade steps with 10 points per decade.

9-2. A shunt-shunt feedback is applied to the amplifier circuit in Fig. 9-8. This is shown in Fig. P9-2. Calculate and print the frequency responses of the output voltage and the input current. The frequency is varied from 10 Hz to 10 MHz in decade steps with 10 points per decade. The model parameters are IS=1N, VTO=−4, BETA=0.5M, CGDO=5.85P, CGSO=3.49P, and LAMBDA=2.395E−3.

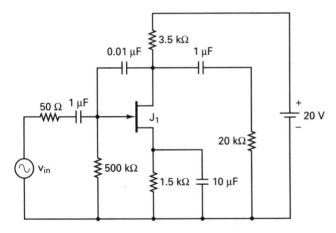

Figure P9-2

9-3. Repeat Example 9-5 if the transistor is an *n*-channel JFET. The model parameters are IS=100E−14, RD=10, RS=10, BETA=1E−3, and VTO=−2. Assume $R_{S2} = 0$.

9-4. For the *n*-channel enhancement-type MOSFET in Fig. P9-4, plot the output characteristics if V_{DS} is varied from 0 to 15 V in steps of 0.1 V and V_{GS} is varied from 0 to 6 V in steps of 1 V. The model parameters are L=10U, W=20U, VTO=2.5, KP=6.5E−3, RD=5, RS=2, RB=0, RG=0, and RDS=1MEG.

9-5. For Problem 9-4, plot the input characteristics if V_{GS} is varied from 0 to 6 V in steps of 0.1 V and $V_{DS} = 15$ V.

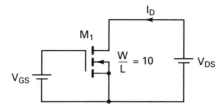

Figure P9-4

9-6. An inverter circuit is shown in Fig. P9-6. For the input voltage as shown in Fig. 9-23(b), plot the transient response of the output voltage from 0 to 80 μs in steps of 2 μs. If the input voltage is 5 V dc, calculate the voltage gain, the input resistance, and the output resistance. Print the small-signal parameters of the PMOS. The model parameters of the PMOS are VTO=-2.5, KP=4.5E-3, CBD=5PF, CBS=2PF, CGSO=1PF, CGDO=1PF, and CGBO=1PF.

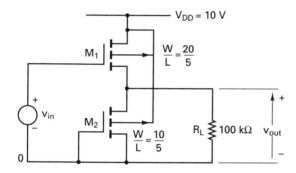

Figure P9-6

9-7. For the NMOS AND-logic circuit in Fig. P9-7, plot the transient response of the output voltage from 0 to 100 μs in steps of 1 μs. The model parameters of the *p*-channel depletion-type MOSFETs are VTO=2, KP=4.5E-3, CBD=5PF, CBS=2PF, RD=5, RS=2, RB=0, RG=0, RDS=1MEG, CGSO=1PF, CGDO=1PF, and CGBO=1PF.

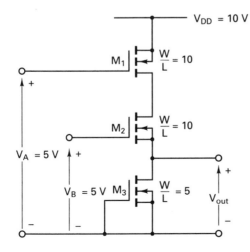

Figure P9-7

9-8. For the NMOS NAND-logic gate circuit in Fig. P9-8, plot the transient response of the output voltage from 0 to 100 μs in steps of 1 μs. The model parameters of the PMOS are VTO=-2.5, KP=4.5E-3, CBD=5PF, CBS=2PF, RD=5, RS=2, RB=0, RG=0, RDS=1MEG, CGSO=1PF, CGDO=1PF, and CGBO=1PF. The model parameters of the NMOS are VTO=2.5, KP=4.5E-3, CBD=5PF, CBS=2PF, RD=5, RS=2, RB=0, RG=0, RDS=1MEG, CGSO=1PF, CGDO=1PF, and CGBO=1PF.

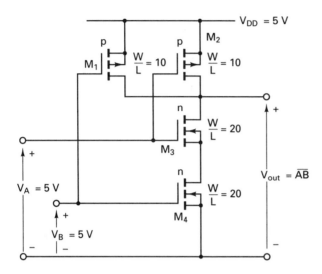

Figure P9-8

9-9. A MOSFET amplifier with active load is shown in Fig. P9-9. Plot the magnitudes of the output voltage and the input current. The frequency is varied from 10 Hz to 100 MHz with a decade increment and 10 points per decade. The peak input voltage is 200 mV. The model parameters of the NMOS are VTO=2.5, KP=4.5E-2, CBD=5PF, CBS=2PF, RD=5, RS=2, RB=0, RG=0, RDS=1MEG, CGSO=1PF, CGDO=1PF, and CGBO=1PF. Print the details of the bias point and the small-signal parameters of the NMOS.

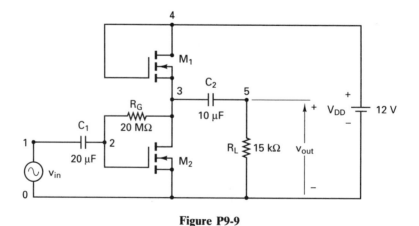

Figure P9-9

9-10. Use PSpice to perform a Monte Carlo analysis for six runs and for the frequency of Problem 9-2. The model parameter is $R=1$ for resistors. The lot deviation for all resistances is $\pm 15\%$. The transistor parameter having uniform deviations is

$$VTO = -4 \pm 1.5$$

(a) The greatest difference from the nominal run is to be printed.
(b) The maximum value of the output voltage is to be printed.
(c) The minimum value of the output voltage is to be printed.

9-11. Use PSpice to perform the worst-case analysis for Problem 9-10.

9-12. Use PSpice to perform a Monte Carlo analysis for six runs and for the transient response of Problem 9-6. The transistor parameter having uniform deviations is

$$VTO = -2.5V \pm 1.2V$$

(a) The greatest difference from the nominal run is to be printed.
(b) The maximum value of the output voltage is to be printed.
(c) The minimum value of the output voltage is to be printed.

9-13. Use PSpice to perform the worst-case analysis for Problem 9-12.

9-14. Use PSpice to perform a Monte Carlo analysis for five runs and for the transient response of Problem 9-7. The transistor parameter having uniform deviations is

$$VTO = 2V \pm 1.5V$$

(a) The greatest difference from the nominal run is to be printed.
(b) The maximum value of the output voltage is to be printed.
(c) The minimum value of the output voltage is to be printed.

9-15. Use PSpice to perform the worst-case analysis for Problem 9-14.

9-16. Use PSpice to perform a Monte Carlo analysis for five runs and for the transient response of Problem 9-8. The transistor parameter having uniform deviations is

$$VTO = -2.5V \pm 1.5V$$

(a) The greatest difference from the nominal run is to be printed.
(b) The maximum value of the output voltage is to be printed.
(c) The minimum value of the output voltage is to be printed.

9-17. Use PSpice to perform the worst-case analysis for Problem 9-16.

9-18. Use PSpice to perform a Monte Carlo analysis for five runs and for the frequency response of Problem 9-9. The transistor parameter having uniform deviations is

$$VTO = 2.5V \pm 1.5V$$

(a) The greatest difference from the nominal run is to be printed.
(b) The maximum value of the output voltage is to be printed.
(c) The minimum value of the output voltage is to be printed.

9-19. Use PSpice to perform the worst-case analysis for Problem 9-18.

Op-Amp Circuits

10-1 INTRODUCTION

An **op-amp** may be modeled as a linear amplifier to simplify the design and analysis of op-amp circuits. The linear models give reasonable results, especially for determining the approximate design values of op-amp circuits. However, the simulation of the actual behavior of op-amps is required in many applications to obtain accurate responses for the circuits. PSpice does not have any model for op-amps. However, an op-amp can be simulated from the circuit arrangement of the particular type of op-amp. The μA741 type of op-amp consists of 24 transistors, and it is beyond the capability of the student (or demo) version of PSpice. However, a macromodel, which is a simplified version of the op-amp and requires only two transistors, is quite accurate for many applications and can be simulated as a subcircuit or library file. Some manufacturers often supply macromodels of their op-amps [1]. In the absence of a complex op-amp model, the characteristics of op-amp circuits may be determined approximately by one of the following models:

Dc linear model
Ac linear model
Nonlinear macromodel

10-2 DC LINEAR MODELS

An op-amp may be modeled as a voltage-controlled voltage source, as shown in Fig. 10-1(a). The input resistance is high, typically 2 MΩ, and the output resistance is very low, typically 75 Ω. For an ideal op-amp, the model in Fig. 10-1(a)

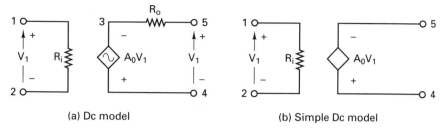

(a) Dc model   (b) Simple Dc model

Figure 10-1   Dc linear models.

can be reduced to that of Fig. 10-1(b). These models do not take into account the saturation effect and slew rate, which do exist in actual op-amps. The gain is also assumed to be independent of the frequency, but the gain of actual practical op-amps falls with the frequency. These simple models are normally suitable for dc or low-frequency applications.

10-3 AC LINEAR MODEL

The frequency response of an op-amp can be approximated by a single break frequency, as shown in Fig. 10-2(a). This characteristic can be modeled by the circuit of Fig. 10-2(b). This is a high-frequency model of op-amps. If an op-amp has more than one break frequency, it can be represented by using as many capacitors as the number of breaks. R_i is the input resistance and R_o is the output resistance.

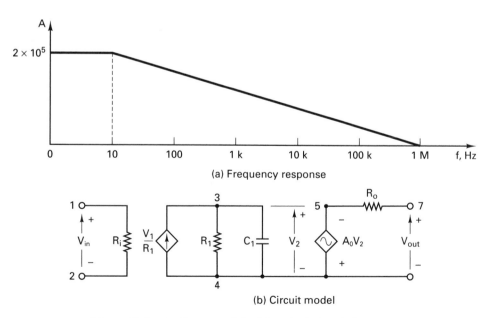

(a) Frequency response

(b) Circuit model

Figure 10-2   Ac linear model with a single break frequency.

The dependent sources of the op-amp model in Fig. 10-2(b) have a common node. Without this, PSpice will give an error message because there is no dc path from the nodes of the dependent current source. The common node could be either with the input stage or with the output stage. This model does not take into account the saturation effect and is suitable only if the op-amp operates within the linear region.

The output voltage can be expressed as

$$V_{\text{out}} = -A_0 V_2 = \frac{-A_0 V_{\text{in}}}{1 + R_1 C_1 s}$$

Substituting $s = j2\pi f$ yields

$$V_{\text{out}} = \frac{-A_0 V_{\text{in}}}{1 + j2\pi f R_1 C_1} = \frac{-A_0 V_{\text{in}}}{1 + jf/f_b}$$

where

$f_b = 1/(2\pi R_1 C_1)$ is called the *break frequency*, in hertz.
$A_0 = $ the *large-signal* (or *dc*) *gain* of the op-amp.

Thus, the open-loop voltage gain is

$$A(f) = \frac{V_{\text{out}}}{V_{\text{in}}} = -\frac{A_0}{1 + jf/f_b}$$

For μA741 op-amps, $f_b = 10$ Hz, $A_0 = 2 \times 10^5$, $R_i = 2$ MΩ, and $R_o = 75$ Ω. Letting $R_1 = 10$ kΩ, $C_1 = 1/(2\pi \times 10 \times 10 \times 10^3) = 1.15619$ μF.

10-4 NONLINEAR MACROMODEL

The circuit arrangement of the **op-amp macromodel** is shown in Fig. 10-3 [1, 2, 3]. The macromodel can be used as a subcircuit with the .SUBCKT command. However, if an op-amp is used in various circuits, it is convenient to have the macromodel as a library file, namely, EVAL.LIB, and it is not required to type the statements of the macromodel in every circuit where the macromodel is employed. The library file EVAL.LIB that comes with the student version of PSpice has macromodels for op-amps, comparators, diodes, MOSFETs, BJTs, and SCRs. The macromodels for the linear operational amplifier of type LM324, the linear operational amplifier of type μA741, and the voltage comparator of type LM111 are included in the EVAL.LIB file. The professional version of PSpice supports library files for many devices.

The macromodel of the μA741 op-amp is simulated at room temperature. The library file EVAL.LIB contains the op-amp macromodel model as a subcircuit definition μA741 with a set of .MODEL statements. This op-amp model contains nominal, not worst-case, devices, and does not consider the effects of temperature.

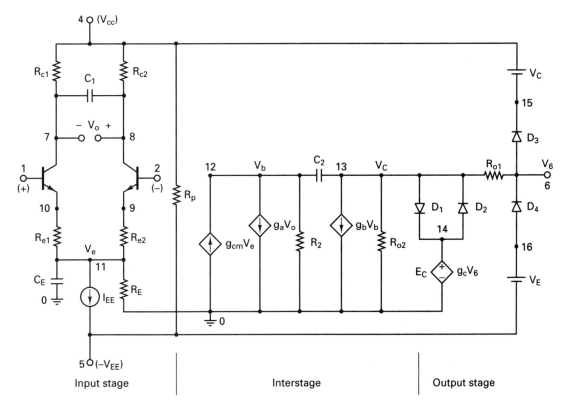

Figure 10-3 Circuit diagram of op-amp macromodel.

The listing of the library file, EVAL.LIB, follows.

```
* Library file "EVAL.LIB" for UA741 op-amp
* connections:    noninverting input
*                 :   inverting input
*                 :   :
*                 :   :       positive power supply
*                 :   :       :   negative power supply
*                 :   :       :   :   output
*                 :   :       :   :   :
.SUBCKT UA741   1   2       4   5   6
*               Vi+ Vi-     Vp+ Vp- Vout
Q1   7   1   10 UA741QA
Q2   8   2   9  UA741QB
RC1  4   7   5.305165D+03
RC2  4   8   5.305165D+03
C1   7   8   5.459553D-12
RE1  10  11  2.151297D+03
RE2  9   11  2.151297D+03
IEE  11  5   1.666000D-05
```

```
CE     11  0    3.000000D-12
RE     11  0    1.200480D+07
GCM    0   12   11  3   5.960753D-09
GA     12  0    8   7   1.884955D-04
R2     12  0    1.000000D+05
C2     12  13   3.000000D-11
GB     13  0    12  0   2.357851D+02
RO2    13  0    4.500000D+01
D1     13  14   UA741DA
D2     14  13   UA741DA
EC     14  0    6   3   1.0
RO1    13  6    3.000000D+01
D3     6   15   UA741DB
VC     4   15   2.803238D+00
D4     16  6    UA741DB
VE     16  5    2.803238D+00
RP     4   5    18.16D+03
*   Models for diodes and transistors
.MODEL UA741DA D (IS=9.762287D-11)
.MODEL UA741DB D (IS=8.000000D-16)
.MODEL UA741QA NPN (IS=8.000000D-16 BF=9.166667D+01)
.MODEL UA741QB NPN (IS=8.309478D-16 BF=1.178571D+02)
*  End of library file
*  End of subcircuit definition
.ENDS
```

Example 10-1

An inverting amplifier is shown in Fig. 10-4. The output is taken from node 5. Calculate and print the voltage gain, the input resistance, and the output resistance. The op-amp, which is modeled by the circuit in Fig. 10-1(a), has $A_0 = 2 \times 10^5$, $R_i = 2$ MΩ, and $R_o = 75$ Ω.

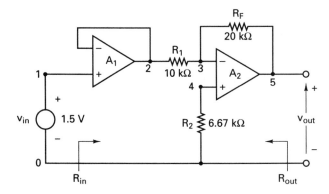

Figure 10-4 Inverting amplifier.

Solution The listing of the subcircuit file follows.

Example 10-1   Inverting amplifier
```
▲ *    Input voltage is 1.5 V dc.
   VIN   1  0   DC   1.5V
```

```
▲▲  R1    2   3   10K
    R2    4   0   6.67K
    RF    3   5   20K
    *  Calling subcircuit OPAMP
    XA1  2   1   2   0   OPAMP
    XA2  3   4   5   0   OPAMP
    *  Subcircuit definition for OPAMP
    .SUBCKT  OPAMP  1   2   5   4
    RI   1   2   2MEG
    RO   3   5   75
    *  Voltage-controlled voltage source with a gain of 2E+5. The polarity of
    *  the output voltage is taken into account by changing the location of
    *  the controlling nodes.
    EA   3   4   2   1   2E+5
    *  End of subcircuit definition
    .ENDS    OPAMP
▲▲▲ *  Transfer-function analysis calculates and prints the dc gain,
    *  the input resistance, and the output resistance.
    .TF  V(5)   VIN
.END
```

The results of the transfer function analysis by the .TF command are given below:

```
****     SMALL-SIGNAL BIAS SOLUTION              TEMPERATURE =  27.000 DEG C
  NODE      VOLTAGE    NODE     VOLTAGE     NODE      VOLTAGE    NODE     VOLTAGE
(    1)     1.5000  (    2)     1.5000   (    3)   15.11E-06  (    4)   50.21E-09
(    5)    -2.9999  (XA1.3)     1.5112   (XA2.3)    -3.0112
    VOLTAGE SOURCE CURRENTS
    NAME           CURRENT
    VIN           -3.778E-12
    TOTAL POWER DISSIPATION  5.67E-12  WATTS


****     SMALL-SIGNAL CHARACTERISTICS
    V(5)/VIN = -2.000E+00
    INPUT RESISTANCE AT VIN = 3.970E+11
    OUTPUT RESISTANCE AT V(5)  = 1.132E+03
        JOB CONCLUDED
        TOTAL JOB TIME          2.42
```

Example 10-2

An integrator circuit is shown in Fig. 10-5(a). For the input voltage as shown in Fig. 10-5(b), plot the transient response of the output voltage for a duration of 0 to 4 ms in steps of 50 μs. The op-amp that is modeled by the circuit in Fig. 10-2(b) has $R_i = 2$ MΩ, $R_o = 75$ Ω, $C_1 = 1.5619$ μF, $R_1 = 10$ kΩ, and $A_0 = 2 \times 10^5$.

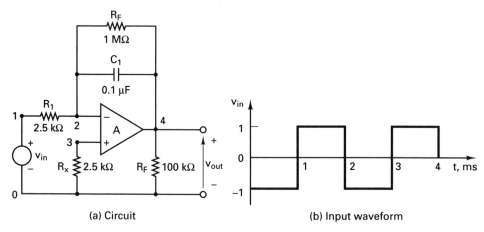

(a) Circuit

(b) Input waveform

Figure 10-5 Integrator circuit.

Solution The listing of the circuit file follows:

Example 10-2 Integrator circuit

▲ * The input voltage is represented by a piecewise linear waveform.

* To avoid convergence problems due to a rapid change, the input

* voltage is assumed to have a finite slope.

```
VIN   1   0   PWL (0  0 1NS −1V  1.0001MS −1V  2MS 1V
+     2.0001MS −1V 3MS −1V  3.0001MS 1V  4MS 1V)
```

▲▲ R1   1   2   2.5K

```
   RF   2   4   1MEG
   RX   3   0   2.5K
   RL   4   0   100K
   C1   2   4   0.1UF
   *  Calling subcircuit OPAMP
   XA1  2   3   4   0   OPAMP
   *  Subcircuit definition for OPAMP
   .SUBCKT OPAMP  1   2   7   4
   RI   1   2   2.0E6
   *  Voltage-controlled current source with a gain of 1
   GB   4   3   1   2   0.1M
   R1   3   4   10K
   C1   3   4   1.5619UF
   *  Voltage-controlled voltage source with a gain of 2E+5
   EA   4   5   3   4   2E+5
   RO   5   7   75
   *  End of subcircuit OPAMP
   .ENDS
```

▲▲▲ * Transient analysis for 0 to 4 ms with 50-μs increment

```
   .TRAN   50US   4MS
   *  Plot the results of transient analysis
   .PLOT   TRAN   V(4)   V(1)
   .PLOT   AC   VM(4)   VP(4)
   .PROBE
.END
```

The transient response for Example 10-2 is shown in Fig. 10-6. The .PLOT statements generate graphical plots in the output file. If the .PROBE command is included, there is no need for the .PLOT commands.

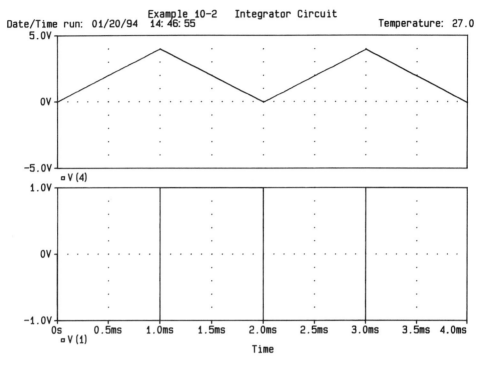

Example 10-2   Integrator Circuit

Figure 10-6   Transient response for Example 10-2.

Example 10-3

A practical differentiator circuit is shown in Fig. 10-7(a). For the input voltage as shown in Fig. 10-7(b), plot the transient response of the output voltage for a duration of 0 to 4 ms in steps of 50 μs. The op-amp, which is modeled by the circuit in Fig. 10-2(b), has $R_i = 2$ MΩ, $R_o = 75$ Ω, $C_1 = 1.5619$ μF, $R_1 = 10$ kΩ, and $A_0 = 2 \times 10^5$.

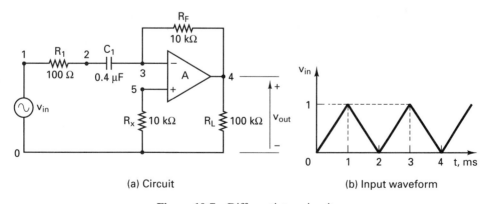

(a) Circuit   (b) Input waveform

Figure 10-7   Differentiator circuit.

Solution The listing of the circuit file follows.

Example 10-3 Differentiator circuit

```
▲ *    The maximum number of points is changed to 410. The default
  *    value is only 201.
  .OPTIONS  NOPAGE  NOECHO LIMPTS=410
  *  Input voltage is a piecewise linear waveform for transient analysis.
  VIN 1  0  PWL (0  0  1MS  1  2MS  0  3MS  1  4MS  0)
▲▲ R1  1  2  100
   RF  3  4  10K
   RX  5  0  10K
   RL  4  0  100K
   C1  2  3  0.4UF
   *  Calling op-amp OPAMP
   XA1  3  5  4  0  OPAMP
   *  Op-amp subcircuit definition
   .SUBCKT  OPAMP  1  2  7  4
   RI  1  2  2.0E6
   *  Voltage-controlled current source with a gain of 0.1M
   GB  4  3  1  2  0.1M
   R1  3  4  10K
   C1  3  4  1.5619UF
   *  Voltage-controlled voltage source with a gain of 2E+5
   EA  4  5  3  4  2E+5
   RO  5  7  75
   *  End of subcircuit OPAMP
   .ENDS  OPAMP
▲▲▲ *  Transient analysis for 0 to 4 ms with 50 µs increment
   .TRAN  10US  4MS
   *  Plot the results of transient analysis 4
   .PLOT  TRAN  V(4)  V(1)
   .PROBE
.END
```

The transient response for Example 10-3 is shown in Fig. 10-8. If the .PROBE command is included, there is no need for the .PLOT command.

Example 10-4

A filter circuit is shown in Fig. 10-9. Plot the frequency response of the output voltage. The frequency is varied from 10 Hz to 100 MHz with an increment of 1 decade and 10 points per decade. For the op-amp modeled by the circuit in Fig. 10-2(b), $R_i = 2$ MΩ, $R_o = 75$ Ω, $C_1 = 1.5619$ μF, $R_1 = 10$ kΩ, and $A_0 = 2 \times 10^5$.

Solution The listing of the circuit file follows.

Example 10-4 A filter circuit

```
▲ *    Input voltage is 1 V peak for ac analysis or frequency response.
   VIN  1  0  AC  1
   R1  1  2  20K
   R2  2  4  20K
   R3  3  0  10K
   R4  1  5  10K
```

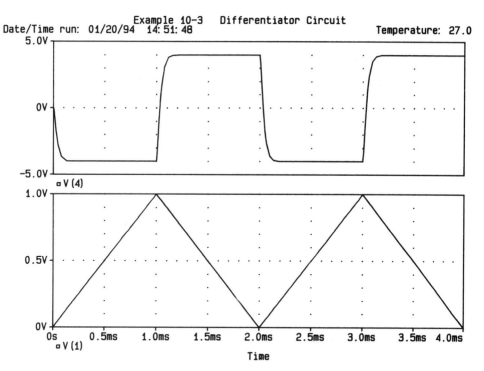

Figure 10-8 Transient response for Example 10-3.

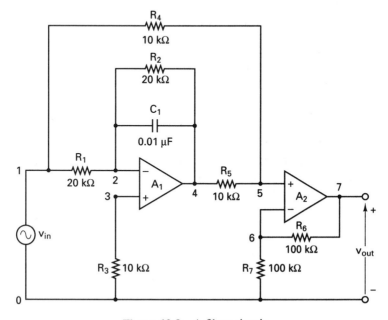

Figure 10-9 A filter circuit.

```
         R5   4   5   10K
         R6   6   7   100K
         RL   7   0   100K
         C1   2   4   0.01UF
▲ ▲  *  Subcircuit call for OPAMP
         XA1  2   3   4   0   OPAMP
         XA2  5   6   7   0   OPAMP
         *  Subcircuit definition for OPAMP
         .SUBCKT  OPAMP  1  2  7  4
         RI   1   2   2.0E6
         *  Voltage-controlled current source with a gain of 0.1M
         GB   4   3   1   2   0.1M
         R2   3   4   10K
         C2   3   4   1.5619UF
         *  Voltage-controlled voltage source of gain 2E+5
         EA   4   5   3   4   2E+5
         RO   5   7   75
         *  End of subcircuit definition
         .ENDS  OPAMP
▲ ▲ ▲ *  AC analysis for 10 Hz to 100 MHz with a decade increment and
         *  10 points per decade
         .AC   DEC   10   10HZ   100MEGHZ
         *  Plot the results of ac analysis
         .PLOT   AC   VM(7)   VP(7)
         .PROBE
     .END
```

The frequency response for Example 10-4 is shown in Fig. 10-10. If the .PROBE command is included, there is no need for the .PLOT command.

Example 10-5

A band-pass active filter is shown in Fig. 10-11. The op-amp can be modeled as a macromodel, as shown in Fig. 10-3. The description of the UA741 macromodel is listed in the library file EVAL.LIB. Plot the frequency response if the frequency is varied from 100 Hz to 1 MHz with an increment of 1 decade and 10 points per decade. The peak input voltage is 1 V.

Solution   The listing of the circuit file follows.

Example 10-5   Band-pass active filter

```
▲  *  Input voltage of 1 V peak for frequency response
     VIN  1   0   AC   1
▲ ▲ R1   1   2   5K
     R2   3   4   1.5K
     R3   2   0   265K
     C1   2   4   0.01UF
     C2   2   3   0.01UF
     RL   4   0   15K
     VCC  6   0   DC   12V
     VEE  0   7   DC   12V
     *  Subcircuit call for UA741
```

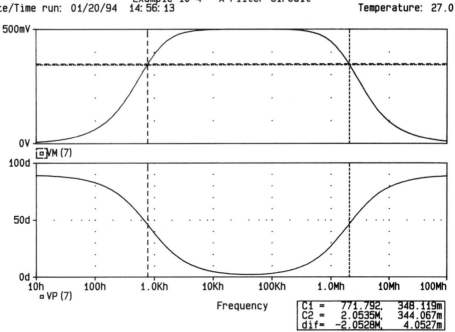

C1 =	771.792,	348.119m
C2 =	2.0535M,	344.067m
dif=	-2.0528M,	4.0527m

Figure 10-10   Frequency response for Example 10-4.

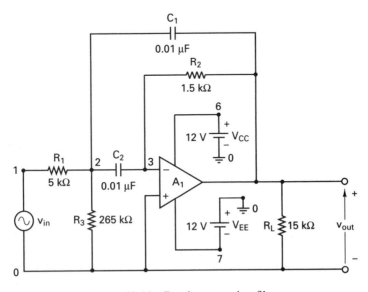

Figure 10-11   Band-pass active filter.

```
    X1   0    3    6    7    4    UA741
    *  Vi+  Vi-  Vp+  Vp-  Vout
    *  Call library file EVAL.LIB
    .LIB  EVAL.LIB
▲ ▲ ▲ *    AC analysis for 100 Hz to 1 MHz with a decade increment and 10
    *    points per decade
    .AC  DEC  10  100HZ  1MEGHZ
    *  Plot the results of the ac analysis: magnitude of voltage at node 4
    .PLOT  AC  VM(4)
    .PROBE
.END
```

The frequency response for Example 10-5 is shown in Fig. 10-12. If the .PROBE command is included, there is no need for the .PLOT command.

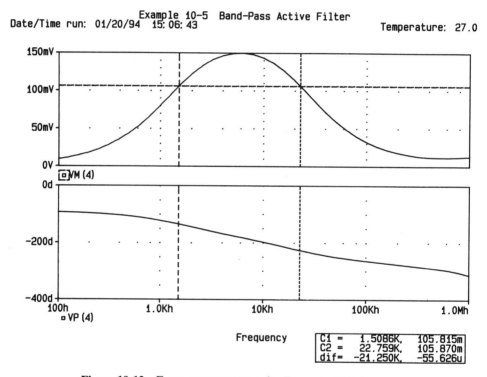

Figure 10-12   Frequency response for Example 10-5.

Example 10-6

A free-running multivibrator circuit is shown in Fig. 10-13. Plot the transient response of the output voltage for a duration of 0 to 4 ms in steps of 20 μs. The op-amp can be modeled as a macromodel as shown in Fig. 10-3. The description of the UA741 macromodel is listed in library file EVAL.LIB. Assume the initial voltage of the capacitor $C_1 = -5$ V.

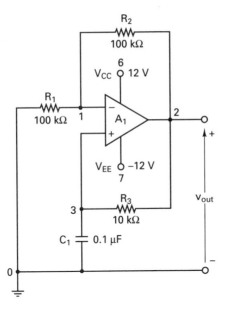

Figure 10-13 Free-running multi-vibrator.

Solution The listing of the circuit file follows.

Example 10-6   Free-running multivibrator

```
▲ VCC   6   0   DC   12V
  VEE   0   7   DC   12V
▲▲ R1   1   0   100K
   R2   1   2   100K
   R3   2   3   10K
   C1   3   0   0.1UF   IC=-5V
   *   Subcircuit call for UA741
   XA1   1     3     6   7   2     UA741
   *      Vi+  Vi-  Vp+  Vp-  Vout
   *   Call library file EVAL.LIB
   .LIB   EVAL.LIB
▲▲▲ *   Transient analysis from 0 to 4 ms in steps of 20 μs
    .TRAN   10US   4MS   UIC
    .PROBE
.END
```

The transient response for Example 10-6 is shown in Fig. 10-14.

Example 10-7

The circuit diagram of a differential amplifier with a transistor current source is shown in Fig. 10-15. Calculate the dc voltage gain, the input resistance, and the output resistance. The input voltage is 0.1 V. The model parameters of the bipolar transistors are BF=50, RB=70, and RC=40.

Solution The listing of the circuit file follows.

Example 10-7   Differential amplifier

```
▲ VCC   11   0     12V
  VEE   0    10    12V
```

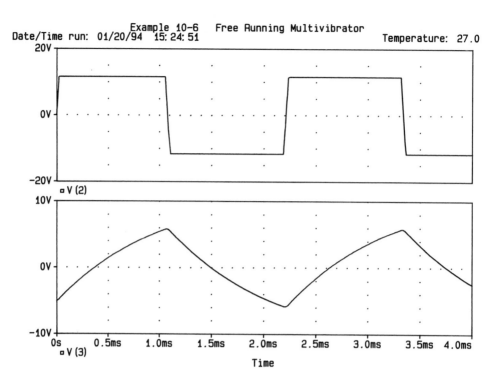

Figure 10-14   Transient response for Example 10-6.

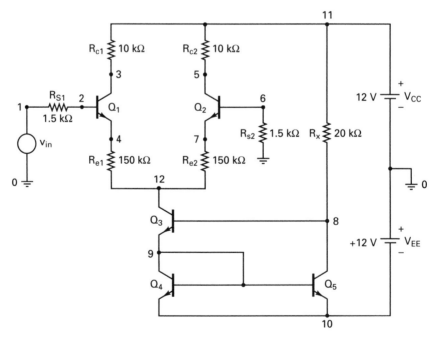

Figure 10-15   Differential amplifier.

```
        VIN  1   0   DC  0.25V
▲ ▲ RC1  11  3   10K
     RC2  11  5   10K
     RE1  4   12  150
     RE2  7   12  150
     RS1  1   2   1.5K
     RS2  6   0   1.5K
     RX   11  8   20K
     *   Model for NPN BJTs with model name QN
     .MODEL QN NPN (BF=50 RB=70 RC=40)
     Q1   3   2   4   QN
     Q2   5   6   7   QN
     Q3   12  8   9   QN
     Q4   9   9   10  QN
     Q5   8   9   10  QN
▲ ▲ ▲ *  DC transfer function analysis
     .TF  V(3,5)  VIN
 .END
```

The results of the transfer-function analysis by the .TF commands are given below:

```
****    SMALL-SIGNAL BIAS SOLUTION              TEMPERATURE =  27.000 DEG C
  NODE    VOLTAGE    NODE    VOLTAGE    NODE    VOLTAGE    NODE    VOLTAGE
(   1)     .2500  (    2)     .2190  (    3)   1.6609  (    4)     -.5575
(   5)   11.3460  (    6)    -.0020  (    7)    -.7057  (    8)  -10.4430
(   9)  -11.2220  (   10)  -12.0000  (   11)  12.0000  (   12)     -.7157
      VOLTAGE SOURCE CURRENTS
      NAME         CURRENT
      VCC        -2.221E-03
      VEE        -2.243E-03
      VIN        -2.068E-05
      TOTAL POWER DISSIPATION  5.36E-02  WATTS

****    SMALL-SIGNAL CHARACTERISTICS
      V(3,5)/VIN = -2.534E+01
      INPUT RESISTANCE AT VIN = 3.947E+04
      OUTPUT RESISTANCE AT V(3,5) = 2.000E+04
      JOB CONCLUDED
      TOTAL JOB TIME          4.01
```

REFERENCES

1. *Linear Circuits—Operational Amplifier Macromodels.* Dallas, Texas: Texas Instruments, 1990.

2. G. Boyle, B. Cohn, D. Pederson, and J. Solomon, ''Macromodeling of integrated circuit operational amplifiers,'' *IEEE Journal of Solid-State Circuits,* Vol. SC-9, No. 6, December 1974, pp. 353–364.

3. I. Getreu, A. Hadiwidjaja, and J. Brinch, "An integrated-circuit comparator macro-model," *IEEE Journal of Solid-State Circuits*, Vol. SC-11, No. 6, December 1976, pp. 826–833.

4. S. Progozy, "Novel applications of SPICE in engineering education," *IEEE Transactions on Education*, Vol. 32, No. 1, February 1990, pp. 35–38.

PROBLEMS

10-1. Plot the frequency response of the integrator in Fig. 10-5 if the frequency is varied from 10 Hz to 100 kHz with a decade increment and 10 points per decade. The peak input voltage is 1 V.

10-2. Plot the frequency response of the differentiator in Fig. 10-7 if the frequency is varied from 10 Hz to 100 kHz with a decade increment and 10 points per decade. The peak input voltage is 1 V.

10-3. Repeat Example 10-2 if the macromodel of the op-amp in Fig. 10-3 is used. The supply voltages are $V_{CC} = 15$ V and $V_{EE} = -15$ V.

10-4. Repeat Example 10-3 if the macromodel of the op-amp in Fig. 10-3 is used. The supply voltages are $V_{CC} = 15$ V and $V_{EE} = -15$ V.

10-5. A full-wave precision rectifier is shown in Fig. P10-5. If the input voltage is $v_{in} = 0.1 \sin(2000\pi t)$, plot the transient response of the output voltage for a duration of 0 to 1 ms in steps of 10 μs. The op-amp can be modeled by the circuit of Fig. 10-2(b), and has $R_i = 2$ MΩ, $R_o = 75$ Ω, $C_1 = 1.5619$ μF, $R_1 = 10$ kΩ, and $A_o = 2 \times 10^5$. Use the default values for the diode model. The supply voltages are $V_{CC} = 12$ V and $V_{EE} = 12$ V.

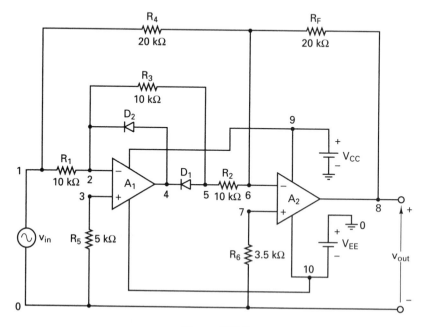

Figure P10-5

10-6. For Fig. P10-5, plot the dc transfer characteristics. The input voltage is varied from −1 V to 1 V in steps of 0.01 V.

10-7. For Fig. P10-7, plot the dc transfer characteristics. The input voltage is varied from −10 V to 10 V in steps of 0.1 V. The op-amp can be modeled as a macromodel, as shown in Fig. 10-3. The description of the macromodel is listed in library file EVAL.LIB. Use the default values for the diode model.

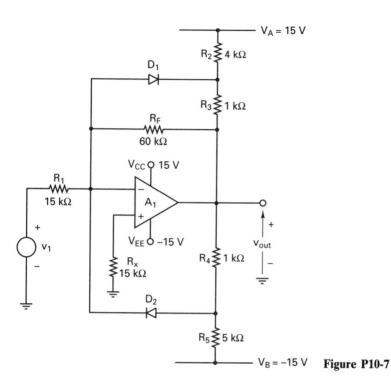

Figure P10-7

10-8. For Fig. P10-8, plot the dc transfer function. The input voltage is varied from −10 V to 10 V in steps of 0.1 V. The Zener voltages are $V_{Z1} = V_{Z2} = 6.3$ V. The op-amp can be modeled as a macromodel, as shown in Fig. 10-3. The description of the macromodel is listed in library file EVAL.LIB. The dc supply voltages of the op-amp are $V_{CC} = |V_{EE}| = 12$ V.

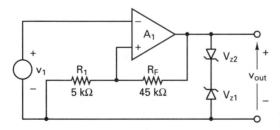

Figure P10-8

10-9. An integrator circuit is shown in Fig. P10-9(a). For the input voltage as shown in Fig. P10-9(b), calculate the slew rate of the amplifier by plotting the transient response of the output voltage for a duration of 0 to 200 μs in steps of 2 μs. For the op-amp modeled by the circuit in Fig. 10-2(b), $R_1 = 2$ MΩ, $R_0 = 75$ Ω, $C_1 = 1.5619$ μF, $R_1 = 10$ kΩ, and $A_0 = 2 \times 10^5$.

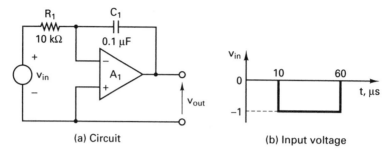

(a) Circuit   (b) Input voltage

Figure P10-9

10-10. Repeat Problem 10-9 if the macromodel of the op-amp in Fig. 10-3 is used. The supply voltages are $V_{CC} = 12$ V and $V_{EE} = -12$ V.

10-11. A sine-wave oscillator is shown in Fig. P10-11. Plot the transient response of the output voltage for a duration of 0 to 2 ms in steps of 0.1 ms. The op-amp can be modeled by the circuit of Fig. 10-2(b), and it has $R_i = 2$ MΩ, $R_o = 75$ Ω, $C_1 = 1.5619$ μF, $R_1 = 10$ kΩ, and $A_0 = 2 \times 10^5$.

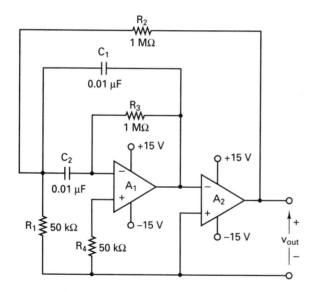

Figure P10-11

10-12. For the gyrator in Fig. P10-12, plot the frequency response of the input impedance. The frequency is varied from 10 Hz to 10 MHz with a decade increment and 10 points per decade. For the op-amp modeled by the circuit in Fig. 10-2(b), $R_i = 2$ MΩ, $R_o = 75$ Ω, $C_1 = 1.5619$ μF, $R_1 = 10$ kΩ, and $A_0 = 2 \times 10^5$.

10-13. Use PSpice to perform a Monte Carlo analysis for five runs and for the dc analysis of Example 10-7. The output voltage is taken between nodes 3 and 5. The model parameter is $R=1$ for resistors. The lot deviation for all resistances is $\pm 15\%$. The transistor parameter having uniform deviations is

$$BF = 50 \pm 20$$

(a) The greatest difference of the output voltage from the nominal run is to be printed.

(b) The maximum value of the output voltage is to be printed.

(c) The minimum value of the output voltage is to be printed.

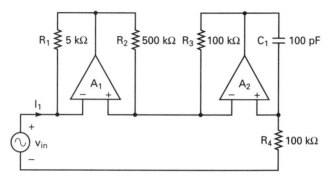

Figure P10-12

10-14. Use PSpice to perform the worst-case analysis for Problem 10-13.

10-15. Use PSpice to perform a Monte Carlo analysis for six runs and for the transient response of Problem 10-5. The model parameter is $R=1$ for resistors. The lot deviation for all resistances is $\pm 20\%$.

(a) The greatest difference of the output from the nominal run is to be printed.

(b) The maximum value of the output voltage is to be printed.

(c) The minimum value of the output voltage is to be printed.

10-16. Use PSpice to perform the worst-case analysis for Problem 10-15.

10-17. Use PSpice to perform a Monte Carlo analysis for five runs and for the dc response of Problem 10-7. The model parameter is $R=1$ for resistors. The lot deviation for all resistances is $\pm 15\%$. The diode parameters having uniform deviations are

$$V_{Z1} = V_{Z2} = 6.3V \pm 1.3V$$

(a) The greatest difference of the output voltage from the nominal run is to be printed.

(b) The maximum value of the output voltage is to be printed.

(c) The minimum value of the output voltage is to be printed.

10-18. Use PSpice to perform the worst-case analysis for Problem 10-17.

10-19. Use PSpice to perform a Monte Carlo analysis for five runs and for the dc response of Problem 10-8. The model parameter is $R=1$ for resistors. The lot deviation for all resistances is $\pm 15\%$. The diode parameters having uniform deviations are

$$V_{Z1} = V_{Z2} = 6.3V \pm 1.3V$$

(a) The greatest difference of the output voltage from the nominal run is to be printed.

(b) The maximum value of the output voltage is to be printed.

(c) The minimum value of the output voltage is to be printed.

10-20. Use PSpice to perform the worst-case analysis for Problem 10-19.

10-21. Use PSpice to perform a Monte Carlo analysis for five runs and for the transient response of Problem 10-11. The model parameter is $R=1$ for resistors, and $C = 1$ for capacitors. The lot deviations for all resistances and capacitances are $\pm 15\%$.

(a) The greatest difference of the output voltage from the nominal run is to be printed.
(b) The maximum value of the output voltage is to be printed.
(c) The minimum value of the output voltage is to be printed.

10-22. Use PSpice to perform the worst-case analysis for Problem 10-21.

Digital Logic Circuits $\quad$

11-1 INTRODUCTION

Digital circuits consist of internal transistors and other circuit components. The simulation of digital logic devices at the circuit level becomes complex and time-consuming, and may even give inaccurate responses due to parasitic elements. PSpice allows simulating off-the-shelf digital devices from their external input/output characteristics. As an introduction to the application of PSpice for simulating digital logic circuits, we shall use only the digital devices that are available in the library file of the student version of PSpice. Further details on other digital devices can be found in the *PSpice Manual* [1].

11.2 DIGITAL DEVICES AND NODES

A digital *state* is a combination of node "logic level" and digital device output "strengths." Digital nodes uniquely take on the values, or output states, shown in Table 11-1. Digital devices may be classified into four categories, as shown in Table 11-2. The U, N, and O devices can be employed to build subcircuits having the characteristics of off-the-shelf devices. Table 11-3 shows some digital devices that are available in the library file .EVAL.LIB of the student version of PSpice.

We will use the following expressions to describe the arguments and/or parameters of devices:

(text)	Text within ( ) is a comment
[item]	Optional item
[item]*	Zero or more of optional item
⟨item⟩	Required item
⟨item⟩*	Zero or more of required item

TABLE 11-1   DIGITAL STATES

State	Meaning
0	Low, false, no, off
1	High, true, yes, on
R	Rising (will change from 0 to 1 sometime during the R-level)
F	Falling (will change from 1 to 0 sometime during the F-level)
X	Unknown: may be high, low, intermediate, or unstable
Z	High impedance: may be high, low, intermediate, or unstable

TABLE 11-2   DIGITAL DEVICE CLASSES

Device class	Type	Description
Primitives	U	Low-level digital devices (gates, flip-flops, etc.)
Stimuli	U	Digital stimuli
Input/Output	N	Digital input
	O	Digital output
	A-to-D interface	
	D-to-A interface	
Power	Power supplies	

TABLE 11-3   OFF-THE-SHELF DIGITAL DEVICES

Subcircuit name	Description
7402	TTL digital 2-input NOR gate
7404	TTL digital inverter
7405	TTL digital inverter, open collector
7414	TTL digital inverter, Schmitt trigger
7474	TTL digital D-type flip-flop
74107	TTL digital JK-type flip-flop
74393	TTL digital 4-bit binary counter

11-3 DIGITAL PRIMITIVES

Digital primitives are low-level devices. These are mainly used to model subcircuits representing off-the-shelf devices such as those in Table 11-3. The primitives use two models: (1) one timing model specifying the propagation delays and the setup and hold times, and (2) one I/O model specifying the device's input/output characteristics. The timing model describes the characteristic of a device, while the I/O model is specific to a whole device family. The general format for a digital primitive is

```
U⟨name⟩  ⟨primitive type⟩  [(⟨parameters value⟩*)]
+    ⟨digital power node⟩  ⟨digital ground node⟩
+    ⟨node⟩*
```

```
+    ⟨timing model name⟩  ⟨I/O model name⟩
+    [MNTYMXDLY = ⟨delay select value⟩]
+    [IO_LEVEL = ⟨interface subcircuit select value⟩]
```

The ⟨*primitive type*⟩ specifies the type of digital device, such as NAND, AND, or OR. The [(⟨*parameter value*⟩*)] specifies the number of inputs that is specific to the primitive type. Digital primitives are listed in Table 11-4. The details of parameters and nodes for the primitives are described in the *PSpice Manual* [1].

TABLE 11-4 DIGITAL PRIMITIVES

Primitive class	Type	Description
Standard gates	BUF	Buffer
	INV	Inverter
	AND	AND gate
	NAND	NAND gate
	OR	OR gate
	NOR	NOR gate
	XOR	Exclusive OR gate
	NXOR	Exclusive NOR gate
	BUFA	Buffer array
	INVA	Inverter array
	ANDA	AND gate array
	NANDA	NAND gate array
	ORA	OR gate array
	NORA	NOR gate array
	XORA	Exclusive OR gate array
	NXORA	Exclusive NOR gate array
	AO	AND-OR compound gate
	OA	OR-AND compound gate
	AOI	OR-NAND compound gate
	NBTG	n-channel transfer gate
	PBTG	p-channel transfer gate
Tri-state gates	BUF3	Buffer
	INV3	Inverter
	AND3	AND gate
	NAND3	NAND gate
	OR3	OR gate
	NOR3	NOR gate
	XOR3	Exclusive OR gate
	NXOR3	Exclusive NOR gate
	BUF3A	Buffer array
	INV3A	Inverter array
	AND3A	AND gate array
	NAND3A	NAND gate array
	OR3A	OR gate array
	NOR3A	NOR gate array
	XOR3A	Exclusive OR gate array
	NXOR3A	Exclusive NOR gate array

(*continued*)

TABLE 11-4 *Continued*

Primitive class	Type	Description
Flip-flops and latches	JKFF	J-K, negative-edge triggered
	DFF	D-type, positive-edge triggered
	SRFF	S-R gated latch
	DLTCH	D gated latch
Pullup/pulldown resistors	PULLUP	Pullup resistor array
	PULLDN	Pulldown array
Delay lines	DLYLINE	Delay line
Programmable logic arrays	PLAND	AND array
	PLOR	OR array
	PLXOR	Exclusive OR array
	PLNAND	NAND array
	PLNOR	NOR array
	PLNXOR	Exclusive NOR array
	PLANDC	AND array, true and complement
	PLORC	OR array, true and complement
	PLXORC	Exclusive OR array, true and complement
	PLNANDC	NAND array, true and complement
	PLNORC	NOR array, true and complement
	PLNXORC	Exclusive NOR array, true and complement
Memory	ROM	Read-only memory
	RAM	Random access read-write memory
Multibit A/D & D/A Converters	ADC	Multibit A/D converter
	DAC	Multibit D/A converter
Behavioral	LOGICEXP	Logic expression
	PINDLY	Pin delay
	CONSTRAINT	Constraint checking

The ⟨*digital power node*⟩ and ⟨*digital ground node*⟩ are used by an interface subcircuit for connecting analog nodes to digital nodes or vice versa. If both analog and digital nodes are specified in the same circuit file, PSpice calls one of the A-to-D or D-to-A subcircuits specified in the device's I/O model. These nodes play no part when a device is connected only to digital nodes. For any of the A-to-D or D-to-A interfaces, PSpice automatically creates one subcircuit with a digital power supply. PSpice always uses node 0 as the analog reference node, GND. The digital power and ground nodes default to global nodes named $G_DPWR and $G_DGND. The default analog output is 5.0 V.

One could create user-defined digital power and ground node names and the desired voltage by calling the subcircuit DIGIFPWR, as in the following example:

```
XPOWER  0  P_PWR  N_GND  DIGIFPWR PARAMS: VOLTAGE=3.5V
```

The listing of the subcircuit DIGIFPWR in the digital library is given below.

```
.SUBCKT   DIGIFPWR   AGND
+       OPTIONAL DPWR=$G_DPWR DGND=$G_DGND
+       PARAMS:   VOLTAGE=5.0v REFERENCE=0v
*
VDPWR    DPWR   DGND   {VOLTAGE}
R1       DPWR   AGND   1MEG
VDGND    DGND   REF    {REFERENCE}
R2       REF    AGND   1E-6
R3       DGND   AGND   1MEG
.ENDS
```

The ⟨node⟩* expression specifies one or more input and output nodes, depending upon the primitive type and its parameters. If both analog and digital devices are connected to a node, then PSpice automatically inserts an interface subcircuit and converts between logic levels and voltages. The node names shown in Table 11-5 are reserved and cannot be used.

The ⟨timing model name⟩ describes the device's timing characteristics, such as propagation delay and setup and hold times. Each timing parameter has minimum, typical, and maximum values that may be selected with the optional MNTYMXDLY device parameter. Each primitive type has a specific timing model and model parameters, discussed in Section 11-4 on Digital Gates and Timing Models.

One or more types of parameters can be specified in a timing model, such as propagation delay (TP), setup times (TSU), hold times (TH), pulse widths (TW), and switching times (TSW). Each parameter can have three values: minimum (MN), typical (TY), and maximum (MX). The typical high-to-low propagation delay on a gate is specified by TPHLTY, and the maximum data-to-clock setup time on a flip-flop is specified by TSUDCLKMX. For example, the timing model for an edge-triggered flip-flop is described by

```
.MODEL   D_393_1   ueff  (
+    tppcqhlty=18ns      tppcqhlmx=33ns
+    tpclkqlhty=6ns      tpclkqlhmx=14ns
+    tpclkqhlty=7ns      tpclkqhlmx=14ns
+    twclkhmn=20ns       twclklmn=20ns
+    twpclmn=20ns        tsudclkmn=25ns
+    )
```

TABLE 11-5 RESERVED NODE NAMES

Reserved node names	Value	Description
0	0 volts	Analog ground
$D_HI	1	Digital high-level
$D_LO	0	Digital low-level
$D_X	X	Digital unknown level

The ⟨*I/O model name*⟩ describes the device's loading and driving characteristics. The I/O model expression contains the names of up to four D-to-A and A-to-D interfacing subcircuits. The I/O models are common to device families. There are four I/O models for the 74LS family:

IO_LS   For standard inputs and outputs
IO_LS_OC   For standard inputs and open-collector outputs
IO_LS_ST   For Schmitt-trigger inputs and standard outputs
IO_LS_ST_OC   For Schmitt-trigger inputs and open-collector outputs

The statement for an I/O model has the following general form:

```
.MODEL  ⟨I/O model name⟩  UIO [model parameters]
```

where the model parameters are those described in Table 11-6.

TABLE 11-6 DIGITAL I/O MODEL PARAMETERS

UIO model parameters	Description
INLD	Input load capacitance
OUTLD	Output load capacitance
DRVH	Output high-level resistance
DRVL	Output low-level resistance
AtoD1 (Level 1)	Name of A-to-D interface subcircuit
DtoA1 (Level 1)	Name of D-to-A interface subcircuit
AtoD2 (Level 2)	Name of A-to-D interface subcircuit
DtoA2 (Level 2)	Name of D-to-A interface subcircuit
AtoD3 (Level 3)	Name of A-to-D interface subcircuit
DtoA3 (Level 3)	Name of D-to-A interface subcircuit
AtoD4 (Level 4)	Name of A-to-D interface subcircuit
DtoA4 (Level 4)	Name of D-to-A interface subcircuit
DIGPOWER	Name of power supply subcircuit
TSWLH1	Switching time: low to high for D-to-A1
TSWLH2	Switching time: low to high for D-to-A2
TSWLH3	Switching time: low to high for D-to-A3
TSWLH4	Switching time: low to high for D-to-A4
TSWHL1	Switching time: high to low for D-to-A1
TSWHL2	Switching time: high to low for D-to-A2
TSWHL3	Switching time: high to low for D-to-A3
TSWHL4	Switching time: high to low for D-to-A4

INLD and OUTLD are used for calculating the loading capacitance and the propagation delay. DRVH and DRVL measure the strength of the output signals. A-to-D1 through A-to-D4 and D-to-A1 through D-to-A4 hold the names of interface subcircuits. If one of the A-to-D or D-to-A interface subcircuits is specified, DIGPOWER specifies the name of the digital power supply that PSpice should call.

The switching times, TSWLH*n* and TSWHL*n*, can be used to compensate for the time it takes for the D-to-A device to change its output voltage from its current level to that of the switching threshold. TSWLH*n* and TSWHL*n* are subtracted from a device's propagation delay on the outputs. This allows the analog signal to reach the switching threshold at the time of the digital transition. TSWLH*n* and TSWHL*n* are used only if the output drives an analog node. An example of an I/O model statement follows.

```
.MODEL IO_STD uio (
+       drvh=96.4              drvl=104
+       AtoD1="AtoD_STD"       AtoD2="AtoD_STD_NX"
+       AtoD3="AtoD_STD_E"     AtoD4="AtoD_STD_NXE"
+       DtoA1="DtoA_STD"       DtoA2="DtoA_STD"
+       DtoA3="DtoA_STD"       DtoA4="DtoA_STD"
```

The expression [*MNTYMXDLY* = ⟨*delay select value*⟩] is an optional device parameter and selects either the minimum, typical, or maximum delay values from the device timing model. The valid values are

0 = the current value of .OPTIONS DIGMNTYMX (default = 2)

1 = minimum

2 = typical

3 = maximum

If not specified, MNTYMXDLY defaults to 0.

The expression [*IO_LEVEL* = ⟨*interface subcircuit select value*⟩] is an optional parameter and selects one of the following four A-to-D or D-to-A interface subcircuits from the device's I/O model.

0 = the current value of .OPTIONS DIGIOLVL (default = 1)

1 = AtoD1/DtoA1

2 = AtoD2/DtoA2

3 = AtoD3/DtoA3

4 = AtoD4/DtoA4

The default value of IO_LEVEL is 0.

Examples of Statements for Digital Devices

```
U1   NAND(2)   $G_DPWR   $G_DGND   1   2    5   DO-GATE   IO-DFT
U2   JKFF(1)   $G_DPWR   $G_DGND   3   5   20   3   3   10   2   D_293ASTD   IO_STD
U3   INV       $G_DPWR   $G_DGND   IN   OUT   D_INV   MNTYMXDLY = 2   IO_LEVEL = 3
```

The 74393 part for the TTL digital 4-bit binary counter is defined as a subcircuit in the digital library and consists of U devices, as shown below.

```
.SUBCKT  74393   A   CLR   QA   QB   QC   QD
+     OPTIONAL: DPWR=$G_DPWR DGND=$G_DGND
+     PARAMS: MNTYMXDLY=0 IO_LEVEL=0
UINV INV DPWR DGND
+     CLR    CLRBAR
+     DO_GATE IO_STD IO_LEVEL={IO_LEVEL}
U1 JKFF(1) DPWR DGND
+     $D_HI CLRBAR A    $D_HI $D_HI   QA_BUF $D_NC
+     D_393_1 IO_STD MNTYMXDLY={MNTYMXDLY} IO_LEVEL={IO_LEVEL}
U2 JKFF(1) DPWR DGND
+     $D_HI CLRBAR QA_BUF    $D_HI $D_HI   QB_BUF $D_NC
+     D_393_2 IO_STD MNTYMXDLY={MNTYMXDLY}
U3 JKFF(1) DPWR DGND
+     $D_HI CLRBAR QB_BUF    $D_HI $D_HI   QC_BUF $D_NC
+     D_393_2 IO_STD MNTYMXDLY={MNTYMXDLY}
U4 JKFF(1) DPWR DGND
+     $D_HI CLRBAR QC_BUF    $D_HI $D_HI   QD_BUF $D_NC
+     D_393_3 IO_STD MNTYMXDLY={MNTYMXDLY}
UBUFF BUFA(4) DPWR DGND
+     QA_BUF QB_BUF QC_BUF QD_BUF   QA QB QC QD
+     D_393_4 IO_STD MNTYMXDLY={MNTYMXDLY} IO_LEVEL={IO_LEVEL}
.ENDS
```

11-4 DIGITAL GATES AND TIMING MODELS

Logic gates can be simple gates and gate arrays. Simple gates have one or more inputs and only one output. Gate arrays contain one or more simple gates in one component. Simple gates can be standard and tri-state. A standard gate is always enabled, but a tri-state gate requires an enable control.

11-4.1 Standard Gates

The symbols and truth tables of eight digital gates are shown in Table 11-7.

The general format for a standard gate is

```
U⟨name⟩ ⟨gate type⟩ [(⟨parameter value⟩*)]
+    ⟨digital power node⟩ ⟨digital ground node⟩
+    ⟨input node⟩* ⟨output node⟩*
+    ⟨timing model name⟩  ⟨I/O model name⟩
+    [MNTYMXDLY = ⟨delay select value⟩]
+    [IO_LEVEL = ⟨interface subckt select value⟩]
```

The standard gate types and their parameters are listed in Table 11-8.

TABLE 11-7 DIGITAL GATES

Name	Symbol	Bolean function	Truth table		

Name	Symbol	Bolean function			
				B	Y
Inverter	A ▷o Y	$Y = A_-$		0	1
				1	0
				B	Y
Buffer	A ▷ Y	$Y = A$		0	0
				1	1
			A	B	Y
AND		$Y = AB$	0	0	0
			0	1	0
			1	0	0
			1	1	1
			A	B	Y
NAND		$Y = (AB)_-$	0	0	1
			0	1	1
			1	0	1
			1	1	0
			A	B	Y
OR		$Y = A + B$	0	0	0
			0	1	1
			1	0	1
			1	1	1
			A	B	Y
NOR		$Y = (A + B)_-$	0	0	0
			0	1	1
			1	0	1
			1	1	1
			A	B	Y
Exclusive-OR (XOR)		$Y = A + B$	0	0	0
			0	1	1
			1	0	1
			1	1	0
			A	B	Y
Exclusive-NOR (XNOR)		$Y = A \cdot B$	0	0	1
			0	1	0
			1	0	0
			1	1	1

TABLE 11-8 STANDARD GATE TYPES

Type	Parameters	Nodes	Description
BUF		in, out	Buffer
INV		in, out	Inverter
AND	(⟨no. of inputs⟩)	in*, out	AND gate
NAND	(⟨no. of inputs⟩)	in*, out	NAND gate
OR	(⟨no. of inputs⟩)	in*, out	OR gate
NOR	(⟨no. of inputs⟩)	in*, out	NOR gate
XOR		in1, in2, out	Exclusive OR gate
NXOR		in1, in2, out	Exclusive NOR gate
BUFA	(⟨no. of gates⟩)	in*, out*	Buffer array
INVA	(⟨no. of gates⟩)	in*, out*	Inverter array
ANDA	(⟨no. of inputs⟩, ⟨no. of gates⟩)	in*, out*	AND gate array
NANDA	(⟨no. of inputs⟩, ⟨no. of gates⟩)	in*, out*	NAND gate array
ORA	(⟨no. of inputs⟩, ⟨no. of gates⟩)	in*, out*	OR gate array
NORA	(⟨no. of inputs⟩, ⟨no. of gates⟩)	in*, out*	NOR gate array
XORA	(⟨no. of gates⟩)	in*, out*	Exclusive OR gate array
NXORA	(⟨no. of gates⟩)	in*, out*	Exclusive NOR gate array
AO	(⟨no. of inputs⟩, ⟨no. of gates⟩)	in*, out	AND-OR compound gate
OA	(⟨no. of inputs⟩, ⟨no. of gates⟩)	in*, out	OR-AND compound gate
AOI	(⟨no. of inputs⟩, ⟨no. of gates⟩)	in*, out	AND-NOR compound gate
OAI	(⟨no. of inputs⟩, ⟨no. of gates⟩)	in*, out	OR-NAND compound gate

The value in ⟨*no. of inputs*⟩ is the number of inputs per gate; ⟨*no. of gates*⟩ is the number of gates. With the asterisk, "in*" and "out*" can be one or more nodes, but "in" and "out" refer to only one node.

In gate arrays, the input nodes come first, followed by the output nodes. The total number of input nodes is ⟨no. of inputs⟩·⟨no. of gates⟩. The number of output nodes is ⟨no. of gates⟩.

A compound gate has two gate levels: first level and second level. All inputs are connected to the first-level gate, and the second-level gate gives the output. Thus, all of the input nodes are followed by one output node.

The general format for a timing model of a standard gate is

```
.MODEL  ⟨timing model name⟩  UGATE [model parameters]
```

where the model parameters are given in Table 11-9.

TABLE 11-9 TIMING MODEL PARAMETERS OF STANDARD GATES

Model parameters	Descriptions	Units	Default
TPLHMN	Delay: low-to-high, minimum	sec	0
TPLHTY	Delay: low-to-high, nominal	sec	0
TPLHMX	Delay: low-to-high, maximum	sec	0
TPHLMN	Delay: high-to-low, minimum	sec	0
TPHLTY	Delay: high-to-low, nominal	sec	0
TPHLMX	Delay: high-to-low, maximum	sec	0

Examples of Some Standard Gates

```
U5   AND(2)  $G_DPWR  $G_DGND  IN0  IN1  OUT   ; Two-input AND gate
+        T_AND2  IO_STD
U2   INV  $G_DPWR  $G_DGND   3   5           ; Simple inverter
+        T_INV   IO_STD
U13  NANDA(2,4)  $G_DPWR   $G_DGND           ; 4 two-input NAND gates
+        INA0 INA1 INB0 INB1 INC0 INC1
+        IND0 IND1 OUTA OUTB OUTC OUTD
+        T_NANDA IO_STD
.MODEL T_AND2 UGATE (                        ; AND2 timing model
+        TPLHMN=15ns TPLHTY=20ns TPLHMX=25ns
+        TPHLMN=10ns TPHLTY=15ns TPLHMX=20ns
+        )
```

11-4.2 Tri-State Gates

The general format for a tri-state gate is

```
U⟨name⟩  ⟨tri-state gate type⟩  [(⟨parameter value⟩*)]
+        ⟨digital power node⟩  ⟨digital ground node⟩
+        ⟨input node⟩*  ⟨enable node⟩  ⟨output node⟩*
+        ⟨timing model name⟩   ⟨I/O model name⟩
+        [MNTYMXDLY = ⟨delay select value⟩]
+        [IO_LEVEL = ⟨interface subckt select value⟩]
```

The tri-state gate types and their parameters are listed in Table 11-10.

TABLE 11-10 STANDARD GATE TYPES

Type	Parameters	Nodes	Description
BUF3		in, en, out	Buffer
INV3		in, en, out	Inverter
AND3	(⟨no. of inputs⟩)	in*, en, out	AND gate
NAND3	(⟨no. of inputs⟩)	in*, en, out	NAND gate
OR3	(⟨no. of inputs⟩)	in*, en, out	OR gate
NOR3	(⟨no. of inputs⟩)	in*, en, out	NOR gate
XOR3		in1, in2, en, out	Exclusive OR gate
NXOR3		in1, in2, en, out	Exclusive NOR gate
BUF3A	(⟨no. of gates⟩)	in*, en, out*	Buffer array
INV3A	(⟨no. of gates⟩)	in*, en, out*	Inverter array
AND3A	(⟨no. of inputs⟩, ⟨no. of gates⟩)	in*, en, out*	AND gate array
NAND3A	(⟨no. of inputs⟩, ⟨no. of gates⟩)	in*, en, out*	NAND gate array
OR3A	(⟨no. of inputs⟩, ⟨no. of gates⟩)	in*, en, out*	OR gate array
NOR3A	(⟨no. of inputs⟩, ⟨no. of gates⟩)	in*, en, out*	NOR gate array
XOR3A	(⟨no. of gates⟩)	in*, en, out*	Excl. OR gate array
NXOR3A	(⟨no. of gates⟩)	in*, en, out*	Excl. NOR gate array

The value in *⟨no. of inputs⟩* is the number of inputs per gate; *⟨no. of gates⟩* is the number of gates. With the asterisk, "in*" and "out*" can be one or more nodes,

---

but "in" and "out" refer to only one node, and "en" refers to the output enable node.

In gate arrays, the input nodes come first, followed by the output nodes. The total number of input nodes is ⟨no. of inputs⟩·⟨no. of gates⟩ + 1. The number of output nodes is ⟨no. of gates⟩.

The general format for a timing model of a tri-state gate is

```
MODEL  ⟨timing model name⟩  UTGATEA [model parameters]
```

where the model parameters are given in Table 11-11.

TABLE 11-11 TIMING MODEL PARAMETERS OF TRI-STATE GATES

Model parameters	Descriptions	Units	Default
TPLHMN	Delay: low-to-high, min	sec	0
TPLHTY	Delay: low-to-high, nom	sec	0
TPLHMX	Delay: low-to-high, max	sec	0
TPHLMN	Delay: high-to-low, min	sec	0
TPHLTY	Delay: high-to-low, nom	sec	0
TPHLMX	Delay: high-to-low, max	sec	0
TPLZMN	Delay: low-to-Z, min	sec	0
TPLZTY	Delay: low-to-Z, nom	sec	0
TPLZMX	Delay: low-to-Z, max	sec	0
TPHZMN	Delay: high-to-Z, min	sec	0
TPHZTY	Delay: high-to-Z, nom	sec	0
TPHZMX	Delay: high-to-Z, nom	sec	0
TPZLMN	Delay: Z-to-low, min	sec	0
TPZLTY	Delay: Z-to-low, nom	sec	0
TPZLMX	Delay: Z-to-low, max	sec	0
TPZHMN	Delay: Z-to-high, min	sec	0
TPZHTY	Delay: Z-to-high, nom	sec	0
TPZHMX	Delay: Z-to-high, max	sec	0

Examples of Some Tri-State Gates

```
U5    AND3(2)  $G_DPWR  $G_DGND  IN0  IN1  ENABLE  OUT         ; Two-input AND
+     T_TRIAND2  IO_STD
U2    INV3  $G_DPWR  $G_DGND   2  10   5                       ; Inverter
+        T_TRIINV   IO_STD
U13   NAND3A(2,4)  $G_DPWR  $G_DGND                            ; Four two-input NAND
+     INA0   INA1   INB0   INB1   INC0   INC1   IND0   IND1
+     ENABLE  OUTA  OUTB  OUTC  OUTD
+     T_TRINAND  IO_STD
.MODEL   T_TRIAND2   UTGATE(                                   ; TRI-AND2 timing model
+     TPLHMN=15ns  TPLHTY=20ns  TPLHMX=25ns

...

+     TPHLMN=10ns  TPHLTY=15ns  TPLHMX=20ns
+     )
```

11-4.3 Bidirectional Transfer Gate

The bidirectional transfer gate is a passive device that connects or disconnects two nodes. The state of the gate input controls whether the gate connects the two nodes. There are two types of transfer gate: NBTG and PBTG. The type NBTG connects the nodes if the gate is 1 and disconnects the nodes if the gate is 0. The type PBTG connects the nodes if the gate is 0 and disconnects them if the gate is 1. The I/O model DRVH and DRVL parameters are used as a ceiling on the strength of a 1 or a 0 that is passed through a bidirectional transfer gate. This transfer gate has no parameters and no load-independent delay or charge storage.

The general formats for the two types of bidirectional transfer gate follow.

```
U⟨name⟩   NBTG
+    ⟨digital power node⟩   ⟨digital ground node⟩
+    ⟨gate node⟩ ⟨channel node 1⟩   ⟨channel node 2⟩
+    ⟨timing model name⟩   ⟨I/O model name⟩
+    [MNTYMXDLY = ⟨delay select value⟩]
+    [IO_LEVEL = ⟨interface subckt select value⟩]
U⟨name⟩   PBTG
+    ⟨digital power node⟩   ⟨digital ground node⟩
+    ⟨gate node⟩ ⟨channel node 1⟩   ⟨channel node 2⟩
+    ⟨timing model name⟩ ⟨I/O model name⟩
+    [MNTYMXDLY = ⟨delay select value⟩]
+    [IO_LEVEL = ⟨interface subckt select value⟩]
```

Examples of Bidirectional Transfer Gates

```
U4 NBTG $G_DPWR $G_DGND GATE SD1 SD2
+BTG1 IO_BTG
.MODEL BTG1 UBTG (
+TONMN=10NS        TONTY-15NS        TONMX-20NS
+TOFFMN=5NS        TOFFTY=10NS       TOFFMX=15NS)
```

The general format for a timing model of a bidirectional transfer gate is

```
.MODEL   ⟨timing model name⟩   UBTG [model parameters]
```

where the model parameters are given in Table 11-12.

TABLE 11-12 TIMING MODEL PARAMETERS OF BIDIRECTIONAL TRANSFER GATES

Model parameters	Descriptions	Units	Default
TONMN	turn-on time, minimum	sec	0
TONTY	turn-on time, typical	sec	0
TONMX	turn-on time, maximum	sec	0
TOFFMN	turn-on time, minimum	sec	0
TOFFTY	turn-on time, typical	sec	0
TOFFMX	turn-on time, maximum	sec	0

PSpice supports both edge-triggered and gated flip-flops. An edge-triggered flip-flop changes its state when the clock changes. This occurs on the falling edge for JKFFs and on the rising edge for DFFs of the clock signal. The state of gated flip-flops follows the input as long as the clock (gate) is high. The state is "frozen" when the clock (gate) falls. Gated flip-flops are also referred to as *latches*.

PSpice initializes all flip-flops to the unknown state X at the beginning of each simulation. Each flip-flop will remain in the unknown state until explicitly set or cleared. The X startup state can be overridden by setting .OPTIONS DIGINITSTATE to either 0 or 1. If set to 0, all flip-flops and latches in the circuit will be cleared; setting to 1 will preset all devices. It should be noted that with initial setting to 0 or 1, the device can still output an X at the beginning of the simulation if the inputs would normally produce an X on the output.

11-5.1 Edge-Triggered Flip-Flops

The general format for an edge-triggered flip-flop is

```
U⟨name⟩  ⟨flip-flop type⟩  (⟨no. of flip-flops⟩)
+        ⟨digital power node⟩   ⟨digital ground node⟩
+        ⟨preset node⟩  ⟨clear node⟩  ⟨clock node⟩
+        ⟨input node⟩*  ⟨output node⟩*
+        ⟨timing model name⟩   ⟨I/O model name⟩
+        [MNTYMXDLY = ⟨delay select value⟩]
+        [IO_LEVEL = ⟨interface subckt select value⟩]
```

The value in ⟨*no. of flip-flops*⟩ specifies the number of flip-flops in the device. The edge-triggered flip-flop types and their parameters are listed in Table 11-13.

TABLE 11-13 EDGE-TRIGGERED FLIP-FLOP TYPES

Type	Nodes	Description
JKFF	preb clrb clkb j* k* q* qb*	J-K, negative-edge-triggered
DFF	preb clrb clk d* q* qb*	D-type, positive-edge-triggered

In the "Nodes" column, "preb," "clrb," and "clk" ("clkb") are preset, clear, and clock nodes, respectively: "preb," "clkb," and "clkb" are active low; "clk" is active high; "j" and "k" are the inputs, and "q" and "qb" (q-bar) are the outputs for JKFFs; "d" is the input, and "d," "q," and "qb" (q-bar) are the outputs for DFF.

The general format for a timing model for an edge-triggered flip-flop is

```
.MODEL ⟨timing model name⟩   UEFF [model parameters]
```

where the model parameters are given in Table 11-14.

TABLE 11-14 TIMING MODEL PARAMETERS OF EDGE-TRIGGERED FLIP-FLOPS

Model parameters	Descriptions	Units	Default
TPPCQLHMN	Delay: preb/clrb to q/qb low-to-hi, min	sec	0
TPPCQLHTY	Delay: preb/clrb to q/qb low-to-hi, typ	sec	0
TPPCQLHMX	Delay: preb/clrb to q/qb low-to-hi, max	sec	0
TPPCQHLMN	Delay: preb/clrb to q/qb hi-to-low, min	sec	0
TPPCQHLTY	Delay: preb/clrb to q/qb hi-to-low, typ	sec	0
TPPCQHLMX	Delay: preb/clrb to q/qb hi-to-low, max	sec	0
TWPCLMN	Min preb/clrb width low, min	sec	0
TWPCLTY	Min preb/crib width low, nom	sec	0
TWPCLMX	Min preb/clrb width low, max	sec	0
TPCLKQLHMN	Delay: clk/clkb edge to q/qb low-to-hi, min	sec	0
TPCLKQLHTY	Delay: clk/clkb edge to q/qb low-to-hi, typ	sec	0
TPCLKQLHMX	Delay: clk/clkb edge to q/qb low-to-hi, max	sec	0
TPCLKQHLMN	Delay: clk/clkb edge to q/qb hi-to-low, min	sec	0
TPCLKQHLTY	Delay: clk/clkb edge to q/qb hi-to-low, typ	sec	0
TPCLKQHLMX	Delay: clk/clkb edge to q/qb hi-to-low, max	sec	0
TWCLKLMN	Min clk/clkb width low, min	sec	0
TWCLKLTY	Min clk/clkb width low, nom	sec	0
TWCLKLMX	Min clk/clkb width low, max	sec	0
TWCLKHMN	Min clk/clkb width hi, min	sec	0
TWCLKHTY	Min clk/clkb width hi, nom	sec	0
TWCLKHMX	Min clk/clkb width hi, max	sec	0
TSUDCLKMN	Setup: j/k/d to clk/clkb edge, min	sec	0
TSUDCLKTY	Setup: j/k/d to clk/clkb edge, typ	sec	0
TSUDCLKMX	Setup: j/k/d to clk/clkb edge, max	sec	0
TSUPCCLKHMN	Setup: preb/clrb hi to clk/clkb edge, min	sec	0
TSUPCCLKHTY	Setup: preb/clrb hi to clk/clkb edge, typ	sec	0
TSUPCCLKHMX	Setup: preb/clrb hi to clk/clkb edge, typ	sec	0
THDCLKMN	Hold: j/k/d after clk/clkb edge, min	sec	0
THDCLKTY	Hold: j/k/d after clk/clkb edge, typ	sec	0
THDCLKMX	Hold: j/k/d after clk/clkb edge, max	sec	0

Examples of Some Edge-Triggered Flip-Flops

```
U5   JKFF(1) $G_DPWR  $G_DGND  PREBAR  CLRBAR  CLKBAR ; One JK flip-flop
+    J   K   Q   QBAR
+    T_JKFF   IO_STD
U2   DFF(2) $G_DPWR $G_DGND PREBAR CLKBAR CLRBAR CLK   ; Two DFF flip-flops
+    D0  D1  Q0  Q1  QBAR0  QBAR1
+    T_DFF   IO_STD
.MODEL  T_JKFF   UEFF(...)                             ; JK timing model
```

11-5.2 Gated Latches

The general format for a gated latch is

```
U⟨name⟩  ⟨flip-flop type⟩  (⟨no. of flip-flops⟩)
+       ⟨digital power node⟩    ⟨digital ground node⟩
+       ⟨preset node⟩   ⟨clear node⟩   ⟨clock node⟩
+       ⟨input node⟩*    ⟨output node⟩*
+       ⟨timing model name⟩   ⟨I/O model name⟩
+       [MNTYMXDLY = ⟨delay select value⟩]
+       [IO_LEVEL = ⟨interface subckt select value⟩]
```

where ⟨*no. of flip-flops*⟩ specifies the number of flip-flops in the device. The gated latch types and their parameters are listed in Table 11-15.

TABLE 11-15   GATED LATCH TYPES

Type	Nodes	Description
SRFF	preb clrb gate s* r* q* qb*	S-R gated latch
DLTCH	preb clrb gate d* q* qb*	D gated latch

As in Table 11-13, "preb," "clrb," and "clk" ("clkb") are preset, clear, and clock nodes, respectively: "preb," "clkb," and "clkb" are active low; "clk" is active high; "s" and "r" are the inputs, and "q" and "qb" (q-bar) are the outputs for SRFFs; "d" is the input, and "q" and "qb" (q-bar) are the outputs for DLTCHs.

The general format for a timing model for a gated latch is

```
.MODEL ⟨timing model name⟩   UGFF [model parameters]
```

where the model parameters are given in Table 11-16.

Some Examples of Gated Latches

```
U5   SRFF(4)  $G_DPWR  $G_DGND  PRESET  CLEAR  GATE  ; Four S-R flip-flops
+    S0  S1  S2  S3  R0  R1  R2  R3
+    Q0  Q1  Q2  Q3  QB0  QB1  QB2  QB3
+    T_SRFF   IO_STD
U2   DLTCH(8)  $G_DPWR  $G_DGND  PRESET  CLEAR  GATE  ; Eight D flip-flops
+    D0  D1  D2  D3  D4  D5  D6  D7
+    Q0  Q1  Q2  Q3  Q4  Q5  Q6  Q7
+    QB0 QB1 QB2 QB3 QB4 QB5 QB6 QB7
+    T_DLTCH  IO_STD
.MODEL T_SRFF  UGFF(...)                          ; SRFF timing model
```

TABLE 11-16 TIMING MODEL PARAMETERS OF GATED LATCHES

Model parameters	Descriptions	Units	Default
TPPCQLHMN	Delay: preb/clrb to q/qb low-to-hi, min	sec	0
TPPCQLHTY	Delay: preb/clrb to q/qb low-to-hi, typ	sec	0
TPPCQLHMX	Delay: preb/clrb to q/qb low-to-hi, max	sec	0
TPPCQHLMN	Delay: preb/clrb to q/qb hi-to-low, min	sec	0
TPPCQHLTY	Delay: preb/clrb to q/qb hi-to-low, typ	sec	0
TPPCQHLMX	Delay: preb/clrb to q/qb hi-to-low, max	sec	0
TWPCLMN	Min preb/clrb width low, min	sec	0
TWPCLTY	Min preb/crib width low, nom	sec	0
TWPCLMX	Min preb/clrb width low, max	sec	0
TPGQLHMN	Delay: gate to q/qb low-to-hi, min	sec	0
TPGQLHTY	Delay: gate to q/qb low-to-hi, typ	sec	0
TPGQLHMX	Delay: gate to q/qb low-to-hi, max	sec	0
TPGQHLMN	Delay: gate to q/qb hi-to-low, min	sec	0
TPGQHLTY	Delay: gate to q/qb hi-to-low, typ	sec	0
TPGQHLMX	Delay: gate to q/qb hi-to-low, max	sec	0
TPDQLHMN	Delay: s/r/d to q/qb low-to-hi, min	sec	0
TPDQLHTY	Delay: s/r/d to q/qb low-to-hi, typ	sec	0
TPDQLHMX	Delay: s/r/d to q/qb low-to-hi, max	sec	0
TPDQHLMN	Delay: s/r/d to q/qb hi-to-low, min	sec	0
TPDQHLTY	Delay: s/r/d to q/qb hi-to-low, typ	sec	0
TPDQHLMX	Delay: s/r/d to q/qb hi-to-low, max	sec	0
TWGHMN	Min gate width hi, min	sec	0
TWGHTY	Min gate width hi, nom	sec	0
TWGHMX	Min gate width hi, max	sec	0
TSUDGMN	Setup: s/r/d to gate edge, min	sec	0
TSUDGTY	Setup: s/r/d to gate edge, typ	sec	0
TSUDGMX	Setup: s/r/d to gate edge, max	sec	0
TSUPCGHMN	Setup: preb/clrb hi to gate edge, min	sec	0
TSUPCGHTY	Setup: preb/clrb hi to gate edge, typ	sec	0
TSUPCGHMX	Setup: preb/clrb hi to gate edge, max	sec	0
THDGMN	Hold: s/r/d after gate edge, min	sec	0
THDGTY	Hold: s/r/d after gate edge, typ	sec	0
THDGMX	Hold: s/r/d after gate edge, max	sec	0

11-6 PULLUP AND PULLDOWN

PSpice allows pulling an output to a 1 level (pullup) by a pullup resistor or a 0 level (pulldown) by a pulldown resistor. The strength of the output is specified by the I/O model.

The general format for pullup and pulldown is

```
U⟨name⟩  ⟨resistor type⟩  (⟨number of resistors⟩)
+    ⟨digital power node⟩    ⟨digital ground node⟩
+    ⟨output node⟩*
+    ⟨I/O model name⟩
+    [IO_LEVEL = ⟨interface subckt select value⟩]
```

The ⟨resistor type⟩ is one of the following:

PULLUP   Pullup resistor array

PULLDN   Pulldown resistor array

The value in ⟨number of resistors⟩ is the number of resistors. PULLUP and PULLDN do not require timing models.

Some Examples of Pullup and Pulldown

```
U5   PULLUP(4)   $G_DPWR   $G_DGND       ; Four pullup resistors
+    BUS0  BUS1  BUS2  BUS3  R1K
U2   PULLDN(1)   $G_DPWR   $G_DGND       ; One pulldown resistor
+    15   R500
```

11-7 DELAY LINE

The output of a delay line is delayed by an amount specified in the timing model. A delay-line statement has only one input and one output node, but no parameters.

The general format for a delay line is

```
U⟨name⟩  DLYLINE
+    ⟨digital power node⟩   ⟨digital ground node⟩
+    ⟨input node⟩  ⟨output node⟩
+    ⟨timing model name⟩   ⟨I/O model name⟩
+    [MNTYMXDLY = ⟨delay select value⟩]
+    [IO_LEVEL = ⟨interface subckt select value⟩]
```

The general format for a timing model of a delay line is

```
.MODEL   ⟨timing model name⟩   UDLY [model parameters]
```

where the model parameters are given in Table 11-17.

TABLE 11-17   TIMING MODEL PARAMETERS OF DELAY LINE

Model parameters	Descriptions	Units	Default
DLYMN	Delay: minimum	sec	0
DLYTY	Delay: typical	sec	0
DLYMX	Delay: maximum	sec	0

Some Examples of Delay Lines

```
U5   DLYLINE   $G_DPWR   $G_DGND   IN   OUT      ; Delay line
+    DLY20NS   IO_STD
.MODEL   DLY20NS   UDLY(                         ; Delay line timing model
+    DLYMN=20ns   DLYTY=20ns   DLYMX=20ns)
```

Stimulus devices are used to apply digital signals to a node. There are two types of stimulus devices: (1) the stimulus generator (STIM), which uses simple commands to generate a wide variety of waveforms; and (2) the file stimulus (FSTIM), which obtains the waveforms from an external file. The stimulus devices do not have a timing model. We shall discuss only the stimulus generator (STIM).

The general format for a stimulus generator is

```
U⟨name⟩  STIM(⟨width⟩, ⟨format array⟩)
+    ⟨digital power node⟩  ⟨digital ground node⟩
+    ⟨node⟩*
+    ⟨I/O model name⟩
+    [IO_LEVEL = ⟨interface subckt select value⟩]
+    [TIMESTEP = ⟨stepsize⟩]
+    ⟨command⟩*
```

The ⟨*width*⟩ specifies the number of signals (nodes) output by the stimulus generator. The ⟨*format array*⟩ is a sequence of digits. Each digit specifies the number of signals (nodes) and must be either a 1 (for binary), 3 (for octal), or 4 (for hexadecimal). The sum of the digits in ⟨format array⟩ must be equal to ⟨width⟩; a 1 represents one output node, and 3 and 4 represent three and four output nodes, respectively. For example, five output nodes can be represented by 11111, 113, or 14. Table 11-18 shows the numbers in binary, octal, and hexadecimal.

If the output has two nodes, the ⟨format array⟩ could be represented in binary, and the stimuli becomes STIM (2, 11). For three output nodes, the ⟨format array⟩ could be in either binary or octal, and the stimuli becomes STIM

TABLE 11-18 BINARY, OCTAL, AND HEXADECIMAL NUMBERS

Decimal (base 10)	Binary (base 2)	Octal (base 8)	Hexadecimal (base 16)
00	0000	00	0
01	0001	01	1
02	0010	02	2
03	0011	03	3
04	0100	04	4
05	0101	05	5
06	0110	06	6
07	0111	07	7
08	1000	10	8
09	1001	11	9
10	1010	12	A
11	1011	13	B
12	1100	14	C
13	1101	15	D
14	1110	16	E
15	1111	17	F

(3, 111) or STIM (3, 3). For four output nodes, the ⟨format array⟩ could be represented in either binary or hexadecimal, and the stimuli becomes STIM (4, 1111) or STIM (4, 4).

For 7 output nodes, the ⟨format array⟩ can be a combination of binary and hexadecimal. The stimuli becomes STIM (7, 1141), in which 4 represents the states of four output signals. That is, nodes 1, 2, and 7 are in binary, and nodes 3, 4, 5, and 6 are represented by hexadecimal. For 16 output nodes, the stimuli can be STIM (16, 4444) in which each 4 represents the states of four output signals.

The ⟨*digital power node*⟩ and ⟨*digital ground node*⟩ are used by the interfacing subcircuits. The ⟨node⟩* specifies one or more node output names of the stimulus generator. The number of nodes must be the same as ⟨width⟩.

The ⟨*I/O model name*⟩ is the name of an I/O model, and describes the driving characteristics of the stimulus generator. I/O models also contain the names of up to four D-to-A interfacing subcircuits. The I/O model named IO_STM in the DIG_IO.LIB library file can be used in many cases. IO_LEVEL is an optional device parameter, discussed in Section 11-3.

TIMESTEP is the number of seconds per clock cycle, or step. Transition times that are specified in clock cycles (with the "C" suffix) are multiplied by this amount to determine the actual time of the transition. If TIMESTEP is not specified, the default is 0 seconds. TIMESTEP has no effect on ⟨time⟩ values, which are specified in seconds (with the "S" suffix).

The expressions in ⟨*command*⟩* describe the stimuli to be generated, using one or more of the following:

⟨ ⟨time⟩ ⟨value⟩ ⟩
⟨ LABEL = ⟨label name⟩ ⟩
⟨ ⟨time⟩ GOTO ⟨label name⟩ ⟨n⟩ TIMES⟩
⟨ ⟨time⟩ GOTO ⟨label name⟩ UNTIL GT ⟨value⟩ ⟩
⟨ ⟨time⟩ GOTO ⟨label name⟩ UNTIL GE ⟨value⟩ ⟩
⟨ ⟨time⟩ GOTO ⟨label name⟩ UNTIL LT ⟨value⟩ ⟩
⟨ ⟨time⟩ GOTO ⟨label name⟩ UNTIL LE ⟨value⟩ ⟩
⟨ ⟨time⟩ INCR BY ⟨value⟩ ⟩
⟨ ⟨time⟩ DECR BY ⟨value⟩ ⟩

The expression ⟨*time*⟩ specifies the time for the new ⟨value⟩, GOTO, or INCR/DECR command to occur. Time values may be stated in seconds or in clock cycles. To specify a time value in clock cycles, the "C" suffix is used; otherwise, the units default to seconds. Times may be absolute, such as 45ns or 10c, or relative to the previous time. Relative time can be specified by a plus-sign prefix (+), such as +5ns or +2c.

The expression ⟨*value*⟩ specifies the value for each node (0, 1, R, F, X, or Z); ⟨value⟩ is interpreted using the ⟨format array⟩. The ⟨*label name*⟩ specifies the name used in GOTO statements. GOTO ⟨label name⟩ will jump to the next non-label statement after the ⟨LABEL = ⟨label name⟩ ⟩ statement, and ⟨n⟩ specifies the number of times to repeat a GOTO loop. An infinite loop can be specified by a −1.

Notes

1. Transitions with absolute times within a GOTO loop will be converted to relative times based on the time of the previous command and the current stepsize.

2. GOTO ⟨label name⟩ must specify a label that has been defined in a previous LABEL = ⟨label name⟩ statement.

3. Times must be in strictly ascending order, except that the transition after a GOTO may be at the same time as the GOTO.

11-9 DIGITAL INPUT

A digital input device can translate logic levels (such as 1, 0, X, Z, R, and F) into representative voltage levels with series resistances. The device is modeled by two time-varying resistors: one from ⟨low-level node⟩ to ⟨interface node⟩, and another from ⟨high-level node⟩ to ⟨interface node⟩. Two capacitors (CLO and CHI) can be used to let the state of the digital signal change exponentially. The circuit model of a digital input is shown in Fig. 11-1. The digital input can be obtained from PSpice Digital Simulation or from a digital file. The digital file can be an output file from another logic simulator. Further details of digital input can be found in the *PSpice Manual* [1].

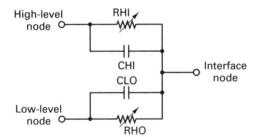

Figure 11-1 Digital input model.

The general format for digital input is

```
N⟨name⟩   ⟨interface node⟩   ⟨low-level node⟩   ⟨high-level node⟩
+         ⟨model name⟩
+         DGTLNET = ⟨digital net name⟩
+         ⟨digital I/O model name⟩
+         [IS = initial state]
```

The general format for digital input files is

```
N⟨name⟩   ⟨interface node⟩ ⟨low-level node⟩ ⟨high-level node⟩
+         ⟨model name⟩
+         [SIGNAME = ⟨digital signal name⟩]
+         [IS = initial state]
```

The general format for the model parameters of digital input is

```
.MODEL ⟨model name⟩  DINPUT [model parameters]
```

where the model parameters are given in Table 11-19.

TABLE 11-19 MODEL PARAMETERS OF DIGITAL INPUT

Model parameters	Descriptions	Units	Default
CLO	Capacitance to low-level node	farad	0
CHI	Capacitance to high-level node	farad	0
SONAME	State "0" character abbreviation		
SOTSW	State "0" switching time	sec	
SORLO	State "0" resistance to low-level node	ohm	
SORHI	State "0" resistance to high-level node	ohm	
S1NAME	State "1" character abbreviation		
S1TSW	State "1" switching time	sec	
S1RLO	State "1" resistance to low-level node	ohm	
S1RHI	State "1" resistance to high-level node	ohm	
S2NAME	State "2" character abbreviation		
S2TSW	State "2" switching time	sec	
S2RLO	State "2" resistance to low-level node	ohm	
S2RHI	State "2" resistance to high-level node	ohm	
.			
.			
.			
S19NAME	State "19" character abbreviation		
S19TSW	State "19" switching time	sec	
S19RLO	State "19" resistance to low-level node	ohm	
S19RHI	State "19" resistance to high-level node	ohm	
FILE	Digital input file name (digital files only)		
FORMAT	Digital input file format (digital files only)		1
TIMESTEP	Digital input file step size (digital files only)	sec	1E−9

Some Examples of Digital Input

```
N1 ANALOG DIGITAL_GND DIGITAL_PWR DIN74 DGTLNET=DIGITAL_NODE IO_STD
NRESET  5  7  12    FROM_TIL
N12     8  0  10    FROM_CMOS SIGNAME=VCO_GATE IS=0
```

11-10 DIGITAL OUTPUT

A digital output device translates analog voltages into digital logic levels (such as 1, 0, X, R, or F). The device is modeled by a resistor and a capacitor, which are connected between ⟨interface node⟩ and ⟨reference node⟩. The circuit model of a digital output is shown in Fig. 11-2. The values of resistors and capacitors are specified in the model statement. Further details of digital input can be found in the *PSpice Manual* [1].

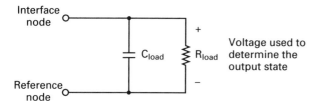

Figure 11-2 Digital output model.

The general format for digital output is

```
O⟨name⟩  ⟨interface node⟩  ⟨reference node⟩  ⟨model name⟩
+       DGTLNET = ⟨digital net name⟩  ⟨digital I/O model name⟩
```

The general format for digital output files is

```
O⟨name⟩  ⟨interface node⟩  ⟨reference node⟩  ⟨model name⟩
+       [SIGNAME = ⟨digital signal name⟩]
```

The general format for the model parameters of digital output is

```
.MODEL ⟨model name⟩  DOUTPUT [model parameters]
```

where the model parameters are given in Table 11-20.

TABLE 11-20 MODEL PARAMETERS OF DIGITAL OUTPUT

Model parameters	Descriptions	Units	Default
RLOAD	Output resistor	ohm	1/GMIN
CLOAD	Output capacitor	farad	0
CHGONLY	0: write each timestep; 1: write upon change		
SONAME	State "0" character abbreviation		
SOVLO	State "0" low-level voltage	volt	
SOVHI	State "0" high-level voltage	volt	
S1NAME	State "1" character abbreviation		
S1RLO	State "1" resistance to low-level voltage	volt	
S1RHI	State "1" resistance to high-level voltage	volt	
S2NAME	State "2" character abbreviation		
S2RLO	State "2" resistance to low-level voltage	volt	
S2RHI	State "2" resistance to high-level voltage	volt	
⋮			
S19NAME	State "19" character abbreviation		
S19RLO	State "19" resistance to low-level voltage	volt	
S19RHI	State "19" resistance to high-level voltage	volt	
FILE	Digital input file name (digital files only)		
FORMAT	Digital input file format (digital files only)		1
TIMESTEP	Digital input file step size (digital files only)	sec	1E−9
TIMESCALE	Scale factor for TIMESTEP (digital files only)		1

```
011   ANALOG_NODE  DIGITAL_GND  DO74  DGTLNET=DIGITAL_NODE  IO_STD
OVCO   7    0   TO_TTL
04    12   10   TO_CMOS  SIGNAME=VCO_OUT
```

11-11 EXAMPLES OF DIGITAL LOGIC CIRCUITS

PSpice allows the creation of library files for many digital devices and the simulation of complex digital logic circuits. The devices in Table 11-3 are available in the library files of the student version of PSpice. We shall use these devices to illustrate the PSpice simulation of digital circuits in the following examples. However, it is possible to develop the characteristics of other digital devices (see the *PSpice Manual* [1]).

Example 11-1

A two-input multiplexer is shown in Fig. 11-3(a). It consists of NOR and INV gates. The digital input and select signals are shown in Fig. 11-3(b). Use PSpice to plot the digital signals A, B, S, and Y from 0 to 50 μs in steps of 0.1 μs.

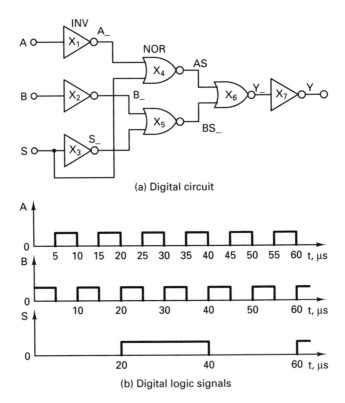

(a) Digital circuit

(b) Digital logic signals

Figure 11-3 Two-input multiplexer.

Solution A multiplexer acts like a rotary switch connecting one of several inputs to a single output. The selection of the input that is to be connected to the output is determined by the binary level of the select (or control) signal. We shall use stimulus devices to generate the logic signals *A*, *B*, and *S*. The listing of the circuit file follows.

Example 11-1   Two-input multiplexer

```
▲ U1 stim(2, 11)   DPWR  D_LO      ; Digital power supply nodes
   +    A     B                    ; Two output signal nodes
   +    IO_STM  ; Use the library digital IO model for a stimulus device.
   +    LABEL = STARTLOOP          ; Loop begins
   +    0US     01                 ; At t= 0, A = 0 and B = 1
   +    +5US    10                 ; 5 μs later, A = 1 and B = 0
   +    +5US  GOTO STARTLOOP -1  TIMES ; 5 μs later, branch to STARTLOOP.
   *    The -1 TIMES causes the loop to repeat indefinitely.
   U2 stim(1, 1)   DPWR  0          ; Digital power supplies
   +    S                          ; One output signal node
   +    IO_STM  ; Use the library digital IO model for a stimulus device.
   +    0US     0                  ; At t=0, S = 0
   +    LABEL = STARTLOOP          ; Loop begins
   +    +20US   1                  ; 20 μs later, S = 1
   +    +20US   0                  ; 20 μs later, S = 0
   +    +20US  GOTO STARTLOOP -1  TIMES  ; 20 μs later, branch to STARTLOOP.
▲▲ X1   A     A_    7404           ; Call statement for inverter 7404
   X2   B     B_    7404           ; Call statement for inverter 7404
   X3   S     S_    7404           ; Call statement for inverter 7404
   X4   A_    S_    AS    7402      ; Call statement for NOR Gate 7402
   X5   B_    S     BS_   7402      ; Call statement for NOR Gate 7402
   X6   AS    BS_   Y_    7402      ; Call statement for NOR Gate 7402
   X7   Y_    Y     7404           ; Call statement for inverter 7404
▲▲▲ .TRAN  0.1US  50US             ; Transient analysis
    .LIB  EVAL.LIB                 ; Calling library file EVAL.LIB
    .PROBE                         ; Graphic post-processor
  .END
```

PSpice plots of digital signals are shown in Fig. 11-4. When the select signal $S = 1$, the input signal *B* will appear at the output; that is, $Y = B$. If the select signal $S = 0$, the input signal *A* will appear at the output; that is, $Y = A$. The logical output *Y* is related to *A* and *B* by $Y = AS + BS_-$.

Note. With the proper choice of the ⟨format array or radices⟩, one could use only one stimulus device to generate *A*, *B*, and *S*. U1 can be modified to have stim(3, 111), and the signal levels at various time intervals should then be represented in three binary digits, with the first digit representing *A*, the second digit *B*, and the third digit *S*.

Example 11-2

A two-bit ripple counter is shown in Fig. 11-5(a). It consists of two J-K flip-flops. The digital clock input and clearbar (clrbar) signals are shown in Fig. 11-5(b). Use PSpice to plot the digital signals: CLK, CLRBAR, *Q1*, and *Q2* from 0 to 80 μs in steps of 0.1 μs.

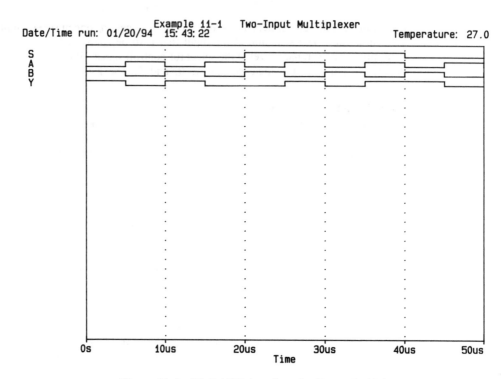

Figure 11-4   Digital PSpice plots for Example 11-1.

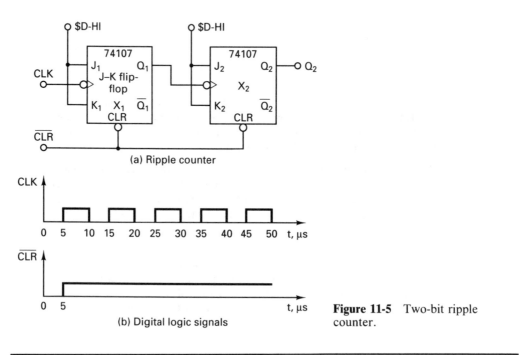

(a) Ripple counter

(b) Digital logic signals

Figure 11-5   Two-bit ripple counter.

Solution   A counter is a sequential circuit that goes through a set of given binary states (e.g., 0 through 1, 0 through 7, 0 through 15, and so on) on successive clock cycles. A counter to sequence from 0 through 7 requires three outputs to represent the three bit positions. Similarly, a counter to sequence from 0 through 15 will require four outputs to represent the four bit positions. We shall use stimulus devices to generate the logic signals CLK and CLRBAR. The listing of the circuit file follows.

Example 11-2   Two-bit ripple counter

```
▲ U1 stim(1, 1) DPWR  D_LO       ; Digital power supply nodes
   +     CLK                      ; Output clock-signal node
   +     IO_STM  ; Use the library digital IO model for a stimulus device.
   +     0US     0                ; At t=0, CLK = 0
   +     LABEL = STARTLOOP        ; Loop begins
   +     +5US    1                ; 5 µs later, CLK = 1
   +     +5US    0                ; 5 µs later, CLK = 0
   +     +5US    GOTO STARTLOOP -1 TIMES ; 5 µs later, branch to STARTLOOP.
   U2 stim(1, 1) DPWR  D_LO       ; Digital power supplies
   +     CLRBAR                   ; Output CLRBAR signal node
   +     IO_STM  ; Use the library digital IO model for a stimulus device.
   +     0NS     0                ; At t=0, CLRBAR = 0
   +     5US     1                ; At t= 5 µs, CLRBAR = 1
   +     10MS    1                ; At t= 10 ms, CLRBAR = 1
▲▲ *  X74107 CLK CLRBAR J K  Q  QBAR  74107 ; Call statement for 74107
      X1  CLK   CLRBAR  $D_HI  $D_HI  Q1  Q1_  74107  ; J-K flip-flop 74107
      X2  Q1    CLRBAR  $D_HI  $D_HI  Q2  Q2_  74107  ; J-K flip-flop 74107
      .LIB  EVAL.LIB            ; Calling library file EVAL.LIB
▲▲▲ .TRAN  0.1US  80US          ; Transient analysis
     .PROBE                     ; Graphic post-processor
   .END
```

PSpice plots of the digital signals are shown in Fig. 11-6. As long as the clearbar (clrbar) is changed from 0 to 1, and remains 1, Q1 switches to 1 at the falling of the clock signal. That is, the frequency of the Q1 signal becomes twice the clock frequency. Since the Q1 signal acts as the clock signal for the flip-flop X2, the Q2 signal switches to 1 at the falling of the Q1 signal. Thus, the frequency of the Q2 signal becomes twice the frequency of the Q1 signal and four times the clock frequency.

Example 11-3

A 3-bit parallel register is shown in Fig. 11-7(a). It consists of two D-type flip-flops. The digital clock inputs and clearbar (clrbar) signals are shown in Fig. 11-7(b). Use PSpice to plot the following digital signals: CLK, CLRBAR, $D1$, $Q1$, $D2$, $Q2$, $D3$, and $Q3$ from 0 to 100 µs in steps of 0.1 µs.

Solution   A register is a sequential logic circuit that can be set to a specific state and retain that state until it is externally changed, usually by a clock signal. The clock input to each flip-flop comes from a common source. On the rising edge of the clock signal, the data on the D input is stored in its flip-flop. Each flip-flop stores a single bit. A register with n bits requires n D-type flip-flops. Since the clock input determines when a flip-flop changes its state, it can be regarded as a load signal. We shall use stimulus devices to generate the logic signals CLK, D-inputs, and CLRBAR.

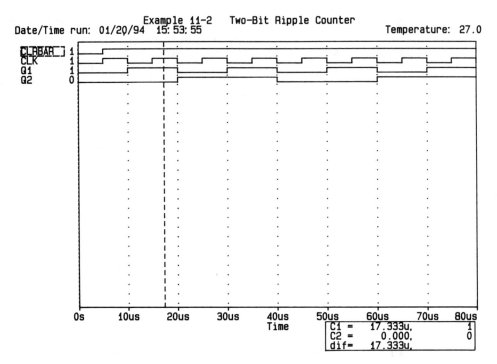

Example 11-2   Two-Bit Ripple Counter

Temperature: 27.0

Figure 11-6   Digital PSpice plots for Example 11-2.

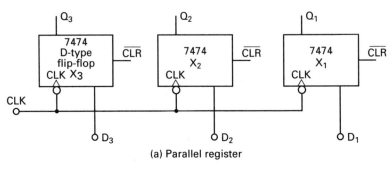

(a) Parallel register

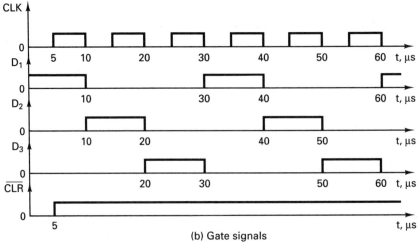

(b) Gate signals

Figure 11-7   Three-bit parallel register.

The listing of the circuit file follows.

Example 11-3   Three-bit parallel register

```
▲ U1 stim(1, 1) DPWR  D_LO        ; Digital power supply nodes
  +    CLK                         ; One output clock-signal node
  +    IO_STM  ; Use the library digital IO model for a stimulus device.
  +    0US      0                  ; At t=0, CLK = 0
  +    LABEL = STARTLOOP           ; Loop begins
  +    +5US     1                  ; 5 μs later, CLK = 1
  +    +5US     0                  ; 5 μs later, CLK = 0
  +    +5US    GOTO STARTLOOP -1  TIMES  ; 5 μs later, branch to STARTLOOP.
  U2 stim(3, 111) DPWR  D_LO       ; Digital power supplies
  +    D3    D2   D1               ; Three output signal nodes
  +    IO_STM  ; Use the library digital IO model for a stimulus device.
  +    LABEL = STARTLOOP           ; Loop begins
  +    0US        001              ; At t = 0, D3 = D2 = 0, and D1 = 1
  +    +10US      010              ; 10 μs later, D3 = D1 = 0, and D2 = 1
  +    +10US      100              ; 10 μs later, D1 = D2 = 0, and D3 = 1
  +    +10US    GOTO STARTLOOP -1  TIMES  ; 10 μs later, branch to STARTLOOP.
  U3 stim(1, 1) DPWR  D_LO         ; Digital power supplies
  +    CLRBAR                      ; Output CLRBAR signal node
  +    IO_STM  ; Use the library digital IO model for a stimulus device.
  +    0NS      0                  ; At t=0, CLRBAR = 0
  +    5US      1                  ; At t = 5 us, CLRBAR = 1
  +    10MS     1                  ; At t = 10 ms, CLRBAR = 1
▲▲ * X7474  1CLRBAR  1D  1CLK  1PREBAR  1Q  1QBAR   7474
     X1  CLRBAR  D1  CLK  $D_HI  Q1  Q1_  7474   ; D-type flip-flop 7474
     X2  CLRBAR  D2  CLK  $D_HI  Q2  D2_  7474   ; D-type flip-flop 7474
     X3  CLRBAR  D3  CLK  $D_HI  Q3  D3_  7474   ; D-type flip-flop 7474
     .LIB  EVAL.LIB                     ; Calling library file EVAL.LIB
▲▲▲ .TRAN  0.1US  100US              ; Transient analysis
     .PROBE                          ; Graphic post-processor
  .END
```

PSpice plots of digital signals are shown in Fig. 11-8. As long as the clearbar (clrbar) is changed from 0 to 1 and remains 1, D-inputs are transferred to the flip-flops, and $Dn = Qn$. The data transfer is done at the rising of the clock signal CLK.

Example 11-4

A ring-oscillator is shown in Fig. 11-9. It consists of an inverter and a Schmitt trigger. Use PSpice to plot the voltage at node 2 and the capacitor voltage across nodes 1 and 3 from 0 to 20 μs in steps of 1 ns.

Solution   The listing of the circuit file follows.

Example 11-4   Ring oscillator

```
▲▲ X1   1    2    7414           ; Subcircuit call for Schmitt trigger 7414
  +    params: IO_LEVEL=3
  *  The use of the IO_LEVEL=3 parameter gives an elaborate IO model,
  *  with clamping diodes on the inputs, and realistic I-V curves.
  *  This is usually important only with capacitively coupled inputs
  *  like this one.
```

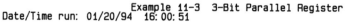

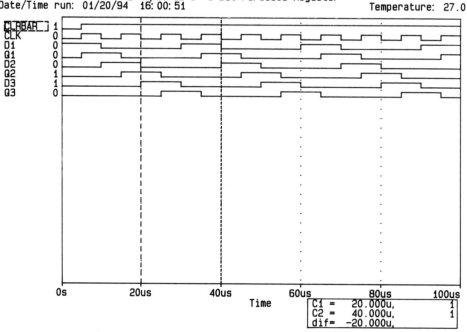

Figure 11-8 Digital PSpice plots for Example 11-3.

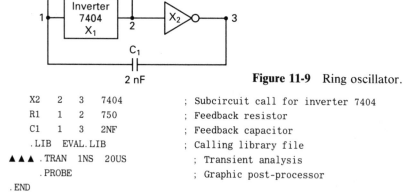

Figure 11-9 Ring oscillator.

```
X2    2   3   7404          ; Subcircuit call for inverter 7404
R1    1   2   750           ; Feedback resistor
C1    1   3   2NF           ; Feedback capacitor
.LIB  EVAL.LIB              ; Calling library file
▲▲▲ .TRAN  1NS  20US         ; Transient analysis
    .PROBE                  ; Graphic post-processor
.END
```

PSpice plots are shown in Fig. 11-10. (V2) is the voltage at node 2, and V(1,3) is the voltage across the capacitor.

Example 11-5

A 4-bit binary counter is shown in Fig. 11-11(a). The digital clock input and set signals are shown in Fig. 11-11(b). Use PSpice to plot the following digital signals: CLK, S, QA, QB, QC, and QD from 0 to 160 μs in steps of 0.1 μs.

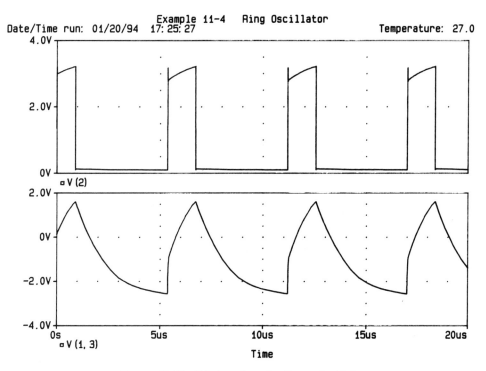

Figure 11-10 PSpice plots for Example 11-4.

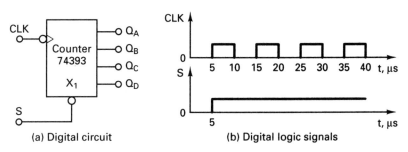

| (a) Digital circuit | (b) Digital logic signals |

Figure 11-11 Four-bit binary counter.

Solution We shall use stimulus devices to generate the logic signals *A* and *S*. The listing of the circuit file follows.

Example 11-5 Four-bit binary counter

```
▲ U1 stim(1, 1)  DPWR  D_LO        ; Digital power supply nodes
  +    CLK                          ; One output clock-signal node
  +    IO_STM  ; Use the library digital IO model for a stimulus device.
  +    0US    0                     ; At t=0, A = 0
  +    LABEL = STARTLOOP            ; Loop begins
```

```
+     +5US    1                         ;  5 µs later, A = 1
+     +5US    0                         ;  5 µs later, A = 0
+     +5US    GOTO STARTLOOP −1  TIMES  ;  5 µs later, branch to STARTLOOP.
U2 stim(1, 1) DPWR  D_LO                ;  Digital power supplies
+     S                                 ;  Output S signal node
+     IO_STM  ;  Use the library digital IO model for a stimulus device.
+     0US     1                         ;  At t=0, S = 1
+     5US     0                         ;  At t = 5 µs, S = 0
+     1MS     0                         ;  At t = 1 ms, S = 0
▲▲ * X74393  A  CLR  QA  QB  QC  QD ;  Call statement for 74393
    X1  A  S  QA  QB  QC  QD  74393  ;  4-bit binary counter 74393
    .LIB  EVAL.LIB                      ;  Calling library file EVAL.LIB
▲▲▲ .TRAN  0.1US  160US                 ;  Transient analysis
    .PROBE                              ;  Graphic post-processor
.END
```

PSpice plots of the digital signals are shown in Fig. 11-12. The states of QA, QB, QC, and QD change at the falling edge of the input signal A, and represent binary numbers corresponding to the number of input pulses.

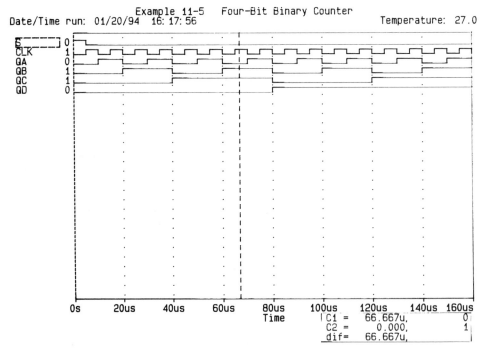

Figure 11-12 Digital PSpice plots for Example 11-5.

Example 11-6

A NAND gate is shown in Fig. 11-13(a). The digital inputs are shown in Fig. 11-13(b). Develop a subcircuit model for the NAND gate, and use PSpice to plot the following digital signals: A, B, and Y from 0 to 50 µs in steps of 0.1 µs.

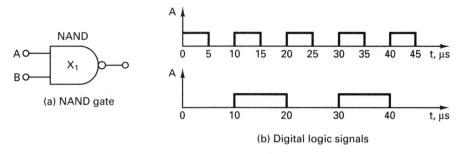

(a) NAND gate

(b) Digital logic signals

Figure 11-13 NAND gate.

Solution We shall use stimulus devices to generate the logic signals *A* and *B*. The listing of the circuit file follows.

Example 11-6 NAND gate

```
▲ U1 stim(1, 1)  DPWR  D_LO      ; Digital power supply nodes
  +    A                         ; Output clock-signal node
  +    IO_STM  ; Use the library digital IO model for a stimulus device.
  +    LABEL = STARTLOOP         ; Loop begins
  +    0US    1                  ; At t = 0, A = 1
  +    +5US   0                  ; 5 µs later, A = 0
  +    +5US   GOTO STARTLOOP -1  TIMES ; 5 µs later, branch to STARTLOOP.
  U2 stim(1, 1)  DPWR  D_LO      ; Digital power supply nodes
  +    B                         ; Output clock-signal node
  +    IO_STM  ; Use the library digital IO model for a stimulus device.
  +    LABEL = STARTLOOP         ; Loop begins
  +    +0US   0                  ; At t = 0, B = 0
  +    +10US  1                  ; 10 µs later, B = 1
  +    +10US  GOTO STARTLOOP -1  TIMES ; 5 µs later, branch to STARTLOOP.
▲▲ X1  A   B   Y   NAND          ; Call two-input NAND gate
   *  Subcircuit definition for NAND gate
   .subckt   NAND   A   B   Y   ;  Two-input NAND gate
   +    optional: DPWR=$G_DPWR DGND=$G_DGND
   +    params: MNTYMXDLY=0 IO_LEVEL=0
   U1  NAND(2)  DPWR   DGND       ; Two-input NAND gate
   +  A   B   Y
   +  T_NAND IO_STD MNTYMXDLY={MNTYMXDLY} IO_LEVEL={IO_LEVEL}
   .MODEL T_NAND UGATE (                        ; NAND timing model
   +    TPLHMN=15ns TPLHTY=20ns TPLHMX=25ns
   +    TPHLMN=10ns TPHLTY=15ns TPLHMX=20ns
   +    )
   .MODEL IO_STD uio (
   +    drvh=96.4    drvl=104
   +    AtoD1="AtoD_STD"    AtoD2="AtoD_STD_NX"
   +    AtoD3="AtoD_STD_E"   AtoD4="AtoD_STD_NXE"
   +    DtoA1="DtoA_STD"   DtoA2="DtoA_STD"
   +    DtoA3="DtoA_STD"   DtoA4="DtoA_STD"
   +    )
   .ENDS NAND                     ; Ends subcircuit definition
```

```
        *
         .LIB  EVAL.LIB                    ; Calling library file EVAL.LIB
    ▲ ▲ ▲ .TRAN  0.1US  50US               ; Transient analysis
         .PROBE                            ; Graphic post-processor
    .END
```

PSpice plots of the digital signals are shown in Fig. 11-14. When both *A* and *B* inputs are high, the output *Y* is low, as expected.

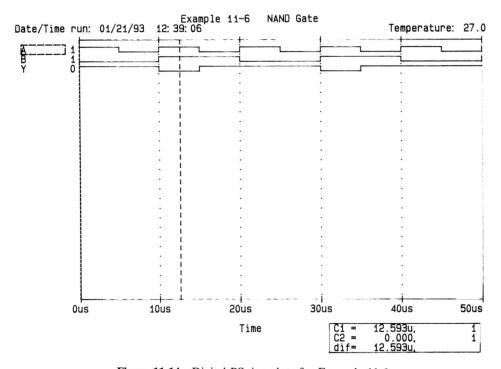

Figure 11-14 Digital PSpice plots for Example 11-6.

SUMMARY

The general format for a digital primitive is

```
U⟨name⟩  ⟨primitive type⟩  [(⟨parameter value⟩*)]
+    ⟨digital power node⟩   ⟨digital ground node⟩
+    ⟨node⟩*
+    ⟨timing model name⟩   ⟨I/O model name⟩
+    [MNTYMXDLY = ⟨delay select value⟩]
+    [IO_LEVEL = ⟨interface subcircuit select value⟩]
```

REFERENCES

1. *PSpice Manual,* Irvine, Calif.: MicroSim Corporation, 1992.
2. J. F. Passafiume and M. Douglas, *Digital Logic Design—Tutorials and Laboratory Exercises.* New York, N.Y.: Harper & Row, 1985.
3. C. H. Roth, Jr., *Fundamentals of Logic Design.* New York, N.Y.: West, 1992.
4. M. M. Mano, *Digital Design.* Englewood Cliffs, N.J.: Prentice Hall, 1984.

PROBLEMS

11-1. The symbols and truth table of eight digital gates are shown in Table 11-7. Use PSpice to verify the truth tables by plotting the digital input and output signals for each gate. Use a stimulus device to generate input signals, and perform transient analysis from 0 to 100 μs in steps of 0.1 μs.

11-2. NOT, AND, or OR can be implemented by NAND gates, as shown in Fig. P11-2. Use PSpice to verify their truth table in Table 11-7 by plotting the digital input and output signals A, B, and Y. Use a stimulus device to generate input signals, and perform transient analysis from 0 to 100 μs in steps of 0.1 μs.

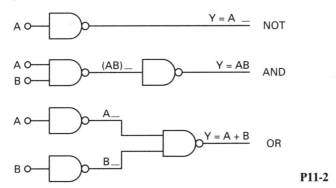

11-3. NOT, OR, or AND can be implemented by NOR gates as shown in Fig. P11-3. Use PSpice to verify their truth table in Table 11-7 by plotting the digital input and output signals A, B, and Y. Use a stimulus device to generate input signals, and perform transient analysis from 0 to 100 μs in steps of 0.1 μs.

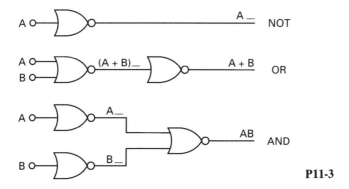

11-4. A half-adder implementation is shown in Fig. P11-4. Use PSpice to verify their truth table by plotting the digital input and output signals A, B, C, and S. Use a stimulus device to generate input signals, and perform transient analysis from 0 to 100 μs in steps of 0.1 μs.

A	B	C	S
0	0	0	0
0	1	0	1
1	0	0	1
1	1	1	0

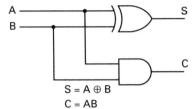

$S = A \oplus B$
$C = AB$

P11-4

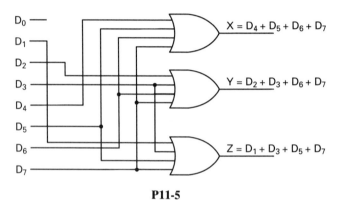

P11-5

11-5. An octal-to-binary encoder is shown in Fig. P11-5. Use PSpice to verify the truth table shown below by plotting the digital input and output signals. Use a stimulus device to generate input signals, and perform transient analysis from 0 to 160 μs in steps of 0.1 μs.

Inputs								Outputs		
D_0	D_1	D_2	D_3	D_4	D_5	D_6	D_7	X	Y	Z
1	0	0	0	0	0	0	0	0	0	0
0	1	0	0	0	0	0	0	0	0	1
0	0	1	0	0	0	0	0	0	1	0
0	0	0	1	0	0	0	0	0	1	1
0	0	0	0	1	0	0	0	1	0	0
0	0	0	0	0	1	0	0	1	0	1
0	0	0	0	0	0	1	0	1	1	0
0	0	0	0	0	0	0	1	1	1	1

11-6. A 4-to-1 line multiplexer is shown in Fig. P11-6. Use PSpice to verify the truth table by plotting the digital input and output signals: S_1, S_0, and Y. Use a stimulus device to generate input signals, and perform transient analysis from 0 to 100 μs in steps of 0.1 μs.

S_1	S_0	Y
0	0	D_0
0	1	D_1
1	0	D_2
1	1	D_3

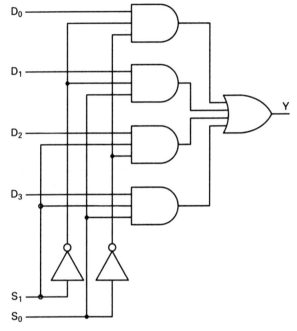

P11-6

11-7. A clocked R-S flip-flop is shown in Fig. P11-7. Use PSpice to verify the truth table by plotting the digital input and output signals: R, S, Q, and $Q(t + 1)$. Use a stimulus device to generate input signals, and perform transient analysis from 0 to 100 μs in steps of 0.1 μs.

Q	R	S	$Q(t + 1)$
0	0	0	0
0	0	1	0
0	1	0	1
0	1	1	X
1	0	0	1
1	0	1	0
1	1	0	1
1	1	1	X

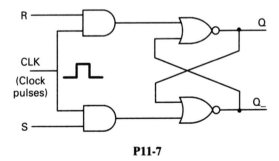

P11-7

11-8. A clocked T flip-flop is shown in Fig. P11-8. Use PSpice to verify the truth table by plotting the digital input and output signals Q, T and $Q(t + 1)$. Use a stimulus device to generate input signals, and perform transient analysis from 0 to 100 μs in steps of 0.1 μs.

Q	T	$Q(t+1)$
0	0	0
0	1	1
1	0	1
1	1	0

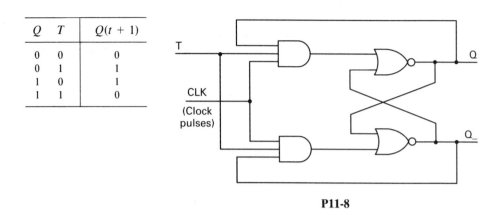

P11-8

11-9. A serial shift-register is shown in Fig. P11-9. The digital clearbar (clrbar) signal is set 1 from 0 after 5 μs. It consists of four D-type flip-flops. Use PSpice to plot the digital signals D, CLK CLRBAR, and S_0 from 0 to 100 μs in steps of 0.1 μs.

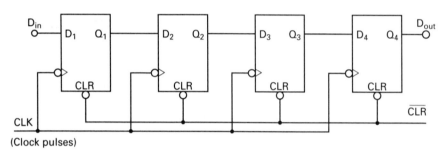

P11-9

11-10. A 3-bit parity generator is shown in Fig. P11-10. It consists of one exclusive-OR gate and one exclusive-NOR gate. Use PSpice to plot the digital signals X, Y, X, and P from 0 to 100 μs in steps of 0.1 μs.

X	Y	Z	P
0	0	0	1
0	0	1	0
0	1	0	0
0	1	1	1
1	0	0	0
1	0	1	1
1	1	0	1
1	1	1	0

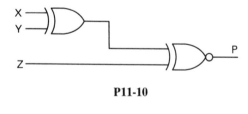

P11-10

Difficulties

12-1 INTRODUCTION

An input file may not run for various reasons, and it is necessary to know what to do when the program does not work. To run a program successfully requires the knowledge of what would not work, why not, and how to fix the problem. There can be many reasons why a program does not work; this chapter covers the problems that are commonly encountered and their solutions. The problems could be due to one or more of the following causes:

Large circuits
Running multiple circuits
Large outputs
Long transient runs
Convergence
Analysis accuracy
Negative component values
Power-switching circuits
Floating nodes
Nodes with fewer than two connections
Voltage source and inductor loops

12-2 LARGE CIRCUITS

The memory requirement (RAM) for the PSpice analysis depends on the size of a circuit file. The peak memory usage (MEMUSE) of a circuit can be found by including the ACCT option in the .OPTIONS statement and by looking at the MEMUSE number in the output file of that circuit. If the circuit does not fit into RAM, PSpice will show "Out of memory."

PSpice sends the results of an analysis (including those for the .PRINT and .PLOT statements) to an output file or one of the temporary files. It should be noted that the results are not sent to RAM.

For DOS, the total available memory can be checked by the command CHKDSK. The possible remedies for a memory problem would be the following:

1. To break the circuit file into pieces and run the pieces separately.
2. To remove other resident software to release enough RAM.

12-3 RUNNING MULTIPLE CIRCUITS

A set of circuits may be placed into one input file and run as a single job. Each circuit should begin with a title statement and end with an .END command. It is important to note that there should not be any blank space or comment line between the .END statement of the preceding circuit and the title line of the following circuit.

PSpice will run all the circuits in the input file and then process each one in sequence. PSpice will store the results in a single output file, which will contain the outputs from each circuit in the order in which they appear in the input file. This feature is most suitable for running a set of large circuits overnight. For example, two circuits are combined into a single input file as follows.

Example 3-4   Transfer-function analysis
```
▲ VIN  1  0  DC  1V   ; Dc input voltage of 1 V
▲▲ R1  1  2  1K
   R2  2  0  20K
   RP  2  6  1.5K
   RE  3  0  250
   F1  4  3  VX    40  ; Current-controlled current-source
   RO  4  3  100K
   RL  4  5  2K
   VX  6  3  DC   0V   ; Measures the current through $R_p$
   VY  5  0  DC   0V   ; Measures the current through $R_L$
▲▲▲ .TF  V(4)  VIN    ; Transfer-function analysis
  .END                ; End of circuit file
```

Example 3-6   Dc sweep
```
▲ VIN   1    0   DC    1V   ; Dc input voltage of 1 V
▲▲ R1   1    2   1K
   R2   2    0   20K
   RP   2    6   1.5K
   RE   3    0   RMOD   250 ; Resistance with model RMOD
  .MODEL  RMOD  RES (R = 1.0) ; Model statement for RE
   F1   4    3   VX     40  ; Current-controlled current-source
```

```
RO     4   3    100K
RL     4   5    2K
VX     6   3    DC    0V     ; Measures the current through R_p
VY     5   0    DC    0V     ; Measures the current through R_L
▲ ▲ ▲ *  Dc sweep for VIN from 0 to 1 V with 0.5 V increment,
      *  using the listed values of parameter R in model RMOD
      .DC  VIN   0   1.5   0.5  RES  RMOD(R)  LIST   0.75  1.0  1.25
      .PRINT  DC  V(1)   V(4)      ; Prints a table in the output file
      .PROBE                       ; Graphical waveform analyzer
  .END                             ; End of circuit file
```

12-4 LARGE OUTPUTS

A large output file will be generated if an input file is run with several circuits, for several temperatures, or with the sensitivity analysis. This will not be a problem with a hard disk. For a PC with floppy disks, the diskette may be filled with the output file. The best solutions for this problem are

1. Direct the output to the printer instead of a file.
2. Direct the output to an empty diskette in a second drive instead of the one containing PSPICE1.EXE by specifying that the PSpice programs are on drive A and the input and output files on drive B. The command to run a circuit file would be: A:PSPICE B:EX2-1.CIR B:EX2-1.OUT.

12-5 LONG TRANSIENT RUNS

Long transient runs can be avoided by the following limit options.

1. LIMPTS limits the number of print steps in a run, and has a default value of 0 (meaning no limit). LIMPTS can be specified as a positive value as high as 32,000. The number of print steps is the final analysis time divided by the print interval time (plus 1).
2. ITL5 is the number of total iterations in a run, and has a default value of 5,000. ITL5 can be set at as high as 2×10^9. It is often convenient to turn it off by setting ITL5=0, which is the effect of setting ITL5 to infinity.
3. The user can limit data points to the Probe limit of 8,000 by suppressing a part of the output at the beginning of the run with a third parameter on the .TRAN statement. For a transient analysis from 0 to 10 ms in steps of 10 μs that should print output only from 8 ms to 10 ms, the command would be .TRAN 10US 10MS 8MS

LIMPTS and ITL5 limits can be set from the "Change Options Analysis" menu or typed in the .OPTIONS statement as follows:

```
.OPTIONS  LIMPTS=6000  ITL5=0.
```

PSpice uses iterative algorithms. These algorithms start with a set of node voltages, and each iteration calculates a new set, which is expected to be closer to a solution of Kirchhoff's voltage and current laws. That is, an initial guess is used, and the successive iterations are expected to converge to the solution. Convergence problems may occur in the following processes:

> Dc sweep
> Bias-point calculation
> Transient analysis

12-6.1 Dc Sweep

If a convergence problem occurs, the analysis fails, and PSpice skips the remaining points of the dc sweep. The convergence problem often occurs in analyzing a circuit with regenerative feedback, such as one with Schmitt triggers. While calculating the hysteresis of such circuits, it is necessary to jump discontinuously from one solution to another at the crossover point, and the analysis fails.

A hysteresis characteristic can be obtained by using transient analysis with a piecewise linear (PWL) voltage source with a very slowly rising ramp. A very slow ramp will cause the input voltage to change slowly until the circuit switches, so that the hysteresis characteristics due to upward and downward switching can be calculated.

Example 12-1

An emitter-coupled Schmitt-trigger circuit is shown in Fig. 12-1(a). Plot the hysteresis characteristics of the circuit from the results of the transient analysis. The input voltage, which is varied slowly from 1 V to 3 V and from 3 V to 1 V, is as shown in Fig. 12-1(b). The model parameters of the transistors are IS=1E−16, BF=50, BR=0.1, RB=50, RC=10, TF=0.12NS, TR=5NS, CJE=0.4PF, PE=0.8, ME=0.4, CJC=0.5PF, PC=0.8, MC=0.333, CCS=1PF, and VA=50. Print the job statistical summary of the circuit.

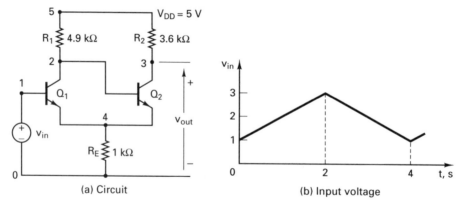

(a) Circuit   (b) Input voltage

Figure 12-1 Schmitt-trigger circuit.

Solution The input voltage is varied very slowly from 1 V to 3 V and from 3 V to 1 V, as shown in Fig. 12-1(b). The listing of the circuit file follows.

Example 12-1   Emitter-coupled trigger circuit
```
*  Dc supply voltage of 5 V
▲ VDD 5  0  DC  5
  *  PWL waveform for transient analysis
  VIN  1  0  PWL (0  1V  2  3V  4  1V)
▲▲ R1  5  2  4.9K
   R2  5  3  3.6K
   RE  4  0  1K
   *  Q1 and Q2 with model QM
   Q1  2  1  4  QM
   Q2  3  2  4  QM
   * Model parameters for QM
   .MODEL  QM NPN (IS=1E-16 BF=50 BR=0.1 RB=50 RC=10 TF=0.12NS TR=5NS
   + CJE=0.4PF PE=0.8 ME=0.4 CJC=0.5PF PC=0.8 MC=0.333 CCS=1PF VA=50)
▲▲▲ * Transient analysis from 0 to 4 s in steps of 0.01 s
    .TRAN  0.01  4
    *  Printing the accounts summary
    .OPTIONS  ACCT
    .PROBE
.END
```

The job statistical summary obtained from the output file is as follows.

```
**** JOB STATISTICS SUMMARY
```

NUNODS	NCNODS	NUMNOD	NUMEL	DIODES	BJTS	JFETS	MFETS	GASFETS
6	6	10	7	0	2	0	0	0
NSTOP	NTTAR	NTTBR	NTTOV	IFILL	IOPS	PERSPA		
12	37	39	13	2	71	72.917		
NUMTTP	NUMRTP	NUMNIT	MEMUSE					
285	55	1293	8906					

	SECONDS	ITERATIONS
MATRIX SOLUTION	20.94	5
MATRIX LOAD	59.29	
READIN	1.54	
SETUP	.05	
DC SWEEP	0.00	0
BIAS POINT	5.54	77
AC and NOISE	0.00	0
TRANSIENT ANALYSIS	122.81	1293
OUTPUT	0.00	
TOTAL JOB TIME	124.90	

The hysteresis characteristics for Example 12-1 are shown in Fig. 12-2.

12-6.2 Bias Point

If the node voltage(s) of a circuit changes very rapidly, PSpice may not find a stable bias point and the calculation fails. This generally occurs in an oscillator

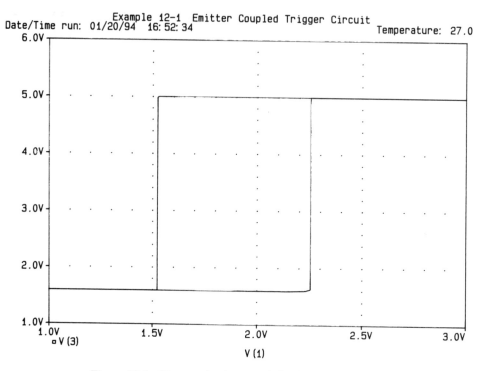

Figure 12-2 Hysteresis characteristics for Example 12-1.

circuit consisting of transistors and capacitors. The failure of the bias point calculation prevents other analyses (e.g., ac analysis, sensitivity, etc.), and the PSpice simulation will stop.

The problems in calculating the bias point can be minimized by giving PSpice initial guesses for node voltages, so that it starts out much closer to the solution. The initial guesses for node voltages can be assigned by the .NODESET statement, e.g., .NODESET V(1)=2.5V. It requires a little judgment in assigning appropriate node voltages. Without any initial guesses, PSpice may not find the bias point of a circuit; but with a carefully selected initial node voltage, it will find the bias point very promptly.

A convergence problem in the bias-point calculation does not generally occur. This is because PSpice contains an algorithm for automatically scaling the power supplies if it is having trouble finding a solution. This algorithm first tries with the full-power supplies. If there is no convergence, and the program cannot find the bias point, then it cuts the power supplies to $\frac{1}{4}$ strength and tries again. If there is still no convergence, then the supplies are cut by another factor of 4 to $\frac{1}{16}$ strength, and so on.

The program will definitely find a solution for some values of the supplies scaled back to 0 V. Once it finds a solution, it then slowly adjusts the power supplies until it finds the bias point at full strength. While this algorithm is in effect, a message such as

Power supplies cut back to 25%
(or some other percentage) appears on the screen.

12-6.3 Transient Analysis

In case of failure due to a convergence problem, the transient analysis skips the remaining time. The remedies available for transient analysis follow:

1. Change the relative accuracy, RELTOL, from 0.001 to 0.01.
2. Set the iteration limits at any point during transient analysis by the ITL4 option. Setting ITL4 = 50 (by the statement .OPTIONS ITL4=50) will allow 50 iterations at each point. More iteration points require a longer simulation time. This is not recommended for circuits that do not have a convergence problem in transient analysis.

12-7 ANALYSIS ACCURACY

The accuracy of PSpice's results is controlled by the following parameters in the .OPTIONS statement:

1. RELTOL controls the relative accuracy of all the voltage and currents that are calculated. The default value of RELTOL is 0.001 (0.1%), which is more accurate than necessary for many applications. The speed can be increased by setting RELTOL = 0.01 (1%), which would increase the average speed by a factor of 1.5.
2. VNTOL can limit the accuracy of all voltages to a finite value; the default value is 1 μV.
3. ABSTOL can limit the accuracy of all currents to a finite value; the default value is 1 μA.
4. CHGTOL can limit the accuracy of capacitor-charges/inductor-fluxes to a finite value.

RELTOL, VNTOL, ABSTOL, and CHGTOL limits can be set from the "Change Options Analysis" menu or typed in the .OPTIONS statement as follows:

```
.OPTIONS   ABSTOL = 5.00U  RELTOL = 0.01   VNTOL = 0.1
```

12-8 NEGATIVE COMPONENT VALUES

PSpice allows negative values for resistors, capacitors, and inductors. If a .NOISE analysis is performed on a circuit with negative values, the noise contribution will be calculated from the absolute values of resistors and will not generate negative noise. However, negative components may cause instabilities during transient analysis.

Example 12-2

A circuit with negative resistances is shown in Fig. 12-3. Calculate the voltage gain, the input resistance, and the output resistance.

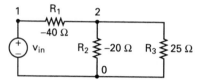

Figure 12-3 A circuit with negative resistances.

Solution The listing of the circuit file follows.

Example 12-2 Circuit with negative components

```
▲ * Dc input voltage of 1 V
  VIN  1  0  DC  1V
  * Negative resistances
  R1   1  2  -40
  R2   2  0  -20
  R3   2  0   25
  * Dc transfer-function analysis
  .TF  V(2)  VIN
.END
```

```
****      SMALL-SIGNAL BIAS SOLUTION         TEMPERATURE =  27.000 DEG C
NODE   VOLTAGE       NODE   VOLTAGE      NODE   VOLTAGE     NODE   VOLTAGE
(   1)   1.0000  (    2)     .7143
       VOLTAGE SOURCE CURRENTS
       NAME           CURRENT
       VIN            7.143E-03
       TOTAL POWER DISSIPATION -7.14E-03  WATTS

****      SMALL-SIGNAL CHARACTERISTICS
       V(2)/VIN =  7.143E-01
       INPUT RESISTANCE AT VIN = -1.400E+02
       OUTPUT RESISTANCE AT V(2) = -2.857E+01
          JOB CONCLUDED
          TOTAL JOB TIME           1.98
```

12-9 POWER-SWITCHING CIRCUITS

The switching period of a power-switching circuit may consist of many switching intervals of rapidly changing voltages and currents. The transient response of power-switching circuits may extend over many switching cycles. PSpice will try to keep the internal time step relatively short compared to the switching period, which may cause long transient runs. This problem can be solved by transforming the switching circuit into an equivalent circuit, which can represent a "quasi-steady state" of the actual circuit and can accurately model the actual circuit's response.

Example 12-3

A single-phase full-bridge resonant inverter is shown in Fig. 12-4(a). The transistors and diodes can be considered as switches whose on-state resistance is 10 mΩ and whose on-state voltage is 0.2 V. Plot the transient response of the capacitor voltage and the current through the load from 0 to 2 ms in steps of 10 μs. The output frequency of the inverter is $f_{out} = 4$ kHz.

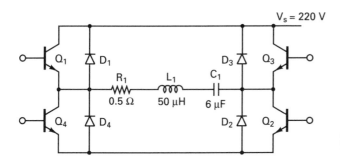

Figure 12-4 Single-phase full-bridge resonant inverter.

Solution When transistors Q_1 and Q_2 are turned on, the voltage applied to the load will be V_s, and the resonant oscillation will continue for the whole resonant period, first through Q_1 and Q_2 and then through diodes D_1 and D_2. When transistors Q_3 and Q_4 are turned on, the load voltage will be $-V_s$, and the oscillation will continue for another whole period, first through Q_3 and Q_4 and then through diodes D_3 and D_4. The resonant period of a series RLC circuit is approximately calculated as

$$\omega_r = \left(\frac{1}{LC} - \frac{R^2}{4L^2}\right)^{1/2}$$

For $L = L_1 = 50$ μH, $C = C_1 = 6$ μF, and $R = R_1 + R_{1(sat)} + R_{2(sat)} = 0.5 + 0.1 + 0.1 = 0.52$ Ω, $\omega_r = 57572.2$ rad/s, and $f_r = \omega_r/2\pi = 9162.9$ Hz. The resonant period is $T_r = 1/f_r = 1/9162.9 = 109.1$ μs. The period of the output voltage is $T_{out} = 1/f_{out} = 1/4000 = 250$ μs.

The switching action of the inverter can be represented by two voltage-controlled switches, as shown in Fig. 12-5(a). The switches are controlled by the voltages, as shown in Fig. 12-5(b). The on-time of switches, which should be approximately equal to the resonant period of the output voltage, is assumed to be 112 μs. The switch S_2 is delayed by 115 μs to take into account overlap. The model parameters of the switches are RON=0.01, ROFF=10E+6, VON=0.001, and VOFF=0.0.

The listing of the circuit file follows.

Example 12-3 Full-bridge resonant inverter

```
▲ *    The controlling voltage for switch S1:
   V1  1  0  PULSE (0  220V  0  1US  1US  110US  250US)
   *    The controlling voltage for switch S2 with a delay time of 115 µs
   V2  3  0  PULSE (0  −220V  115US  1US  1US  110US  250US)
   *    Voltage-controlled switches with model SMOD
```

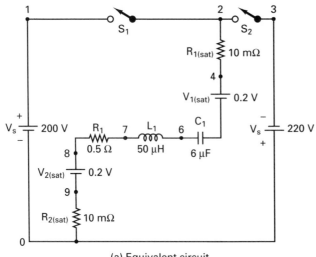

(a) Equivalent circuit

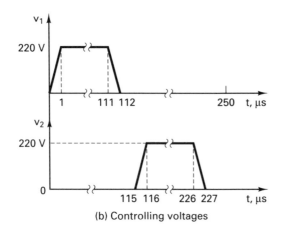

(b) Controlling voltages

Figure 12-5 Equivalent circuit for Fig. 12-4.

```
▲▲ S1   1   2   1   0   SMOD
   S2   2   3   0   3   SMOD
   *  Switch model parameters for SMOD
   MODEL  SMOD  VSWITCH (RON=0.01  ROFF=10E+6  VON=0.001 VOFF=0.0)
   RSAT1   2   4   10M
   VSAT1   4   5   DC   0.2V
   RSAT2   9   0   10M
   VSAT2   8   9   DC   0.2V
   *  Assuming an initial capacitor voltage of −250 V to reduce settling time
   C1   5   6   6UF  IC=−250V
   L1   6   7   50UH
   R1   7   8   0.5
   *  Switch model parameters for SMOD
   MODEL  SMOD  VSWITCH (RON=0.01  ROFF=10E+6  VON=0.001 VOFF=0.0)
```

```
▲ ▲ ▲  *  Transient analysis with UIC condition
       .TRAN  2US  500US  UIC
       .PROBE
  . END
```

The transient response for Example 12-3 is shown in Fig. 12-6.

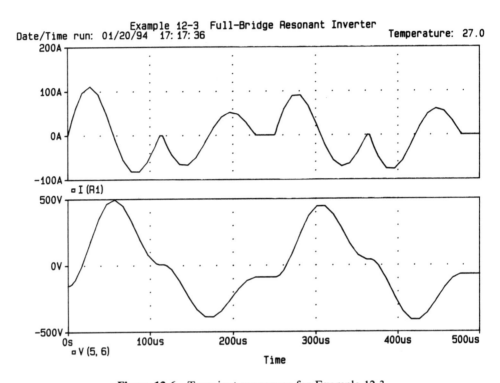

Figure 12-6   Transient responses for Example 12-3.

12-10 FLOATING NODES

PSpice does not allow floating nodes. A floating node is present if there is no dc path from a particular node to the ground. In general, PSpice requires that there must be a dc path from every node to the ground. Resistors, inductors, diodes, and transistors provide dc paths. If the circuit file contains a floating node, PSpice will indicate a read-in error on the screen, and the output file will contain a message indicating the node number. For example, if node 25 is floating, the output file will contain the following message:

```
ERROR: Node 25 is floating
```

Floating nodes can occur in many circuits, as shown in Fig. 12-7. Node 4 in Fig. 12-7(a) is floating because it does not have a dc path. The floating-node condition can be avoided by connecting node 4 to node 0, as shown by dotted lines (or by connecting node 3 to node 2). A similar situation can occur in voltage-controlled and current-controlled sources, as shown in Fig. 12-7(b) and 12-7(c). The model of an op-amp shown in Fig. 12-7(d) has five floating nodes. Nodes 0, 3, and 5 can be connected (or, alternatively, nodes 1, 2, and 4 may be connected) to avoid floating nodes.

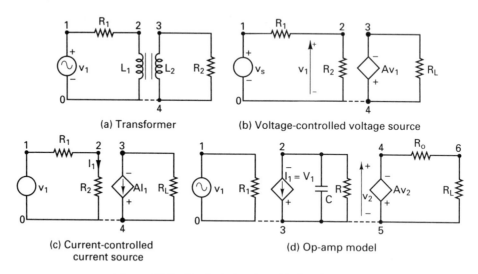

(a) Transformer

(b) Voltage-controlled voltage source

(c) Current-controlled current source

(d) Op-amp model

Figure 12-7 Typical circuits with floating nodes.

Capacitors can also cause floating nodes, because there is no dc path between the two sides of a capacitor. Let us consider the circuit of Fig. 12-8. Nodes 3 and 5 do not have dc paths; however, dc paths can be provided by connecting a very large resistance R_3 (say, 100 MΩ) across capacitor C_3 as shown by the dotted lines.

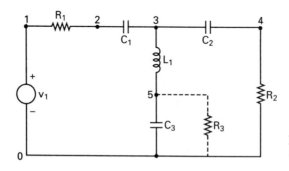

Figure 12-8 Typical circuit without a dc path.

Example 12-4

A passive filter is shown in Fig. 12-9. The output is taken from node 9. Plot the magnitude and phase of the output voltage separately against the frequency. The frequency should be varied from 100 HZ to 10 kHz in steps of 1 decade and 10 points per decade.

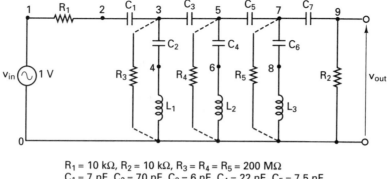

$R_1 = 10\ k\Omega,\ R_2 = 10\ k\Omega,\ R_3 = R_4 = R_5 = 200\ M\Omega$
$C_1 = 7\ nF,\ C_2 = 70\ nF,\ C_3 = 6\ nF,\ C_4 = 22\ nF,\ C_5 = 7.5\ nF$
$C_6 = 12\ nF,\ C_7 = 10.5\ nF,\ L_1 = 1.5\ mH$
$L_2 = 1.75\ mH,\ L_3 = 2.5\ mH$

Figure 12-9 A passive filter.

Solution The nodes between C_1 and C_3, C_3 and C_5, and C_5 and C_7 do not have dc paths to the ground. Therefore, the circuit cannot be analyzed without connecting resistors R_3, R_4, and R_5, as shown in Fig. 12-9 by dotted lines. If the values of these resistances are very high, say 200 MΩ, their influence on the ac analysis would be negligible.

The listing of the circuit file follows.

Example 12-4 A passive filter

```
 *    Input voltage is 1 V peak for ac analysis or frequency response.
 VIN  1  0  AC  1
 R1   1  2  10K
 R2   9  0  10K
 *   Resistances R3, R4, and R5 are connected to provide dc paths
 R3   3  0  200MEG
 R4   5  0  200MEG
 R5   7  0  200MEG
 C1   2  3  7NF
 C2   3  4  70NF
 C3   3  5  6NF
 C4   5  6  22NF
 C5   5  7  7.5NF
 C6   7  8  12NF
 C7   7  9  10.5NF
 L1   4  0  1.5MH
 L2   6  0  1.75MH
 L3   8  0  2.5MH
 *   Ac analysis for 100 Hz to 10 kHz with a decade increment and
 *   10 points per decade
 .AC  DEC  10  100   10KHZ
 *   Plot the results of ac analysis for the magnitude of voltage
 *   at node 9.
 .PLOT  AC  VM(9)  VP(9)
```

```
.PLOT    AG    VP(9)
.PROBE
.END
```

The frequency response for Example 12-4 is shown in Fig. 12-10.

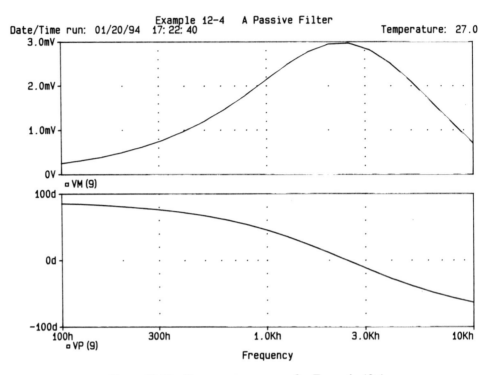

Figure 12-10   Frequency response for Example 12-4.

12-11 NODES WITH LESS THAN TWO CONNECTIONS

PSpice requires that each node must be connected to at least two other nodes. Otherwise, PSpice will give an error message similar to the following:

```
ERROR: Less than two connections at node 10
```

This means that node 10 must have at least one more connection. A typical situation is shown in Fig. 12-11(a), where node 3 has only one connection. This problem can be solved by short-circuiting resistance R_2, as shown by the dotted lines.

An error message may be indicated in the output file for a circuit with voltage-controlled sources as shown in Fig. 12-11(b). The input to the voltage-controlled source will not be considered to have connections during the check by

PSpice. This is because the input draws no current and it has infinite impedance. A very high resistance (say, $R_i = 10$ GΩ) may be connected from the input to the ground, as shown by the dotted lines.

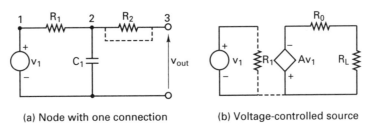

(a) Node with one connection   (b) Voltage-controlled source

Figure 12-11 Typical circuits with less than two connections at a node.

12-12 VOLTAGE SOURCE AND INDUCTOR LOOPS

If there is a voltage source E with zero resistance ($R = 0$), PSpice will try to divide E by 0, which is impossible. It should be noted that a zero value for voltage source E will not cause problems. PSpice always looks for zero-resistance loops, and if there are any, PSpice indicates a read-in error on the screen. For example, if loop voltage V15 is involved, the output file will contain the following message:

`ERROR: Voltage loop involving V15`

The zero-resistance components are independent voltage sources (V), inductors (L), voltage-controlled voltage sources (E), and current-controlled voltage sources (H). Typical circuits with such loops are shown in Fig. 12-12.

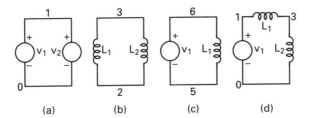

(a)   (b)   (c)   (d)

Figure 12-12 Typical circuits with zero-resistance loops.

The problem of a zero-resistance loop can be solved by adding a very small series resistance to at least one component in the loop. The resistor's value should be small enough so that it does not disturb the operation of the circuit, but it should not be less than 1 μΩ.

12-13 RUNNING A PSpice FILE ON SPICE

PSpice should give the same results as SPICE-2G (or SPICE). However, there could be small differences due to different methods used for handling convergence problems. If a circuit file is developed for running on PSpice, it may be necessary to make some changes before running it on SPICE, because the following features are *not* available in SPICE:

1. Extended output variables for .PRINT and .PLOT statements are not available. SPICE allows only voltages between nodes $V(x)$ or $V(x,y)$ and currents through voltage sources.
2. Group delay is not available.
3. The gallium-arsenide model is not available.
4. The nonlinear magnetic (transformer) model is not available.
5. Voltage-controlled and current-controlled switches are not available.
6. The temperature coefficients for capacitors and inductors and exponential temperature coefficients for resistors are not available.
7. The model parameters RG, RDS, L, W, and WD are not available in the MOSFET's .MODEL statement.
8. The sweep variable is limited to an independent current or voltage source.
9. Sweeping of model parameters or temperature is not allowed.
10. The .LIB and .INCLUDE statements are not available.
11. The current description of the input file must be typed in uppercase rather than in lowercase.

12-14 RUNNING A SPICE FILE ON PSpice

If a circuit file is developed for running on SPICE-2G (or SPICE), it may be necessary to make some changes before running it on PSpice, because the following features are *not* available in PSpice:

1. .DISTO (small-signal distortion) analysis and distortion output variables (HD2, DIM3, etc.) are not available. The .DISTO analysis should be replaced by a .TRAN analysis and a .FOUR analysis.
2. The IN = option in the .WIDTH statement is not available.
3. Temperature coefficients for resistors must be put into a .MODEL statement instead of in the resistor statement.
4. The voltage coefficients for capacitors and the current coefficients for inductors are placed in the .MODEL statements instead of in their own statements.

REFERENCES

1. *PSpice Manual.* Irvine, Calif.: MicroSim Corporation, 1992.
2. Wolfram Blume, "Computer circuit simulation," *BYTE*, Vol. 11, No. 7, July 1986, p. 165.

3. M. H. Rashid, *SPICE For Power Electronics and Electric Power*. Englewood Cliffs, N.J.: Prentice Hall, 1993.

PROBLEMS

12-1. For the inverter circuit in Fig. P9-6, plot the hysteresis characteristics.

12-2. For the circuit in Fig. P12-2, plot the hysteresis characteristics from the results of the transient analysis. The input voltage is varied slowly from -4 V to 4 V and from 4 V to -4 V. The op-amp can be modeled as a UA741 macromodel, as shown in Fig. 10-3. The description of the macromodel is listed in library file EVAL.LIB. The supply voltages are $V_{CC} = 12$ V and $V_{EE} = -12$ V.

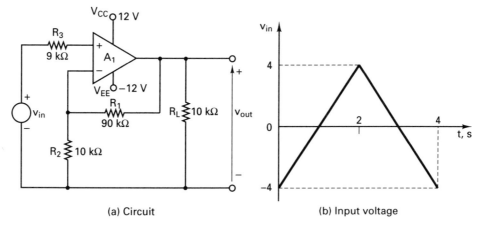

(a) Circuit (b) Input voltage

Figure P12-2

Running PSpice on PCs A

PSpice programs are available in high-density (1.2-megabyte) diskettes or normal (360-kilobyte) diskettes. The first step is to have the directory listing of the files on the program diskettes. The next step is to print and read the README.DOC file. It contains a brief description of the type of display and hard copy that are allowed by PSpice and Probe. It also contains the system requirements and instructions for running PSpice programs.

PSpice will run on any IBM PC, the Macintosh II, or any compatible computer. The student version of PSpice does not require the coprocessor for running Probe. The display could be on monochrome or color graphics monitors. There is no requirement for special printer features. The types of printers and display can be set by editing the PROBE.DEV file, or from the "Display/Printer Setup" menu. The simulation of a circuit requires

Installing PSpice on the PC
Creating input files
Run command
DOS (disk operating system) commands

A-1 INSTALLING PSpice ON PCs

The steps to be followed to install the PSpice program on a PC with a hard disk follow.

1. Create the directory PSPICE by typing

```
MD PSPICE  or  MD\PSPICE
```

2. Copy all the PSpice files from the diskettes onto the hard disk by typing

COPY a:*.* (and then press Return)

3. Change your current directory to PSPICE by typing

CD PSPICE or CD\PSPICE

4. Type

PS

5. Now the PSpice menu should appear on your monitor screen.
6. The File menu should be highlighted. If not, highlight the File menu by pressing the arrow keys.
7. Press the Return key.
8. Move to the Display/Printer Setup submenu and press Return.
9. Choose the type of display and printer:

Display (for example IBMVGA)   (Press F4 for options/choices)
Port (for example PRN for printer)   (Press F4 for options/choices)
Printer (for example EPSON printer)   (Press F4 for options/choices)

10. Press ESC to get out.
11. Move to Current File menu and press Return.
12. Type the name of your new circuit file or the location and name of your existing circuit file. Press Return.
13. You should be in editing mode, and you can type the circuit descriptions in order to create (or edit) your circuit file.
14. Move to the Analysis menu, and run the circuit simulation.
15. Move to the Probe menu, and run Probe for a graphical display of output plots on the monitor.
16. Use the Browse Output menu to look through your output file.

Note. Move to a menu with the arrow keys and press the Return key to work on the menu. Always press ESC to get out of a menu.

A-2 CREATING INPUT FILES

The PSpice program has a built-in-editor with shell. The input files can also be created by text editors. The text editor that is always available is EDLIN. It comes with DOS and is described in the DOS user's guide. There are other editors, such as Program Editor (from WordPerfect Corporation). Word processing programs such as WordStar, WordStar 2000, and Word may also be used to create the input file. The word processor normally creates a file that is not a text file. It contains embedded characters to determine margins, paragraph boundaries, pages, and the like. However, most word processors have a command or mode to create a text file without these control characters. For example, WordStar 200 creates text files with UNIFORM format.

A-3 RUN COMMAND

PSpice is run from the menu, but it can be run by typing

```
PSPICE   ⟨input file⟩   ⟨output file⟩
```

By default, the input file has the extension of .CIR, and the output file has the extension of .OUT. The name of the output file defaults to the name of the input file. If the input file, EX2-1.CIR, is on the default drive, the following commands are equivalent:

```
PSPICE  EX2-1
PSPICE  EX2-1.CIR
PSPICE  EX2-1.CIR  EX2-1
PSPICE  EX2-1.CIR  EX2-1.OUT
```

The output file can be assigned to the printer that is connected to the PC by

```
PSPICE  EX1  PRN
```

The commands that will instruct PSpice about the location of the input file and the program files will depend on the type of disk drives. There are two types of disks: fixed (or hard) disk and floppy disk, which is not fixed.

With a fixed disk.   Running PSpice with a fixed disk is straightforward. The PSPICE.BAT file must be in the default drive. Running PSpice will call the PSPICE.BAT file, which in turn will call the PSPICE1.EXE file and then the PROBE.EXE file, if required. PSPICE1.EXE creates one temporary file for storing intermediate results and deletes this temporary file when it finishes. If the circuit file EX2-1.CIR is on a diskette in drive A, the command for running PSpice is

```
PSPICE  A:EX2-1.CIR  A:EX2-1.OUT
```

Without a fixed disk.   The input file and the PSPICE.BAT file must be on Diskette 1 in drive A. The PSPICE.BAT file is searched by DOS for programs and commands. Running PSpice will cause PSPICE.BAT to call PSPICE1.EXE from drive B. PSPICE1.EXE creates one temporary file for storing intermediate results and automatically deletes the temporary file after writing to the output file when it finishes. The command for running the input file EX2-1.CIR is

```
PSPICE  A:EX2-1.CIR
```

A-4 DOS COMMANDS

The DOS commands that are frequently used are

To format a new diskette in drive A, type

```
FORMAT A:
```

To list the directory of a diskette in drive A, type

```
DIR A:
```

To delete the file EX2-1.CIR in drive A, type

```
Delete A:EX2-1.CIR   (or Erase A:EX2-1.CIR)
```

To copy the file EX2-1.CIR in drive A to the file EX2-2.CIR in drive B, type

```
COPY A:EX2-1.CIR B:EX2-2.CIR
```

To copy all the files on a diskette in drive A to a diskette in drive B, type

```
COPY A:*.*  B:
```

To type the contents of the file EX2-1.OUT on the diskette in drive A, type

```
TYPE A:EX2-1.OUT
```

To print the contents of the file EX2-1.CIR in drive A to the printer, first activate the printer by pressing the Ctrl (Control) and Prtsc (Print Screen) keys together, and then type

```
TYPE A:EX2-1.CIR
```

The printer can be deactivated by pressing the Ctrl (Control) and Prtsc (Print Screen) keys again.

Noise Analysis

Noise is generated in electronic circuits. That is, an electronic circuit will have output even without any input signal. Noise can be classified as one of five types:

Thermal noise
Shot noise
Flicker noise
Burst noise
Avalanche noise

B-1 THERMAL NOISE

Thermal noise is generated in resistors due to random motion produced by the thermal agitation of electrons. This noise is dependent on temperature. The equivalent circuit for thermal noise in resistors is shown in Fig. B-1. The mean square value of the noise generator is expressed as

$$V_t^2 = 4kTR\Delta f \qquad V^2$$

$$I_t^2 = 4kT\Delta f/R \qquad A^2$$

where k = Boltzmann's constant $(1.38 \times 10^{-23}$ J/K$)$
T = absolute temperature, kelvin
R = resistance, ohms
Δf = noise bandwidth, hertz

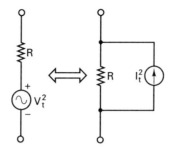

Figure B-1 Equivalent circuit for noise in resistors.

B-2 SHOT NOISE

Shot noise is generated by random fluctuations in the number of charged carriers that are emitted from a surface or diffused from a junction. This noise is always associated with a direct current flow and is present in bipolar transistors. The mean square value of the noise current is expressed by

$$I_s^2 = 2qI_D \, \Delta f \qquad A^2$$

where Δf = noise bandwidth, hertz
q = electron charge (1.6×10^{-19} C)
I_D = dc current, amps

B-3 FLICKER NOISE

Flicker noise is generated due to surface imperfections resulting from the emission. This noise is associated with all active devices and some discrete passive elements such as carbon resistors. The mean square value of the noise current is expressed by

$$I_f^2 = K_f \frac{I_D^a}{f} \, \Delta f \qquad A^2$$

where Δf = noise bandwidth, hertz
I_D = direct current
K_f = flicker constant for a particular device
a = flicker-exponent constant in the range of 0.5 to 2

B-4 BURST NOISE

Burst noise is generated due to the presence of heavy metal ion contamination and is found in some integrated circuits and discrete transistors. The repetition rate of noise pulses is in the audio frequency range (a few kilohertz or less) and produces a ''popping'' sound when played through a speaker. This noise is also known as *popcorn noise*. The mean square value of the noise current is expressed as

$$I_b^2 = K_b \frac{I_D^c}{1 + (f/f_c)} \Delta f \qquad A^2$$

where Δf = noise bandwidth, hertz
I_D = direct current
K_b = burst constant for a particular device
c = burst-exponent constant
f_c = a particular frequency for a given noise

B-5 AVALANCHE NOISE

Avalanche noise is produced by Zener or avalanche breakdown in *p-n* junctions. The holes and electrons in the depletion region of a reverse-biased *p-n* junction acquire sufficient energy to create hole-electron pairs by collision. This process is cumulative, resulting in the production of a random series of large noise spikes. This noise is associated with direct current and is much greater than shot noise for the same current. Zener diodes are normally avoided in circuits requiring low noise.

B-6 NOISE IN DIODES

The equivalent circuit for noise in diodes is shown in Fig. B-2. There are two generators. The voltage generator is due to thermal noise in the resistance of the

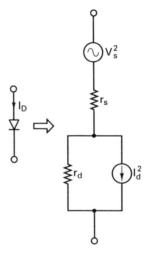

Figure B-2 Equivalent circuit for noise in diodes.

silicon. The current source is due to the shot noise and flicker noise. The noise voltage is given by

$$V_s^2 = 4kTr_s\,\Delta f \qquad \text{V}^2$$

$$I_d^2 = 2qI_D\Delta f + K_f\frac{I_D^a}{f}\,\Delta f \qquad \text{A}^2$$

where I_D = forward diode current, amps
  r_s = resistance of the silicon, ohms
  K_f = flicker constant for a particular device
  a = flicker-exponent constant in the range of 0.5 to 2

B-7 NOISE IN BIPOLAR TRANSISTORS

The equivalent circuit for noise in bipolar transistors is shown in Fig. B-3. The current generator in the collector is due to shot noise. The noise voltage generator in the base circuit is due to thermal noise in the base resistance. The current generator in the base circuit consists of shot noise, flicker noise, and burst noise. The noise is expressed by

$$V_b^2 = 4kTr_b\Delta f \qquad \text{V}^2$$

$$I_c^2 = 2qI_C\Delta f \qquad \text{A}^2$$

$$I_b^2 = 2qI_B\Delta f + K_f\frac{I_B^a}{f}\,\Delta f + K_b\,\frac{I_B^c}{1 + (f/f_c)}\,\Delta f \qquad \text{A}^2$$

where I_B = base bias current, amps
  I_C = collector bias current, amps
  r_b = resistance at the transistor base, ohms
  K_b = burst constant for a particular device
  c = burst-exponent constant in the range of 0.5 to 2
  f_c = a particular frequency for a given noise
  K_f = flicker constant for a particular device
  a = flicker-exponent constant in the range of 0.5 to 2

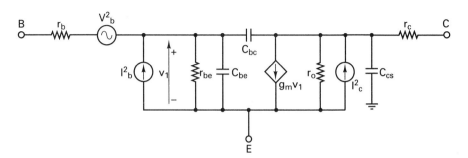

Figure B-3 Equivalent circuit for noise in bipolar transistors.

The equivalent circuit for noise in field-effect transistors (FETs) is shown in Fig. B-4. The current generator in the gate is due to shot noise, which is very small. The current generator in the drain circuit consists of thermal and flicker noise. The noise current is expressed by

$$I_g^2 = 2qI_G\Delta f \qquad A^2$$

$$I_d^2 = 4kT\left(\frac{2}{3}g_m\right)\Delta f + K_f\frac{I_D^a}{f}\Delta f \qquad A^2$$

where I_D = drain bias current, amps
I_G = gate leakage current, amps
g_m = transconductance at bias point, amps per volt
r_b = resistance of the transistor base, ohms
K_f = flicker constant for a particular device
a = flicker-exponent constant in the range of 0.5 to 2

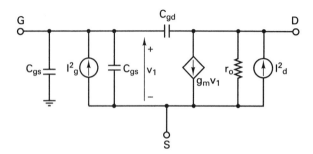

Figure B-4 Equivalent circuit for noise in field-effect transistors.

B-9 EQUIVALENT INPUT NOISE

Each noise generator contributes to the output of a circuit. The effect of all the noise generators can be found by summing the mean square value of individual noise contributions. Once the total mean square noise output voltage is found, all the noises can be represented by an equivalent input noise at a desired source, as shown in Fig. B-5. This input noise is found by dividing the output voltage by the gain. The gain is the output noise voltage with respect to a defined input. PSpice calculates the output noise and the equivalent input noise by the .NOISE command, which is covered in Section 6-13. It may be noted that PSpice calculates the noise in $V/\sqrt{Hz}$ or $A/\sqrt{Hz}$. Dividing the mean square value of the noise output voltage $V_{out(noise)}$ by the noise bandwidth gives the output noise spectrum V_{out}. That is,

$$V_{out} = \frac{V_{out(noise)}}{\Delta f} \qquad V/\sqrt{Hz}$$

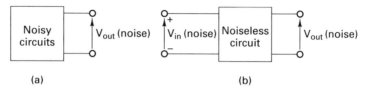

Figure B-5 Equivalent noise input.

Dividing the output noise spectrum by the gain yields the equivalent input noise spectrum:

$$V_{\text{in}} = \frac{V_{\text{out}}}{G_v} \qquad \text{V}/\sqrt{\text{Hz}}$$

where G_v is the voltage gain with respect to the equivalent input source.

If the equivalent input is a current source, the equivalent input current noise spectrum becomes

$$I_{\text{in}} = \frac{V_{\text{out}}}{R_t} \qquad \text{V}/\sqrt{\text{Hz}}$$

where R_t is the transresistance with respect to the equivalent input source.

Example B-1

For the TTL inverter circuit in Fig. 8-28, calculate and print the equivalent input and output noise. The frequency is varied from 1 Hz to 100 kHz with a decade increment and 1 point per decade. The input voltage is 1 V for ac analysis and 3.5 V for dc analysis.

Solution The listing of the circuit file follows.

Example B-1   TTL inverter
```
▲ *  Input voltage of 3.5 V for dc analysis and 1 V for ac analysis
  VIN  1   0  DC  3.5V  AC  1V
  VCC  13  0  5V
▲▲ RS  1   2  50
  RB1 13   3   4K
  RC2 13   5   1.4K
  RE2  6   0   1K
  RC3 13   7   100
  RB5 13  10   4K
  *  BJTs with model QNP and substrate connected to ground by default
  Q1   4   3   2  QNP
  Q2   5   4   6  QNP
  Q3   7   5   8  QNP
  Q4   9   6   0  QNP
  Q5  11  10   9  QNP
  *  Diodes with model DIODE
  D1   8   9  DIODE
  D2  11  12  DIODE
  D3  12   0  DIODE
```

```
* Model of NPN transistors with model QNP
.MODEL QNP NPN (BF=50 RB=70  RC=40 TF=0.1NS TR=10NS VJC=0.85 VAF=50
+KF=6.5E-16 AF=1.2)
* Diodes with model DIODE
.MODEL DIODE D (RS=40 TT=0.1NS)
```
▲▲▲
```
* Ac sweep from 1Hz and 100 kHz with a decade increment
* and 1 point per decade
.AC  DEC 1  1HZ  100KHZ
* Noise analysis between output voltage, V(9) and input voltage, VIN
.NOISE  V(9)  VIN
* Printing the results of noise analysis
.PRINT  NOISE  ONOISE INOISE
```
.END

The results of the noise analysis are given next.

****   SMALL-SIGNAL BIAS SOLUTION   TEMPERATURE = 27.000 DEG C

NODE	VOLTAGE	NODE	VOLTAGE	NODE	VOLTAGE	NODE	VOLTAGE
(1)	3.5000	(2)	3.4724	(3)	2.7564	(4)	1.9134
(5)	1.1702	(6)	1.0205	(7)	4.9998	(8)	.5585
(9)	.0635	(10)	.9265	(11)	.0816	(12)	.0408
(13)	5.0000						

VOLTAGE SOURCE CURRENTS

NAME	CURRENT
VIN	−5.524E−04
VCC	−4.317E−03

TOTAL POWER DISSIPATION 2.35E−02 WATTS

****   AC ANALYSIS   TEMPERATURE = 27.000 DEG C

FREQ	ONOISE	INOISE
1.000E+00	8.146E−10	7.205E+02
1.000E+01	8.146E−10	7.205E+02
1.000E+02	8.146E−10	7.205E+02
1.000E+03	8.146E−10	7.205E+02
1.000E+04	8.146E−10	7.205E+02
1.000E+05	8.146E−10	7.204E+02

JOB CONCLUDED

TOTAL JOB TIME   22.58

Nonlinear Magnetic Model

The nonlinear magnetic model uses MKS (metric) units. However, the results for Probe are converted to Gauss and Oersted and may be displayed using $B(Kxx)$ and $H(Kxx)$. The $B\text{-}H$ curve can be drawn by a transient run with a slowly rising current through a test inductor and then by displaying $B(Kxx)$ against $H(Kxx)$.

Characterizing core materials may be done by trial by using PSpice and Probe. The procedures for setting parameters to obtain a particular characteristic are the following:

1. Set domain wall pinning constant, $K = 0$. The curve should be centered in the $B\text{-}H$ loop, like a spine. The slope of the curve at $H = 0$ should be approximately equal to that when it crosses the x-axis at $B = 0$.

2. Set the magnetic saturation, $MS = B\text{max}/0.01257$.

3. Set the slope, ALPHA. Start with the mean field parameter, ALPHA = 0, and vary its values to get the desired slope of the curve. It may be necessary to change MS slightly to get the desired saturation value.

4. Change K to a nonzero value to create hysteresis; K affects the opening of the hysteresis loop.

5. Set C to obtain the initial permeability. Probe displays the permeability, which is $\Delta B/\Delta H$. Probe calculates differences, not derivatives, so the curves will not be smooth. The initial value of $\Delta B/\Delta H$ is the initial permeability.

Example C-1

The coupled inductors in Fig. 5-8(a) are nonlinear. (See Fig. C-1.) The parameters of the inductors are $L_1 = L_2 = 500$ turns, $k = 0.9999$. Plot the $B\text{-}H$ characteristic of the core from the results of transient analysis if the input current is varied very slowly from 0 to -15 A, -15 A to 15 A, and from 15 A to -15 A. The load resistance of $R_L = 1$ KΩ is connected to the secondary of the transformer. The model parame-

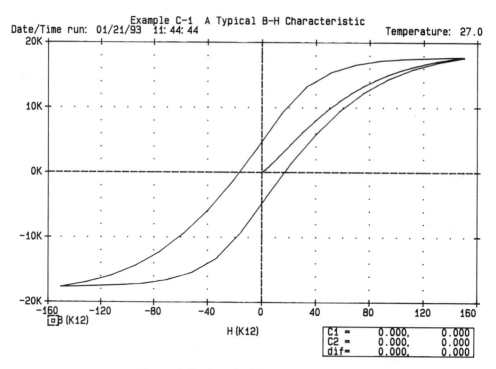

Figure C-1   A typical *B-H* characteristic.

ters of the core are AREA=2.0, PATH=62.73, GAP=0.1, MS=1.6E+6, ALPHA=1E−3, A=1E+3, C=0.5, and K=1500.

Solution   The circuit life for the coupled inductors in Fig. 5-8(a) would be as follows.

Example C-1   A typical *B-H* characteristic

```
▲ *   PWL waveform for transient analysis
   IN  1  0  PWL (0  0  1  −15  2  15   3  −15)
▲▲ *  Inductors represent the number of turns
      L1  1  0  500
      L2  2  0  500
      R2  2  0  1000
      *   Coupled inductors with k = 0.9999 and model CMOD
      K12  L1  L2  0.9999   CMOD
      *   Model parameters for CMOD
      .MODEL CMOD  CORE (AREA=2.0 PATH=62.73 GAP=0.1 MS=1.6E+6 ALPHA=1E−3 A=1E+3
      +   C=0.5 K=1500)
▲▲▲ *  Transient analysis from 0 to 3 s in steps of 0.03 s
       .TRAN  0.05  3
       .PROBE
   .END
```

BIBLIOGRAPHY

1. Allen, Phillip E., *CMOS Analog Circuit Design*. New York: Holt, Rinehart and Winston, 1987.

2. Antognetti, Paolo, and Guiseppe Massobrio, *Semiconductor Device Modeling with SPICE*. New York: McGraw-Hill, 1988.

3. Banzhaf, Walter, *Computer-Aided Circuit Analysis Using SPICE*. Englewood Cliffs, N.J.: Prentice Hall, 1989.

4. Bugnola, Dimitri S., *Computer Programs for Electronic Analysis and Design*. Reston, Va.: Reston, 1983.

5. Chattergy, Rahul, *SPICEY Circuits*. Boca Raton, Fla.: CRC Press, 1992.

6. Chua, Leon O., and Pen-Min Lin, *Computer-Aided Analysis of Electronic Circuits—Algorithms and Computational Techniques*. Englewood Cliffs, N.J.: Prentice Hall, 1975.

7. Ghandi, S. K., *Semiconductor Power Devices*. New York: Wiley, 1977.

8. Gray, Paul R., and Robert G. Meyer, *Analysis and Design of Analog Integrated Circuits*. New York: Wiley, 1984.

9. Grove, A. S., *Physics and Technology of Semiconductor Devices*. New York: Wiley, 1967.

10. Hodges, D. A., and H. G. Jackson, *Analysis and Design of Digital Integrated Circuits*. New York: McGraw-Hill, 1988.

11. Keown, John, *PSpice and Circuit Analysis*. New York: Macmillan, 1991.

12. McCalla, William J., *Fundamentals of Computer-Aided Circuit Simulation*. Norwell, Mass.: Kluwer Academic, 1988.

13. MicroSim Corporation, *PSpice Manual*. Irvine, Calif.: MicroSim, 1992.

14. Nagel, Laurence, W., *SPICE2—A computer program to simulate semiconductor circuits*, Memorandum no. ERL-M520, May 1975, Electronics Research Laboratory, University of California, Berkeley.

15. Nashelsky, Louis, and Robert Boylestad, *BASIC for Electronics and Computer Technology*. Englewood Cliffs, N.J.: Prentice Hall, 1988.

16. Rashid, M. H., *Power Electronics—Circuits, Devices and Applications*. Englewood Cliffs, N.J.: Prentice Hall, 1993.

17. Rashid, M. H., *SPICE For Power Electronics and Electric Power*. Englewood Cliffs, N.J.: Prentice Hall, 1993.

18. Spence, Robert, and John P. Burgess, *Circuit Analysis by Computer—from Algorithms to Package*. London: Prentice Hall International (UK), 1986.

19. Tuinenga, Paul W., *SPICE: A guide to circuit simulation and analysis using PSPice*, Englewood Cliffs, N.J.: Prentice Hall, 1992.

20. Ziel, Aldert van der, *Noise in Solid State Devices*. New York: Wiley, 1986.

Index

Metal oxide field effect transistor (*continued*)
    model parameters of, 242
    model statement of, 243
    Schichman and Hodges model of, 241
    small-signal model of, 241
Model parameters, 25
.MODEL statement, 25
Models:
    behavior, 119
    BJT, 186
    capacitor, 59
    diode, 163
    Gummel-Poon, 186
    I/O, 294
    inductor, 60
    JFET, 227
    Jiles-Atherton magnetics, 108
    MOSFET, 240
    resistor, 23
    Schichman-Hodges MOSFET, 241
    timing, 298
Mutual coupling, coefficient of, 106
Mutual inductance, 106
Names:
    element, 8
    model, 25

N

Negative component values, 333
Nodes, 8
    circuit, 8
    floating, 337
    ground, 8
.NODESET statement, 127
Noise:
    avalanche, 350
    BJT, 351
    burst, 348
    diode, 350
    equivalent input, 352
    flicker, 349
    JFET, 352
    resistor, 348
    shot, 348
    thermal, 348
Noise analysis, 134
    .NOISE statement, 134

Nonlinear inductor, 106
Nonlinear magnetics, 355

O

ONOISE, 135, 354
Op-amp macromodel, 270
Op-amps, 268
    ac linear model of, 269
    dc linear model of, 268
    nonlinear macromodel of, 270
Operating point, 42
Options, 127
    list of, 127
.OP statement, 42
.OPTIONS statement, 127
Output:
    current, 34, 68, 96
    voltage, 34, 68, 96

P

Parameters:
    BJT, 188
    diode, 163
    JFET, 229
    MOSFET, 242
Passive elements:
    capacitor, 59
    inductor, 60
    resistor, 23
Phase angle, 96
Piecewise linear source, 64
.PLOT statement, 38
Plus (+) sign, 12
Polup, 305
Polynomial source, 28
Primitives, 290
    digital, 290
.PRINT statement, 37
Probe, 38
PROBE.DEV, 42
PROBE.DAT, 38
Probe output, 39
.PROBE statement, 38
PSpice, 2, 3
    limitations of, 4
    running, 344, 346

Pulldown, 305
Pulse source, 63

R

Real part of complex values, 96
Resistors, models of, 23
Run command, 346

S

Scaling element values, 7
Semiconductor devices, type names of, 25
Semiconductor diode, 156
Sensitivity analysis, dc, 136
.SENS statement, 136
Single-frequency FM source, 64
Sinusoidal source, 65
Small-signal dc analysis, 48
Small-signal transfer function, 45
Source (s), 26, 27, 61
   current-controlled current, 32
   current-controlled voltage, 33
   dependent, 27
   exponential, 61
   independent current, 27, 67, 98
   independent voltage, 27, 67, 98
   modeling, 10, 61
   piecewise linear, 64
   polynomial, 28
   pulse, 63
   single-frequency frequency-modulation, 64
   sinusoidal, 65
   voltage-controlled current, 31
   voltage-controlled voltage, 30
SPICE, 1
   types of, 2
   running, 344
SPICE2, 2
SPICE3, 2
Stimulus devices, 307
Subcircuits, 123
   .ENDS statement, 123
   SUBCKT statement, 123
   X (subcircuit call) device, 123
Suffixes:
   scale, 7
   units, 7

Sweeping temperature, 26
Switch, 79
   current-controlled, 83
   voltage-controlled, 80
Symbols:
   devices, 34
   elements, 9

T

TABLE, 120
Temperature:
   operating, 26
   sweeping, 26
   .TEMP statement, 138
.TF statement, 45
Thermal noise, 248
Tolerances, DEV/LOT, 146
Transfer function:
   analysis, 45
   circuit gain, 45
   input and output resistance, 45
   .TF statement, 45
Transient response, 71
   .TRAN statement, 71
Transistor:
   bipolar junction, 186
   junction field effect, 227
   MOS field effect, 240
Transmission lines:
   lossless, 110
   lossy, 110
.TRAN statement, 71

U

UIC (Use Initial Conditions), 70

V

VALUE, 119
Values:
   element, 7
     scaling, 7
Variables:
   ac analysis, 96
   dc sweep, 34

LIBRARY
ST. LOUIS COMMUNITY COLLEGE
AT FLORISSANT VALLEY

Tear out this card and fill in all necessary information. Then enclose this card with your check or money order *only* in an envelope and mail to:

Book Distribution Center
PRENTICE HALL
Route 59 at Brook Hill Drive
West Nyack, NY
10995

SPICE FOR POWER ELECTRONICS AND ELECTRIC POWER
Muhammad H. Rashid

Hardware requirements: IBM PC or compatible, 640K minimum memory, a hard drive, 1.2mb diskette drive, MS DOS 3.0 or later. (Floating point coprocessor optional.)

Please send the item(s) checked below. PAYMENT ENCLOSED (check or money order only). The Publisher will pay all shipping and handling charges.

____ PSpice® Student Version Disks (two 5¼" disks) IBM PC compatible. $7.50 each set. (73476-4) Release 5.0

____ PSpice® Student Version Disks (two 3½" disks) IBM PC compatible. $9.00 each set. (73475-6) Release 5.0

____ PSpice® Student Version Disks (three 3½" disks) MAC II compatible. $15.50 each set. (73474-9) Release 5.0

NAME _____

DEPT. _____

SCHOOL _____

CITY _____ STATE _____ ZIP _____

NOTE: PROFESSIONAL/REFERENCE BOOKS ARE TAX DEDUCTIBLE.
Prices subject to change without notice. Please add sales tax for your area.

Tear out this card and fill in all necessary information. Then enclose this card with your check or money order *only* in an envelope and mail to:

Book Distribution Center
PRENTICE HALL
Route 59 at Brook Hill Drive
West Nyack, NY
10995

SPICE FOR POWER ELECTRONICS AND ELECTRIC POWER
Muhammad H. Rashid

Hardware requirements: IBM PC or compatible, 640K minimum memory, a hard drive, 1.2mb diskette drive, MS DOS 3.0 or later. (Floating point coprocessor optional.)

Please send the item(s) checked below. PAYMENT ENCLOSED (check or money order only). The Publisher will pay all shipping and handling charges.

____ PSpice® Student Version Disks (two 5¼" disks) IBM PC compatible. $7.50 each set. (73476-4) Release 5.0

____ PSpice® Student Version Disks (two 3½" disks) IBM PC compatible. $9.00 each set. (73475-6) Release 5.0

____ PSpice® Student Version Disks (three 3½" disks) MAC II compatible. $15.50 each set. (73474-9) Release 5.0

NAME _____

DEPT. _____

SCHOOL _____

CITY _____ STATE _____ ZIP _____

NOTE: PROFESSIONAL/REFERENCE BOOKS ARE TAX DEDUCTIBLE.
Prices subject to change without notice. Please add sales tax for your area.